Lecture Notes in Electrical Engineering

Volume 722

The book series *Lecture Notes in Electrical Engineering* (LNEE) publishes the latest developments in Electrical Engineering - quickly, informally and in high quality. While original research reported in proceedings and monographs has traditionally formed the core of LNEE, we also encourage authors to submit books devoted to supporting student education and professional training in the various fields and applications areas of electrical engineering. The series cover classical and emerging topics concerning:

- Communication Engineering, Information Theory and Networks
- Electronics Engineering and Microelectronics
- Signal, Image and Speech Processing
- Wireless and Mobile Communication
- Circuits and Systems
- Energy Systems, Power Electronics and Electrical Machines
- Electro-optical Engineering
- Instrumentation Engineering
- Avionics Engineering
- Control Systems
- Internet-of-Things and Cybersecurity
- Biomedical Devices, MEMS and NEMS

For general information about this book series, comments or suggestions, please contact leontina.dicecco@springer.com.

To submit a proposal or request further information, please contact the Publishing Editor in your country:

China

Jasmine Dou, Editor (jasmine.dou@springer.com)

India, Japan, Rest of Asia

Swati Meherishi, Editorial Director (Swati.Meherishi@springer.com)

Southeast Asia, Australia, New Zealand

Ramesh Nath Premnath, Editor (ramesh.premnath@springernature.com)

USA, Canada:

Michael Luby, Senior Editor (michael.luby@springer.com)

All other Countries:

Leontina Di Cecco, Senior Editor (leontina.dicecco@springer.com)

**** This series is indexed by EI Compendex and Scopus databases. ****

More information about this series at http://www.springer.com/series/7818

T. Laxminidhi · Jyoti Singhai · Sreehari Rao Patri ·
V. V. Mani
Editors

Advances in Communications, Signal Processing, and VLSI

Select Proceedings of IC2SV 2019

Springer

Editors
T. Laxminidhi
National Institute of Technology Karnataka
Mangalore, India

Sreehari Rao Patri
National Institute of Technology Warangal
Warangal, Telangana, India

Jyoti Singhai
Department of Electronics
and Communication Engineering
Maulana Azad National Institute
of Technology
Bhopal, Madhya Pradesh, India

V. V. Mani
Department of Electronics
and Communication Engineering
National Institute of Technology Warangal
Warangal, India

ISSN 1876-1100   ISSN 1876-1119 (electronic)
Lecture Notes in Electrical Engineering
ISBN 978-981-33-4057-2   ISBN 978-981-33-4058-9 (eBook)
https://doi.org/10.1007/978-981-33-4058-9

This Springer imprint is published by the registered company Springer Nature Singapore Pte Ltd.
The registered company address is: 152 Beach Road, #21-01/04 Gateway East, Singapore 189721, Singapore

Organization

Program Chairs

Dr. Venkata Mani Vakamulla, NIT, Warangal, India
Dr. Sreehari Rao Patri, NIT, Warangal, India

Preface

This book presents the advances in communications, signal processing, and VLSI taken from the excerpts of the proceedings of the International Conference on Communications, Signal Processing and VLSI held by the ECE Department, National Institute of Technology Warangal during October 23–24, 2019. This conference offered an excellent forum for exchange of ideas among interested researchers, students, peers, and practitioners in the upcoming areas of signal processing, 5G communications, chip design, image processing, and machine learning. It focused on the congregation of the deft designers across academy and industry, aiming to improve the living standards of mankind.

The selected proceedings of IC2SV 2019 included in this book were peer reviewed and thoroughly checked for plagiarism using Turnitin software. We thank the administration, NIT Warangal, for their support and all the authors for sharing their research outcomes that helped to publish this book.

Mangalore, India — T. Laxminidhi
Bhopal, India — Jyoti Singhai
Warangal, India — Sreehari Rao Patri
Warangal, India — V. V. Mani

Contents

Gibbs-Shannon Entropy and Related Measures: Tsallis Entropy 1
G. Ramamurthy and T. Jagannadha Swamy

Recognition of Natural and Computer-Generated Images Using Convolutional Neural Network 11
K. Rajasekhar and Gopisetti Indra Sai Kumar

Design of a Compact 2 × 2 MIMO Diversity Antenna for X-band Applications 23
Tathababu Addepalli and V. R. Anitha

Performance Analysis of Activation Functions on Convolutional Neural Networks Using Cloud GPU 35
Vamshi Krishna Kayala and Prakash Kodali

Pareto Optimal Approach for Contrast Adjustment in Gray Level Images Using IDS Algorithm 49
Sowjanya Kotte

Emotion Recognition from Speech Biometric System Using Machine Learning Algorithms 63
Chevella Anil Kumar and Kancharla Anitha Sheela

Implementation of Single Image Histogram Equalization and Contrast Enhancement on Zynq FPGA 75
Prathap Soma, Chundru Sravanthi, Palaparthi Srilakshmi, and Ravi Kumar Jatoth

Design and Simulation of a Dual-Band Radiometer for Humidity and Temperature Profiling 83
Akash Tyagi, Bijit Biswas, G. Arun Kumar, and Biplob Mondal

IOT-Based Domestic Aid Using Computer Vision for Specially Abled Persons 91
G. Sahitya, C. Kaushik, Vasagiri Krishnasree, and Ch. Yajna Mallesh

Aeroload Simulation of Aerospace Vehicle Using Fin Load Simulator 103
MVKS Prasad, Sreehari Rao Patri, and Jagannath Nayak

Design and Analysis of Wearable Step-Shaped Sierpinski Fractal Antenna for WBAN Applications 115
Sandhya Mallavarapu and Anjaneyulu Lokam

Evaluation of Real-Time Embedded Systems in HILS and Delay Issues 123
L. A. V. Sanhith Rao, Sreehari Rao Patri, I. M. Chhabra, and L. Sobhan Kumar

Acoustic Feedback Cancellation Using Optimal Step-Size Control of the Partition Block Frequency-Domain Adaptive Filter 139
S. Siva Prasad and C. B. Rama Rao

A Compact IPD Based on-Chip Bandpass Filter for 5G Radio Applications 153
M. V. Raghunadh and N. Bheema Rao

An 18-Bit Incremental Zoom ADC with Reduced Delay Overhead Data Weighted Averaging for Battery Measurement Systems 165
Manoj Katta Venkata and Veeresh Babu Vulligaddala

A Non-Uniform Digital Filter Bank Hearing Aid for Aged People with Hearing Disability 179
K. Ayyappa Swamy, C. Sushma, Zachariah C. Alex, and S. Prathima

Optimal Resource Allocation Based on Particle Swarm Optimization 199
Naidu Kalpana, Hemanth Kumar Gai, Amgothu Ravi Kumar, and Vanlin Sathya

Design and Simulation of a W-Band FMCW Radar for Cloud Profiling Application 213
Akash Tyagi, Bijit Biswas, G. Arun Kumar, and Biplob Mondal

Single Layer Asymmetric Slit Circularly Polarized Patch Antenna for WLAN Application 219
Swetha Ravikanti and L. Anjaneyulu

Contour and Texture-Based Approaches for Dental Radiographic and Photographic Images in Forensic Identification 227
G. Jaffino, J. Prabin Jose, and M. Sundaram

An Efficient Low Latency Router Architecture for Mesh-Based NoC ... 241
Aruru Sai Kumar and T. V. K. Hanumantha Rao

High-Performance Image Visual Feature Detection and Matching 249
N. Sambamurthy and M. Kamaraju

Systolic-Architecture-Based Matrix Multiplications and Its Realization for Multi-Sensor Bias Estimation Algorithms 263
B. Gopala Swamy, U. Sripati Acharya, P. Srihari, and B. Pardhasaradhi

Fast Encoding Using X-Search Pattern and Coded Block Flag Fast Method 273
S. Karthik Sairam and P. Muralidhar

Energy Efficient and Accurate Hybrid CSA-CIA Adders 283
Gayathri Devi Tatiknoda, Kalyani Thummala, Aruna Kumari Neelam, and Sarada Musala

Machine Learning Oriented Dynamic Cost Factors-Based Routing in Communication Networks 293
Rohit Misra and Rahul Jashvantbhai Pandya

Design of Look-Up Table with Low Leakage Power Enhanced Stability 10 T SRAM Cell 301
R. Manoj Kumar and P. V. Sridevi

Conformal Omni Directional Antenna for GPS Applications 315
Shravya Krishna Komati and Damera Vakula

Total Internal Reflection Quasi-Phase-Matching-Based Broadband Second Harmonic Generation in Isotropic Semiconductor: A Comparative Analysis 323
Minakshi Deb Barma, Sumita Deb, and Satyajit Paul

Automated Bayesian Drowsiness Detection System Using Recurrent Convolutional Neural Networks 335
Mahima Jeslani and Naidu Kalpana

Enhancement of Bandwidth and VSWR of Double Notch E-Shaped Inset-Fed Patch Antenna 349
Pavada Santosh and Prudhvi Mallikarjuna Rao

Multiple Adjacent Bit Error Detection and Correction Codes for Reliable Memories: A Review 357
K. Neelima and C. Subhas

Optimisation of Cloud Seeding Criteria Using a Suite of Ground-Based Instruments 373
P. Sudhakar, K. Anitha Sheela, and M. Satyanarayana

Periodic Octagon Split Ring Slot Defected Ground Structure for MIMO Microstrip Antenna 379
F. B. Shiddanagouda, R. M. Vani, and P. V. Hunagund

Multipath Delay Cell-Based Coupled Oscillators for $\Sigma\Delta$ Time-to-Digital Converters 391
R. S. S. M. R. Krishna, Amrita Dikshit, Ashis Kumar Mal, and Rajat Mahapatra

Low-Hardware Digit-Serial Sequential Polynomial Basis Finite Field GF(2^m) Multiplier for Trinomials 401
Siva Ramakrishna Pillutla and Lakshmi Boppana

Author Index 411

About the Editors

Dr. T. Laxminidhi is currently working as a Professor at the National Institute of Technology, (Surathkal) Karnataka India. He has over 20 years of teaching and research experience in his specific areas of interests, which include analog and mixed signal IC design and signal processing. He has published several reputed international journal papers and international conference papers. He received his B.Tech degree from Mangalore University and ME degrees in Industrial Electronics from NIT Karnataka. He obtained his PhD from IIT Madras.

Dr. Jyoti Singhai is currently a Professor at the department of Electronics and Communication Engineering, Maulana Azad National Institute of Technology Bhopal. She obtained her B.E. (Electronics), M.Tech (Digital Communication) and PhD from MANIT Bhopal. Her major areas of research interests include Image and video processing, wireless communication and routing protocols. She has published 50 papers in respected international journals. She received the DST BOYSCAST Fellowship in 2010, MP Young Scientist award and AICTE Career Award to Young Teachers.

Dr. Sreehari Rao Patri is currently working as an Associate Professor at the National Institute of Technology in Warangal, India. He has over 20 years of teaching and research experience in his specific areas of interests, which include analog and mixed signal IC design, power management integrated circuit design and communications. He has published 19 international journal papers and 18 international conference papers. He received his B.Tech degree from Nagarjuna University and ME degrees in communication systems from IIT Roorkee. He obtained his PhD from NIT Warangal.

Dr. V. V. Mani is currently working as an Associate Professor at the National Institute of Technology in Warangal, India. She has over 15 years of teaching and research experience in her specific areas of interest, which include signal processing for communication and smart antenna design. She has published 20 international journal papers and 20 international conference papers. She received her BE and ME degrees in electronics and communications engineering from Andhra University in Vishakhapatnam and earned her PhD from the Indian Institute of Technology Delhi, India.

Gibbs-Shannon Entropy and Related Measures: Tsallis Entropy

G. Ramamurthy and T. Jagannadha Swamy

Abstract In this research paper, it is proved that an approximation to the Gibbs-Shannon entropy measure naturally leads to the Tsallis entropy for the real parameter q = 2. Several interesting measures based on the input as well as the output of a discrete memoryless channel are provided and some of the properties of those measures are discussed. It is expected that these results will be of utility in Information Theoretic research.

Keywords Entropy · Approximation · Memoryless channel · Entropy-like measures

1 Introduction

From the considerations of statistical physics, J. Willard Gibbs proposed an interesting entropy measure. Independently, motivated by the desire to capture the uncertainity associated with a random variable, C. E. Shannon proposed an entropy measure. It is later realized that Gibbs and Shannon entropy measures are very closely related. From the considerations of the statistical communication theory, defining mutual information (based on the definition of conditional entropy measure), Shannon successfully proved the channel coding Theorem (which defined the limits of reliable communication from a source to destination over a noisy channel) [1]. Thus, from the point of view of information theory, Shannon entropy measure became very important and useful.

Also, in recent years, Tsallis proposed an entropy measure generalizing the Boltzmann-Gibbs entropy measure. The authors became interested in the relationship between the Tsallis entropy and the Shannon entropy. Under some conditions,

G. Ramamurthy (✉)
Department of CSE, MECHYD, Hyderabad, India
e-mail: rama.murthy@mechyd.ac.in; rammurthy@iiit.ac.in

T. Jagannadha Swamy
Department of ECE, GRIET, Hyderabad, India
e-mail: tatajagan@gmail.com

T. Laxminidhi et al. (eds.), *Advances in Communications, Signal Processing, and VLSI*, Lecture Notes in Electrical Engineering 722,
https://doi.org/10.1007/978-981-33-4058-9_1

the authors proved that the Shannon entropy leads to the Tsallis entropy. As a natural generalization, the authors proposed interesting measures defined on probability mass functions and the channel matrix of a Discrete Memoryless Channel (DMC).

This research paper is organized as follows. In Sect. 2, an approximation to the Shannon entropy is discussed and the relationship to the Tsallis entropy is proved. In Sect. 3, interesting measures on probability mass functions are proposed. Finally, conclusions are reported in Sect. 4.

2 Approximation to Gibbs-Shannon Entropy Measure: Tsallis Entropy

It is well known that the Gibbs-Shannon entropy measure associated with a discrete random variable (specified by the probability mass function $\{\{p_i\}$ for $1 \leq i \leq M\}$ is given by

$$H(X) = -\sum_{i=1}^{M} p_i \log_2 p_i \tag{1}$$

Also, in recent years, Tsallis proposed another entropy measure which in the case of a discrete random variable is given by

$$S_q(p) = \frac{1}{q-1}\left(1 - \sum\nolimits_x p(x)\right)^q \tag{2}$$

where S denotes the entropy, p (.) is the probability mass function of interest and "q" is a real parameter. In the limit as q approaches 1, the normal Boltzmann—Gibbs entropy is recovered. The parameter "q" is a measure of the non-extensivity of the system of interest. The authors became interested in knowing whether there is any relationship between the Gibbs-Shannon entropy measure and the Tsallis entropy under some conditions. This question naturally led to a discovery which is summarized in the following Lemma.

Lemma 1 Consider a discrete random variable X with finite support for the probability mass function. Under reasonable assumptions, we have that

$$H(X) \approx \left(1 - \sum\nolimits_{i=1}^{M} p_i^2\right) \log_2^e = S_2(p) \log_2^e \tag{3}$$

Proof From the basic theory of infinite series [2], we have the following: for $|x| < 1$, we have that

$$\log_e(1-x) = -x + \frac{x^2}{2} - \frac{x^2}{3} + \cdots + (-1)^{n+1}\frac{(-x)^n}{n} + \cdots$$

Let $p_i = (1 - q_i)$ with $0 < p_i < 1$. Then we have $0 < q_i < 1$.
Thus, we have

$$\log_e(1-q_i) = -q_i + \frac{q_i^2}{2} - \frac{q_i^3}{3} + \cdots \tag{4}$$

Now let us consider the entropy H(X) of a discrete random variable, where X assumes finitely many values. We have that

$$\begin{aligned} H(x) &= -\sum\nolimits_{i=1}^{M} p_i \log_2^{p_i} \ bits \\ &= -\sum\nolimits_{i=1}^{M} (1-q_i)\log_2(1-q_i) \ bits \\ &= -\sum\nolimits_{i=1}^{M} (1-q_i)\log_e(1-q_i)\log_2 e \end{aligned} \tag{5}$$

Now using the above infinite series and neglecting the terms $\frac{q_i^2}{2}, \frac{q_i^3}{3}$, and so on, we have that

$$\begin{aligned} H(x) &\approx -\sum\nolimits_{i=1}^{M} (1-q_i)(-q_i)\log_2^e \\ &\approx \left(\sum\nolimits_{i=1}^{M} p_i\right)(1-p_i)\log_2^e \\ &\approx \left(1 - \sum\nolimits_{I=1}^{M} p_i^2\right)\log_2^e \end{aligned} \qquad \text{(Q.E.D.)}$$

Remark Thus, the square of the $L^2 - norm$ of the vector corresponding to the probability mass function (of a discrete random variable) is utilized to approximate the entropy of the discrete random variable. In summary, we have that

$$H(x) \approx f(p_1, p_2, \ldots p_M) = (1 - \sum\nolimits_{i-1}^{M} p_i^2)\log_2^e \tag{6}$$

Thus, an approximation to the Gibbs-Shannon entropy naturally leads to the scaled Tsallis entropy for the real parameter q = 2. The quantity H(X) with the above approximation is rounded-off to the nearest integer. For the continuous case, i.e. for probability density functions associated with continuous random variables, similar result can easily be derived and is avoided for brevity.

Note If the logarithm is taken to a different base, a scaling constant should be included.

We would like to study the properties satisfied by the function f (.,.,…,) approximating the entropy. The following claim can easily be proved.

Lemma 2 The maximum value of $f(p_1, p_2, \ldots, p_M)$ is attained when $\{p_i\}_{i=1}^{M}$ are all equal, i.e. $p_i = \frac{1}{M}$ for $1 \leq i \leq M$ Q.E.D.

Proof The proof follows by the application of the Lagrange Multipliers method. Detailed proof is avoided for brevity Q.E.D.

Thus, the maximum value of approximation to the entropy of a discrete random variable assuming "M" values is $\left(1 - \frac{1}{M}\right) \log_2^e$.

It is easy to see that this approximation to the Gibbs-Shannon entropy satisfies only two (out of four) axioms satisfied by the Shannon entropy functional.

Remark As in the proof of the above Lemma, it is possible to provide higher order approximations to the Gibbs-Shannon entropy measure. Also, in the spirit of Lemma 1, Renyi and other types of entropies can easily be approximated. The details are avoided for brevity.

3 Novel Measures on Probability Distributions

Shannon's entropy of a discrete random variable constitutes an important scalar-valued measure defined on the class of probability mass functions (of the discrete random variables). In contrast to the moments of discrete random variables, the entropy does not depend on the actual values assumed by the discrete random variable. Thus, one is naturally led to the definition of other measures associated with discrete random variables which depend only on the probability mass function (and not the values assumed by it).

3.1 L^q-Norm of Probability Vectors: Tsallis Entropy

- We first treat the probability mass function of a discrete random variable as a vector of probabilities. It should be kept in mind the M-dimensional probability vector (corresponding to "M" values assumed by the discrete random variable) lies on a hyperplane in the "positive orthant" (of the M-dimensional Euclidean space) only. Also, as a natural generalization, we can also conceptualize an "infinite-dimensional" probability vector corresponding to a discrete random variable which assumes infinitely many values.
- Consider a "probability vector" (corresponding to the associated probability mass function—finite or infinite dimensional) and define the L^q-norm of the vector (in the same manner as done in pure mathematics). Let

$$M_q(\overline{p}) = \left[\sum_{j=1}^{\infty} [p_x(j)]^q\right]^{\frac{1}{q}} \text{ for } q \geq 1 \tag{7}$$

As discussed in [3], some interesting properties are satisfied by M_q. Also, all the results associated with L^q-norm (in pure mathematics) such as the Holder and Minkowski inequalities can be readily invoked with the measure M_q.

It is elementary to see that such a measure can easily be related to the Tsallis entropy. Specifically, we have that

$$M_q^q = 1 - (q-1)S_q(p) \tag{8}$$

Now let us define the following function naturally associated with the L^q-norm, i.e. M_q.

$$g_q(\overline{p}) = 1 - M_q\overline{(p)}$$

It is easy to see that for any two real numbers q_1, q_2 such that $q_1 > q_2$, we have that

$$g_{q_1}(\overline{p}) > g_{q_2}(\overline{p})$$

In view of this, it is easy to reason that $\lim_{q\to\infty} g_q(\overline{p}) = 1$. In a similar spirit, it is possible to derive an inequality associated with $(q-1)S_q(p)$, i.e. scaled Tsallis entropy (for different values of the real parameter "q").

- Based on the properties of M_q, it is easy to see that the probability mass function-based infinite-dimensional probability vectors always belong to discrete Hilbert space.
- Let us first consider the case where the support of the probability mass function is finite. The L^2-norm of the associated probability vector is

$$M_2 = \left[\sum_{j=1}^{M} [px(j)]^2\right]^{\frac{1}{2}} \tag{9}$$

We reasoned in the previous section that such a measure naturally arises in approximating the Gibbs-Shannon entropy functional/measure (to a good degree of accuracy).

- Using a similar approach, the conditional entropy can be approximated. Also, using the approximation for H(X/Y)/H(Y/X), H(X)/H(Y), the mutual information between the input and output of a Discrete Memoryless Channel (DMC) can be approximated. The details are avoided for brevity.

3.2 Quadratic Forms Associated with Probability Mass Functions

- Clearly, the expression in (3) is an interesting measure defined over the Vector, $\overline{p_X}$ representing the probability mass function. Thus, one is naturally led to the definition of a quadratic form defined over the vector $\overline{p_X}$. Specifically, let us define quadratic forms associated with the channel matrix, Q (of a DMC),
  i.e. $\overline{p_X^T}Q\overline{p_X}$.
  Since $\overline{p_X^T}Q = \overline{p_Y}$, we readily have that

$$\overline{p_X^T}Q\overline{p_X} = \overline{p_Y^T}\overline{p_X} = \langle \overline{p_Y}, \overline{p_X} \rangle \tag{10}$$

Claim Thus, the quadratic form associated with the channel matrix of a DMC represents the inner producy between the probability vectors $\overline{p_Y}and\overline{p_X}$.

- It readily follows that in the case of a "noiseless channel", we have that Q = I and thus the quadratic form becomes the "square of the Euclidean length" (L^2-norm) of the probability vector. It is thus always positive.

- In view of the relationship of the Tsallis entropy to the Gibbs-Shannon entropy measure, we define the following measure associated with the stochastic matrix W and the probability vector $\overline{p_X}$, i.e.

$$\overline{S_2} = 1 - \overline{p_X^T}W\overline{p_X}.$$

If W is the channel matrix of a discrete memoryless channel, the above measure has an interesting interpretation (discussed previously). In this case,

$$\overline{S_2} = 1 - \overline{p_X^T}W\overline{p_X} = 1 - \overline{p_X^T}\overline{p_Y}.$$

It is easy to reason that this measure is non-negative. Also using Lemma 2, the above entropy-type measure can be bounded.

It is interesting to see its interpretation when W is the state transition matrix of a homogeneous Markov chain. In this case, the state of the dynamical system captured through an associated probability vector evolves through the associated Markov chain. (We can capture the idea of initial entropy, transient entropy and equilibrium entropy of the associated Markov chain modeling the physical phenomena.)

Furthermore, W could be a doubly stochastic matrix. It is immediate that when W happens to be an identity matrix, i.e. W = I, then the above measure is the Tsallis entropy for parameter q = 2 (i.e. an approximation to the Gibbs-Shannon entropy measure).

- Hence, we would like to study the properties of the quadratic form using the Inner product between two probability vectors (namely the input and output probability vectors of a DMC). In that effort, we would like to address the following question:

Q: How does the inner product of two probability vectors summarize the "similarity/dissimilarity" of probability mass functions?

In this effort, we invoke the Cauchy-Schwartz inequality associated with bounding the inner product between two vectors:

$$\left[\overline{p_Y^T}\,\overline{p_Y}\right]^2 \le \left[\sum_{i=1}^{M} p_X^2(i)\right]\left[\sum_{i=1}^{M} p_Y^2(i)\right] \tag{11}$$

- It is easily seen that the following holds true:

$$\sum_{i=1}^{M} p_X^2(i) = \begin{cases} \{1 \text{ if } \overline{p_X} \text{ is degenerate} \\ \{< 1 \text{ if } \overline{p_X} \text{ is non - degenerate} \end{cases} \tag{12}$$

- Furthermore, the minimum possible value of $\sum_{i=1}^{M} p_X^2(i)$ (i.e. value of 1/M) occurs when $p_X(i) = 1/M$ for all $1 \le i \le M$.
- Also, it should be noted that the inequality in (11) reduces to equality only when $p_X(i) = p_y(i)$ for all $1 \le i \le M$.

That is, the inner product between probability vectors $\overline{p_X}$ and $\overline{p_Y}$ attains the "maximum" value when they are both the same (equal).

- Suppose $\overline{p_X}$ is the invariant probability distribution (also called the steady-state probability distribution) of the homogeneous Discrete Time Markov Chain (DTMC) associated with the Channel matrix Q (a stochastic matrix). In this case, we have that

$$\overline{p_X^T} Q = \overline{p_X^T} \tag{13}$$

Then the quadratic form associated with $\overline{p_X}$ becomes

$$\overline{p_X^T} Q \overline{P_X} = \overline{p_X^T}\,\overline{p_X} > 0$$

Thus, the quadratic form attains the maximum value. Equivalently, we have that in this case, the value of the quadratic form is the same as that in the case of a noiseless channel.

- In the same spirit of the definition of mutual information, let us define the following scalar "measure" between the input and output of a Discrete Memoryless Channel (DMC).

$$\text{J(X; Y)} = \overline{p_X^T} Q \overline{p_Y} = \overline{p_Y^T} \overline{p_Y} \tag{14}$$

where $\overline{p_X}$ corresponds to the input probability vector (i.e. the input probability mass function) and $\overline{p_Y}$ corresponds to the output probability vector. Let us investigate some of the properties of the scalar measure J(X; Y).

(i) Since $\overline{p_X^T} Q = \overline{P_Y^T}$, we have that $\overline{p_Y^T} \overline{p_Y} > 0$.

Also, we have that J(X; X) $= = \overline{p_X^T} Q \overline{p_X} = \overline{p_X} \geq 0$.
That is, J(X; X) is zero when the probability vectors $\overline{p_X}, \overline{p_Y}$ are orthogonal vectors (as in the case of vector spaces).

(ii) J(Y; X) $= \overline{p_Y^T} Q \overline{p_X}$. Now substituting $\overline{p_Y^T} = \overline{p_X^T} Q$,
we have that J (Y; X) $= \overline{p_Y^T p_Y} =$ J (X; Y). Thus the scalar measure is symmetric.

Now we check whether the scalar-valued measure satisfies the triangular inequality. (The random variable X is the input to a discrete memoryless channel whose output is Y. Y is in turn the input to another discrete memoryless channel whose output is Z.)

$$J(X; Y) = \overline{p_X^T} Q \overline{p_Y} = \overline{p_Y^T} \overline{p_Y},$$

$$J(Y; Z) = \overline{p_Y^T} Q \overline{p_Z} = \overline{p_Z^T} \overline{p_Z},$$

Hence, we necessarily have that

$$J(X; Y) + J(Y; Z) = \overline{p_Y^T} \overline{p_Y} + \overline{p_Z^T} \overline{p_Z}$$

But by definition $J(X; Z) = \overline{p_X^T} Q \overline{P_Z} = \overline{P_Z^T} \overline{P_Z}$
Thus $J(X; Y) + J(Y; Z) \geq J(X; Z)$
Hence the triangular inequality is satisfied. Thus, the following Lemma is established.

Lemma 3 The scalar-valued measure

$$J(X; Y) = \overline{p_X^T} Q \overline{p_Y} = \overline{p_Y^T} \overline{p_Y} \tag{15}$$

between the probability vectors (corresponding to the probability mass functions of the random variables X, Y) is a "Pseudo-Metric" on the space of probability vectors (where the random variable Y is the output of a Discrete Memoryless Channel whose input is X).

In the spirit of the definition of the Tsallis entropy, we can define an Interesting entropy-type measure 1- J(X; Y). It is precisely the Tsallis entropy of the probability mass function of the output of a discrete memoryless channel.

Remark It is well known that higher degree forms (multivariate polynomials) are captured through Tensors. Thus using tensors, new measures can be defined generalizing the above ideas. The details are not provided for brevity.

3.3 *Now We Summarize the Results Discussed so Far in the Following*

$J(X) = \left[\sum_{j=1}^{\infty} [p_X(j)]^2\right]^{\frac{1}{2}}$ is like the "Euclidean Length" of a probability vector.

In Lemma 1, it was shown that $1 - [J(X)]^2$ approximates the Gibbs-Shannon entropy of the random variable X.

- J(X; Y) is a scalar-valued measure on the input X and output Y of a discrete memoryless channel.

4 Conclusions

In this research paper, the relationship between the Gibbs-Shannon entropy measure and the Tsallis entropy (for q = 2) is demonstrated. Based on this result, various interesting measures associated with probability mass functions (defined at the input and output of a Discrete Memoryless Channel) are defined. It is expected that these results will be of utility in Information Theoretic Investigations.

References

1. Ash RB, Information theory. Dover Publications, Inc, New York
2. Knopp K, Theory and applications of infinite series. Dover Publications, Inc, New York
3. Rama Murthy G, Weakly short memory stochastic processes: signal processing perspectives. In: Proceedings of international conference on frontiers of interface between statistics and sciences, December 20, 2009 to January 02, 2010

Recognition of Natural and Computer-Generated Images Using Convolutional Neural Network

K. Rajasekhar and Gopisetti Indra Sai Kumar

Abstract Recognizing Natural Images (NI) and Computer-Generated Images (CGI) by a human is difficult due to the use of new-age computer graphics tools for designing more photorealistic CGI images. Identifying whether an image was captured naturally or if it is a computer generated image is a fundamental research problem. For this problem, we design and implement a new Convolutional Neural Network (ConvNet) architecture along with data augmentation techniques. Experimental results show that our method outperforms existing methods by 2.09 percentage for recognizing NI and CGI images.

Keywords Convolutional neural network · Computer-generated image · Natural image · MATLAB · Image processing · Image forensics

1 Introduction

Identification of Natural Images (NI) and Computer-Generated Images (CGI) has become an important research problem. Computer graphics now a days has evolved into the same photorealism as natural images due to various graphics designing tools. However, recognizing the differences between the images is difficult. The methods that exist now are based on the statistical and intrinsic properties of images, i.e. to design a category-distinctive feature with a proper threshold to separate the

K. Rajasekhar
Department of Electronics and Communication Engineering, University College of Engineering Kakinada (A), Jawaharlal Nehru Technological University Kakinada, Kakinada, Andhra Pradesh, India
e-mail: rajakarumuri87@gmail.com

G. I. S. Kumar (✉)
Computers and Communications (M.Tech), Department of Electronics and Communication Engineering, University College of Engineering Kakinada(A), Jawaharlal Nehru Technological University Kakinada, Kakinada, Andhra Pradesh, India
e-mail: gopisettiindrasaikumar@gmail.com

T. Laxminidhi et al. (eds.), *Advances in Communications, Signal Processing, and VLSI*,
Lecture Notes in Electrical Engineering 722,
https://doi.org/10.1007/978-981-33-4058-9_2

Fig. 1 Computer-generated image

Fig. 2 Naturally captured image

two classes. This method performs well for simple datasets and poorly on complex datasets, for example, Columbia dataset, comprising images of different origins.

Today's methods based on Convolutional Neural Network (ConvNet) have gained more popularity in analyzing visual imagery and the reason for this is to learn automatically multiple levels of representation for a given task in an "end-to-end" manner [1]. This makes ConvNet more suitable for complex datasets in image recognition tasks. Inspired by the recent success of ConvNet, we and several other researchers chose this approach for CGI and NI recognition tasks. Our method gives good performance for the complex dataset of Google Image Search (Google) images versus Photo-Realistic Computer Graphics (PRCG) images from different origins and close to real-world applications.

Image forensics is an active field where images are analyzed. But in recent years, computers are able to generate photorealistic images which have become more challenging for forensics to determine whether the image is real or fake. In multimedia security, a number of approaches use ConvNet for steganalysis and image forensics. From Figs. 1 and 2, we get an impression that the first image is naturally generated and the second image is computer generated which is in contrast to reality.

2 Related Work

ConvNet for Recognition of Computer Generated Image and Natural Images. For the recognition of CGI and NI using ConvNet, there are two different works where one is presented in IEEE WIFS by Rahmouni et al. [2]. In this work, they followed a three-step procedure: filtering, statistical feature extraction, and classification. They considered a relatively simple dataset (Raise versus Level-Design) with homogeneous NIs and CGIs from the green channel. The second work is done by Quan et al. by using deeper and particularly designed cascaded convolutional layer with the Columbia dataset mainly Google versus PRCG and using all three channels of an image, i.e. red, blue, and green [3]. Qian et al. proposed a model on ConvNet for steganalysis and reported promising results [4]. Later, Pibre et al. studied the "shape" of ConvNet and identified the best ConvNet model after numerous experiments [5].

3 Dataset Used for Implementation

We conducted our experiment on Columbia photographic images and photorealistic computer graphics dataset [6, 7] considering

- 800 PRCG images from 40 3D graphic websites (prcg_images)
- 800 natural images from Google image search. (google_images)
- 800 natural images from authors' personal collection (personal images)

We considered all three channels of an image, i.e. Red, Blue, and Green. These images are collected from different sites and are of different origins.

4 Framework Proposed

Recognition of NIs and CGI images can be treated as a binary classification problem because we give an image as input and obtain a binary label as output.

Before giving input to the network, we augment the data using imageDataAugmenter so that the data will be according to the input size of the network and also it provides preprocessing image augmentation options like resizing, rotation, reflection, etc. for the data. These preprocessed images are fed to the ConvNet network architecture which is arranged as shown in Fig. 3 so a trained model (.mat file) of required accuracy is obtained after training the network with the data. The model generated will be used to recognize the NI and CGI images. Here, we split the data into two sets: one is for training and the other is for validation; this is mainly for the purpose of obtaining the generalized model for recognition.

4.1 Network Architecture

Our network architecture consists of four convolution layers, three fully connected layers, one SoftMax layer, and a classification layer. An image is taken as an input using the image input layer. The input size of the image for the network is (233 $\times$ 233 $\times$ 3) Normalization as zerocenter. In our network, a convolutional 2D layer

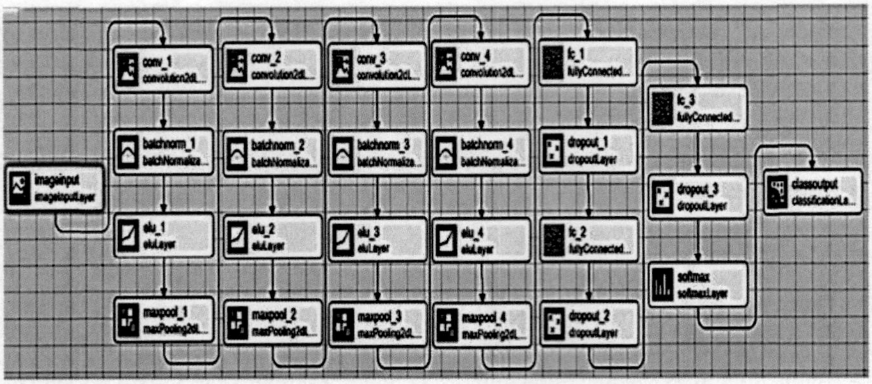

Fig. 3 Network architecture

(Conv_1), batch normalization layer (batchnorm_1), ELU layer (elu_1), and max-pooling 2D layer (maxpool_1) are treated as a single layer as there are 4 layers [8, 9]. All the max-pooling 2D layers have the same pool size of 3 × 3 and a stride of 2,2, and zero padding. The fully connected layers are followed by the Dropout layer each [10]. The dropout probability value is kept default. The SoftMax layer gives a probability vector of class labels. Therefore, the dimension of its output is equal to the number of classes, and the sum of its output is 1. Finally, classification layer is used to produce the output label. Our network layer description is shown in Fig. 4.

4.2 *Optimizer*

The stochastic gradient descent with momentum (SGDM) is the optimizer used in the network for training [11]. The training options are chosen as follows: minibatch size of 36, initial learn rate as 10^{-9}, validation frequency as 9, and the number of epochs to train is based on the accuracy that is needed to get. We actually trained for 300 epochs in order to obtain the required accuracy for the model. We also shuffled the data at every epoch so that the model will be able to generalize the data outside the datasets used. We can also choose the learn rate schedule to be piecewise and the drop factor and the drop period to be any random values. But at the initial stage, we choose the learn rate schedule to be constant and later we manually decreased the learn rate.

4.3 *Training*

We actually split the data into two sets with 75% of data used to train the network and the remaining 25% data used to validate the trained model. Minibatch size is made

```
25x1 Layer array with layers:

 1   'imageinput'    Image Input              233x233x3 images with 'zerocenter' normalization
 2   'conv_1'        Convolution              32 7x7x3 convolutions with stride [1  1] and padding [0  0  0  0]
 3   'batchnorm_1'   Batch Normalization      Batch normalization with 32 channels
 4   'elu_1'         ELU                      ELU with Alpha 1
 5   'maxpool_1'     Max Pooling              3x3 max pooling with stride [2  2] and padding [0  0  0  0]
 6   'conv_2'        Convolution              64 7x7x32 convolutions with stride [1  1] and padding 'same'
 7   'batchnorm_2'   Batch Normalization      Batch normalization with 64 channels
 8   'elu_2'         ELU                      ELU with Alpha 1
 9   'maxpool_2'     Max Pooling              3x3 max pooling with stride [2  2] and padding [0  0  0  0]
10   'conv_3'        Convolution              48 5x5x64 convolutions with stride [1  1] and padding 'same'
11   'batchnorm_3'   Batch Normalization      Batch normalization with 48 channels
12   'elu_3'         ELU                      ELU with Alpha 1
13   'maxpool_3'     Max Pooling              3x3 max pooling with stride [2  2] and padding [0  0  0  0]
14   'conv_4'        Convolution              64 3x3x48 convolutions with stride [1  1] and padding 'same'
15   'batchnorm_4'   Batch Normalization      Batch normalization with 64 channels
16   'elu_4'         ELU                      ELU with Alpha 1
17   'maxpool_4'     Max Pooling              3x3 max pooling with stride [2  2] and padding [0  0  0  0]
18   'fc_1'          Fully Connected          4096 fully connected layer
19   'dropout_1'     Dropout                  50% dropout
20   'fc_2'          Fully Connected          4096 fully connected layer
21   'dropout_2'     Dropout                  50% dropout
22   'fc_3'          Fully Connected          2 fully connected layer
23   'dropout_3'     Dropout                  50% dropout
24   'softmax'       Softmax                  softmax
25   'classoutput'   Classification Output    crossentropyex with classes 'google_images' and 'prcg_images'
```

Fig. 4 Layer description

to be 36 and the number of epochs is chosen to be 200. So the network is trained for 6600 iterations at a constant learn rate of 0.001 got by a validation accuracy of 79 percent for this run as shown in Fig. 5. Though we can choose the learn rate to be piecewise, we actually choosen learn rate manually after obtaining the model with 79 accuracy and varied the learn rate within a range from 10^{-3} to 10^{-6} for 100 epochs so that our model achieved good accuracy. Here, the validation frequency is maintained at 8; this is mainly used to calculate the validation accuracy and loss of the model to generalize the model to other data. For every epoch, it took around 5 min and for every iteration it took 0.135 min; here in MATLAB, there is an option to do it parallely so it uses cores of the CPU to work simultaneously.

One point to be noted here is we split the data according to the label of the folder name, i.e. folder containing natural images are named google_images and the other folder is named as prcg_images. So the classification layer gets the names for their binary labels from these folder names.

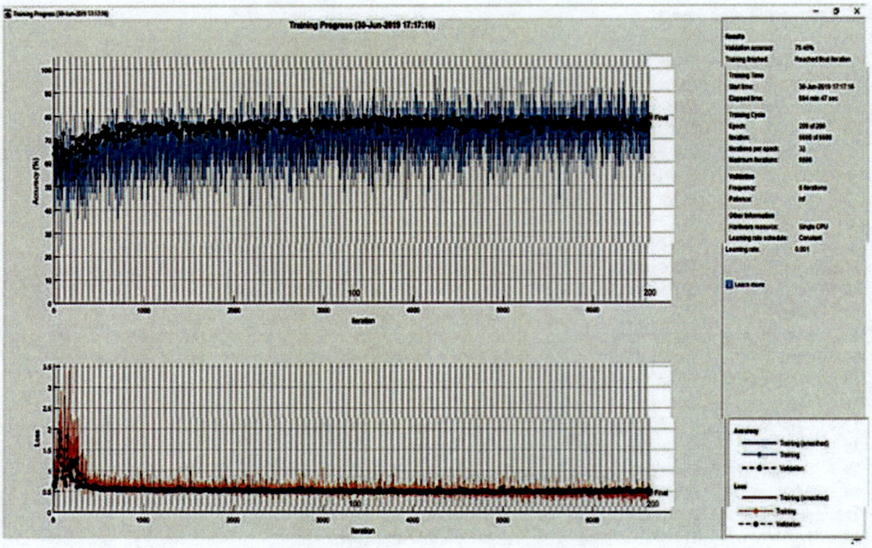

Fig. 5 Training plot for accuracy versus ıteration and loss versus ıteration

5 Results

5.1 Implementation Details

MATLAB R2019a. All of the experiment is done in the MATLAB R2019a Deep Learning package [10, 11] using Deep Learning Toolbox and the Deep Network Designer application. We have designed our network architecture using the Deep Network Designer app and generated code for the network architecture. We used image augmentation techniques to the data before applying them to the network. We split the data into training and validation in a 3:1 ratio.

In Training options, we choose sgdm optimizer and related parameters for this experiment and trained the network. After training, in order to improve the accuracy of the model, we save the checkpoints and retrain the network. Figure 6 shows the classified images by the trained model, and Fig. 7 gives the probabilities for the classification of the images as prcg_image and google_image.

Fig. 6 Classified ımages

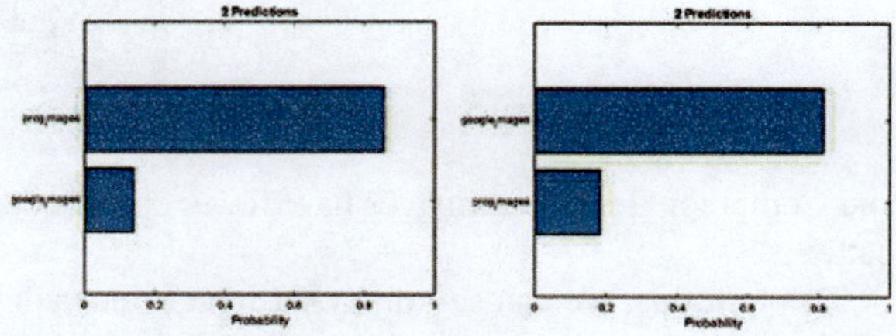

Fig. 7 Classification probabilities

5.2 *Visualization of ConvNet Layers*

After the model with the required accuracy is obtained, we visualize the network layers for the data it gets trained in. For this purpose, in MATLAB the Deep-DreamImage option will be available. Figure 8 gives details about the patterns that are grasped by the hidden layers in the ConvNet.

Figure 9 shows the table for activation strength and the classification layer through this layer visualization; we can say that Natural Images have more complex patterns

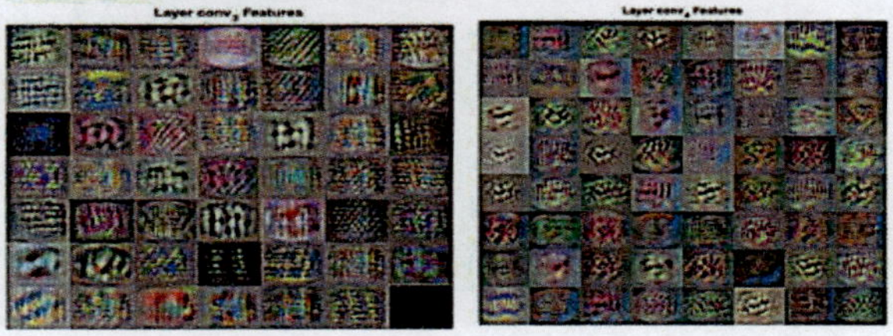

Fig. 8 Visualization of hidden ConvNet layers

Fig. 9 Classification layer activation strength and visualization

Iteration	Activation Strength	Pyramid Level
1	0.58	1
2	5.11	1
3	7.53	1
4	10.16	1
5	12.43	1
6	14.05	1
7	14.58	1
8	15.90	1
9	16.54	1
10	17.90	1

Layer fc_3 Features

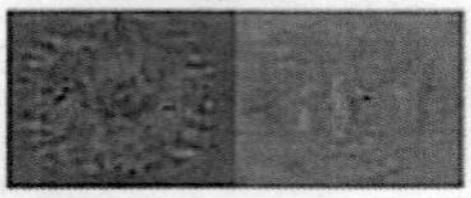

and Computer Generated Images have fewer complex patterns and are mostly simple patterns.

Through this, we can say that CGI images nowadays are getting more photorealistic so the pattern complexity in CGI is increasing. Maybe in the future, their differentiation will be more difficult compared to now.

Visualization of Activations for Test Images. For the testing purpose, we give some Natural and Computer Generated images and visualize the activations of the layers. Figure 10 shows the activations of the layers for the NI and CGI images. Figure 11 shows the strong activation of the layer to recognize the image. These are actually the normalized images which are gray; otherwise, they are black and white in color. After that, we apply this model to the video to detect the frames in the video whether

Fig. 10 Activations of ConvNet layer 1 by CGI and NI

Fig. 11 Strong activations of ConvNet layer 1

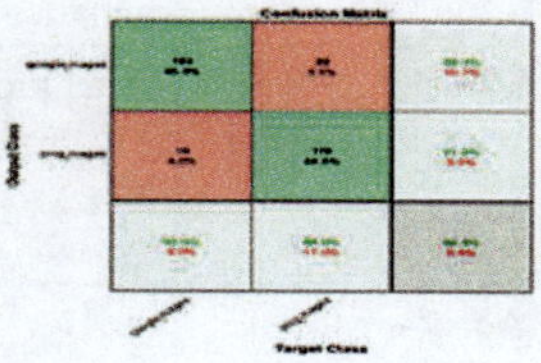

Fig. 12 Confusion Matrix google versus prcg datasets

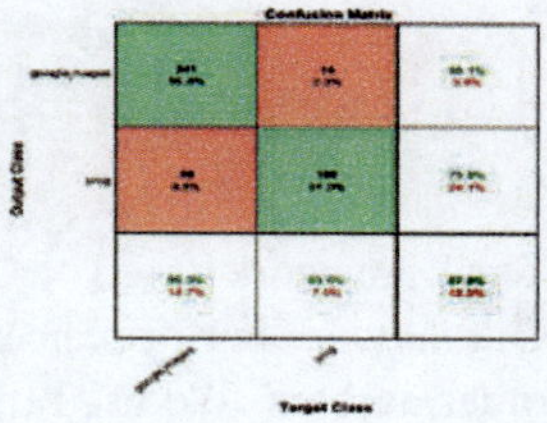

Fig. 13 Confusion Matrix google + personal versus prcg datasets

the image frame is CGI or NI. The final classification layer will be used in a loop to detect the video frames whether the frame is CGI or NI.

Confusion Matrix. Figures 12 and 13 give the details about the accuracy of the model, the diagonal values give the details about the correctly recognized images in the validation set, i.e. the images belong to the same class and are predicted to be of the same class. The off-diagonal values give the details about the wrongly predicted images. Recall and Precision are shown along the first column and first row of the matrix. False positive rate and false omission rate are shown in the second column and second row.

5.3 Comparison

We have not followed the method based on local patches and majority voting to distinguish the whole image to be either NI or CGI as Weize Quan. We only choose the entire image as one and resized it according to the input of the network for the recognition purpose so that the computation cost is reduced. The image cannot be entirely CGI nowadays in a frame; we can see some part to be CGI and the remaining part to be either humans or natural objects. So, through a local patch it is not feasible to decide on the entire image. We actually used simple data augmentation techniques in order to increase the training data and to improve the accuracy. Table 1 shows the accuracies of the Weize Quan model and our model for google versus prcg datasets.

Table 1 Accuracy comparison

Model	Quan model for google versus prcg	Trained Model for google versus prcg	Quan model for google + personal versus prcg	Trained model for google + personal versus prcg
Accuracy for full-sized images in percentage	88.41	90.5	86.43	87.8

6 Conclusion

The framework based on ConvNet to recognize CGI and natural images is proposed. Our network performance is better than the other parallel ConvNets proposed by the other authors. We used the latest MATLAB tools for visualization of the network and improved the accuracy of the model for recognition of Computer Generated Images and Natural Images. For data augmentation, MATLAB in built functions is used. We applied our model to the videos so that the image frames in the video can be recognized whether they are naturally captured or computer-generated and satisfactory results are obtained while CGI recognition.

References

1. Krizhevsky A, Sutskever I, Hinton GE (2012) ImageNet classification with deep convolutional neural networks. Adv Neural Inf Process Syst
2. Rahmouni N, Nozick V, Yamagishi J, Echizen I (2017) Distinguishing computer graphics from natural images using convolution neural networks. In: Proceedings of IEEE ınternational workshop ınformation forensics security, Dec 2017, pp 1–6
3. Quan W, Wang K, Yan D-M, Zhang X (2018) Distinguishing between natural and computer-generated ımages using convolutional neural networks. IEEE Trans Inf Forens Secur 13(11), Nov 2018
4. Qian Y, Dong J, Wang W, Tan T (2015) Deep learning for steganalysis via convolutional neural networks. Proc SPIE 9409: 94090J, Mar 2015
5. Pibre L, Pasquet J, Ienco D, Chaumont M (2016) Deep learning is a good steganalysis tool when embedding key is reused for different images, even if there is a cover source-mismatch. Proc IST Electron Imag, 1–23
6. Ng T-T, Chang S-F, Hsu J, Pepeljugoski M (2004) Columbia photographic images and photo-realistic computer graphics dataset, Columbia University, New York, NY, USA, Tech. Rep. #205-2004-5
7. Ng T-T, Chang S-F (2013) Discrimination of computer synthesized or recaptured images from real images. In: Sencar HT, Memon N (eds) Digital image forensics. Springer, New York, NY, USA, pp 275–309
8. Nagi J, Ducatelle F, Di Caro GA, Ciresan D, Meier U, Giusti FN, Schmidhuber J, Gambardella LM (2011) Max-Pooling Convolutional Neural Networks for Vision-based Hand Gesture Recognition. In: IEEE ınternational conference on signal and ımage processing applications (ICSIPA2011)
9. Ioffe S, Szegedy C (2015) Batch normalization: accelerating deep network training by reducing internal covariate shift. preprint, arXiv:1502.03167

10. Srivastava N, Hinton G, Krizhevsky A, Sutskever I, Salakhutdinov R (2014) Dropout: a simple way to prevent neural networks from overfitting. J Machine Learn Res 15:1929–1958
11. LeCun Y, Bottou L, Bengio Y, Haffner P (1998) Gradient-based learning applied to document recognition. Proc IEEE 86:2278–2324

Design of a Compact 2 × 2 MIMO Diversity Antenna for X-band Applications

Tathababu Addepalli and V. R. Anitha

Abstract A compact 2 × 2 MIMO diversity antenna of size 32 × 32 × 1.6 mm^3 is presented in this paper for portable devices. The proposed structure uses polarization and pattern diversities for isolation enhancement. The orthogonal arrangement of radiating elements, pattern diversity, and defected ground structure is achieved isolation. The whole configuration is designed and fabricated over the FR4 substrate with ε_r 4.4, and fed by a 50Ω microstrip line. The proposed structure operates from 7.9 GHz to 13 GHz, and it resonates at 9 GHz, 12.3 GHz. The isolation between orthogonal elements is above 20 dB and the anti-parallel element is above 15 dB. The radiation efficiency and peak gain values are above 86% and 3.1–5.2 dB, respectively. The diversity metric value of ECC is less than 0.07, DG is above 9.985 dB, and TARC is below −10 dB for the entire desired band. Simulated and measured values are in good concord.

Keywords Multiple-input multiple-output (MIMO) · Isolation · Envelope correlation coefficient (ECC) · Diversity gain (DG) · Total active reflection coefficient (TARC) · X-band applications

1 Introduction

Nowadays, wireless communication technologies are increasing rapidly because of various applications like cordless telephones, satellite television, GPS, WIFI, WLAN, WiMAX, and wireless computer parts, but they require high data rate transmission and reliability. Using the MIMO technology can achieve the above requirements with multiple radio channels placed on the transmitting end and the receiving end. But it

T. Addepalli (✉)
Department of ECE, JNTUA, Ananthapur, Andra Pradesh, India
e-mail: babu.478@gmail.com

V. R. Anitha
Antenna Research Laboratory, Sree Vidyanikethan Engineering College, Tirupathi, Andra Pradesh, India
e-mail: anithavr@ieee.org

T. Laxminidhi et al. (eds.), *Advances in Communications, Signal Processing, and VLSI*, Lecture Notes in Electrical Engineering 722,
https://doi.org/10.1007/978-981-33-4058-9_3

does not require additional bandwidth and input power. In portable devices, antennas are placed very closely. Due to the limited space between the antennas, the radiation from individual antennas collides with each other. Hence, the receiving end can't receive the original information due to coupling. This also leads to degrading system performance and channel capacity. So the coupling between antennas becomes a major problem in the MIMO technology.

Coupling arises due to the collision of radiation from antennas in free space, the current flowing on the metals, and surface waves in the dielectric substrate [1]. The effect of coupling between antennas in the same substrate can't be minimized easily due to the compactness. It degrades system performance and SNR value [2]. Different techniques are there for the reduction of mutual coupling like pattern diversity and polarization diversity, by changing the orientation of the antenna and placing decoupling networks between the antennas. The polarization diversity is achieved through the orthogonal arrangement of antenna elements with different polarization, where one is horizontal and the other one is a vertically polarized antenna. The following literature describes the four antenna elements that are arranged in rows and columns in isolation enhancement, and the size of the MIMO structure is also specified.

The arrangement of four radiating elements on a common substrate with low mutual coupling is a challenging issue for the MIMO design. The polarization diversity of a MIMO antenna with size 270×210 mm^2 and the isolation improvement is 20 dB [3]. Antenna elements are arranged in an orthogonal manner with isolation; they even have the same radiation patterns [4]. The size of the conical-shaped array antenna was 200×200 mm^2 with an isolation of 20 dB [5]. A new wideband slot array with polarization diversity (size = 114×114 mm^2) results in an isolation improvement of higher than 25 dB [6]. The substrate of size 100×120 mm^2 consists of four dipole antennas that are placed in an orthogonal manner for isolation enhancement and also to achieve multi-beam radiation patterns [7]. An inset-fed four-element rectangular-shaped MIMO antenna of size 100×50 mm^2 is designed for ISM band applications, and the isolation of 10 dB is achieved using CSRR structure [8]. The substrate of size 95×60 mm^2 consisting of four radiating elements is placed in its corners for isolation enhancement (S21 < 11.5 dB) [9]. Similarly, four-slot antennas are arranged on the substrate of size 70.11×70.11 mm^2 in an orthogonal manner for isolation [10]. A G-shaped MIMO antenna with an isolation of 21 dB using polarization diversity of size 70×70 mm^2 [11]. The G-shaped antenna elements are arranged in a parallel manner in the MIMO structure of size 55×50 mm^2 is designed for wireless portable applications and the L-shaped stubs with inverted T-shaped slot are used for the isolation enhancement [12]. Four inverted L-shaped antennas with a rectangular-shaped CSRR structure for isolation enhancement better than 17 dB are used and also low ECC values are achieved [13]. A diamond-shaped patch with four orthogonal feeds of structure size 120×140 mm^2 is used and the isolation achieved is better than 15 dB with a ground plane cutting in proper place [14].

Most of the work in the literature is about four-element antennas, which are arranged in an orthogonal manner for isolation and are also larger in size and so it is difficult to design the structures. But, the proposed structure is of compact nature and easy to design. Four antenna elements are arranged in an orthogonal manner for better isolation of structure size 32 × 32 mm^2 and fabricated on FR4 dielectric material with a dielectric constant 4.4. It covers the entire X-band applications, i.e., 7.9–13 GHz. Orthogonal elements in the structure give an isolation of more than 20 dB and the anti-parallel elements give an isolation of above 15 dB.

2 Antenna Design Procedure

The proposed structure, a compact 2 × 2 MIMO antenna, is shown in Fig. 1a–d. The overall structure size is 32 × 32 × 1.6 mm^3. It is simulated and fabricated on a low-cost FR4 substrate material of height 1.6 mm with dielectric constant (€r) 4.4. The shape of the top layer of the proposed structure is three half circles cutting (left, right, and top) the circular patch antenna with inside feed, and the shape of the bottom layer of the proposed antenna has defected ground structure with small

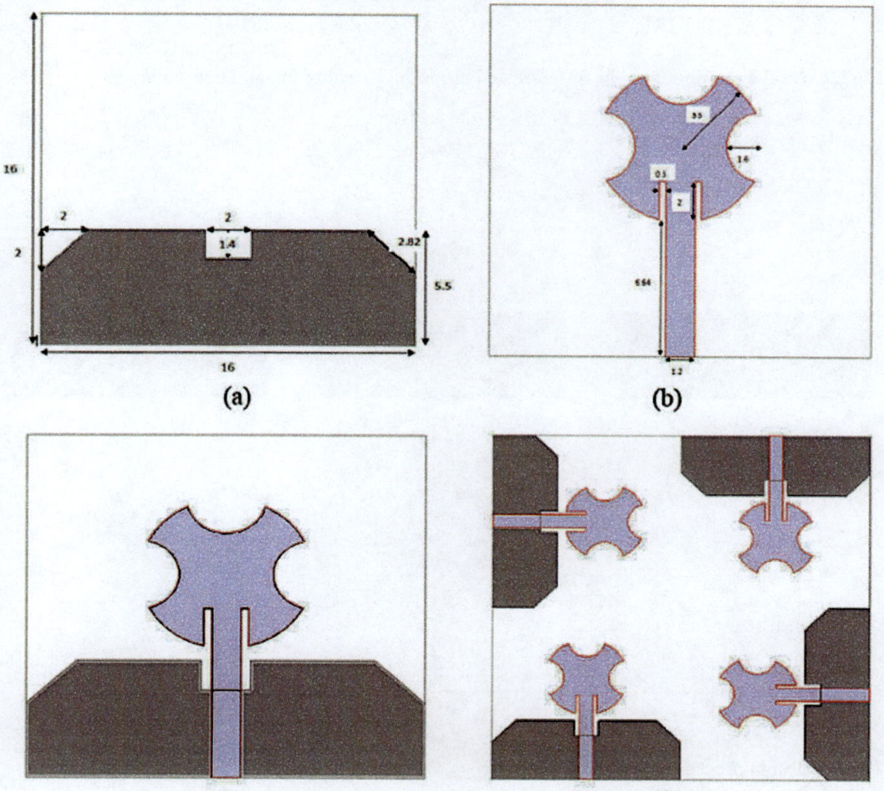

Fig. 1 **a** and **b** Geometry of bottom and top layers; **c** Proposed single-element antenna; **d** Proposed 2 × 2 MIMO structure

rectangular slit and edge cuttings. The proposed structure bottom layer is formed by a small triangular-shaped patch is removed at both edges of the partial ground plane, and a small rectangular-shaped patch removed in the centre of the partial ground plane. The photographs of the fabricated proposed structure top and bottom layers are shown in Fig. 2a, b. The design and evolution procedure of the proposed structure is shown in Fig. 3a–f.

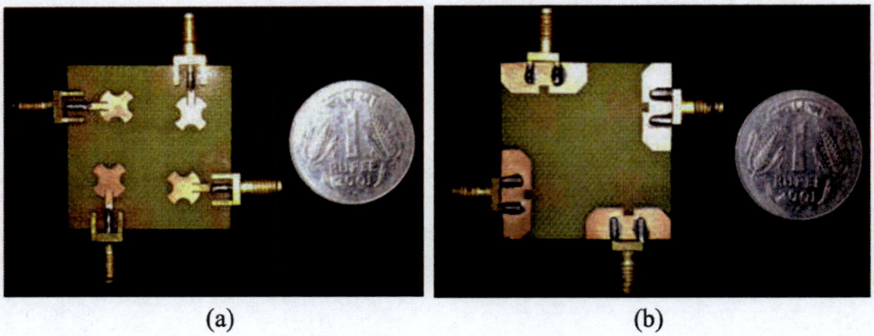

Fig. 2 **a** and **b** are photographs of fabricated proposed structure top and bottom layers

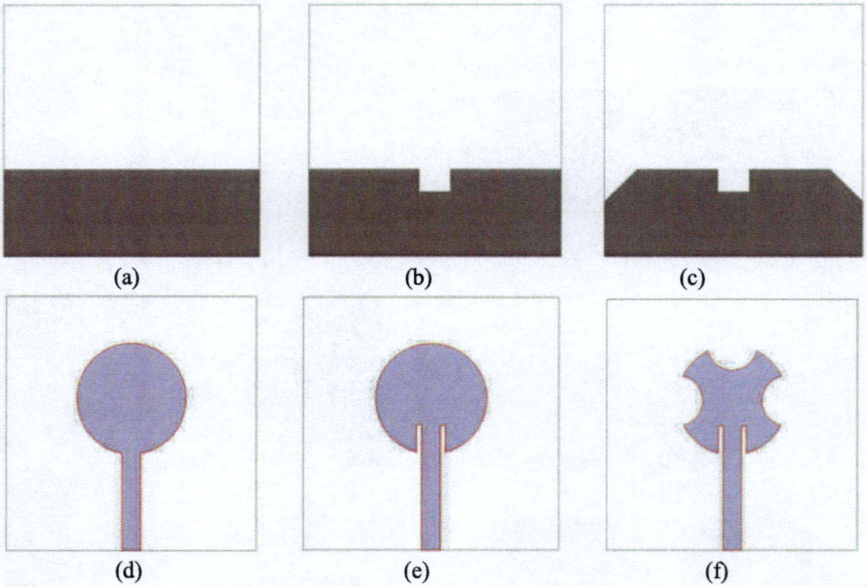

Fig. 3 Design process of proposed structure

3 Simulation and Measurement Results

Figure 4 shows the design procedure of the proposed structure. For getting the desired results, the optimization process is also done. The simulated results of three antenna configurations with the proposed 2 × 2 MIMO structure are shown in Fig. 5a. From the figure, it is observed that the variation in the bandwidth of the proposed single element and 2 × 2 MIMO structure is slightly varying. It is due to the coupling between the orthogonal elements and anti-parallel elements. However, the proposed 2 × 2 structure covers the complete X-band application and also achieves better isolation.

S-parameters are the most important parameters for any antenna design. The reflection coefficient (S11) is "the ratio of the reflected to the incident power", but return loss is the dB's representation of the reflection coefficient. The transmission coefficient (S21) describes the total transmitted power relative to the incident power. The acceptable value of the transmission coefficient of theMIMO design is minimum −15 dB. Figure 5a, b shows the S-parameter comparison of antenna configuration and comparison of simulated and measured S-parameter values when port 1 is excited, respectively. The simulation and measurement values of the proposed structure are slightly varied. In simulation results, the lower band starts at 7.9 GHz but in measured results, the lower band starts at 8.3 GHz; this is due to tolerances at the inside feed cutting and soldering problems. Figure 5c, d shows the photographs of S11- and S21-parametrs of Anritsu MS2073C VNA (vector network analyzer) master.

Figure 6a–d illustrates the comparison of isolation when individual ports are excited. For adjacent radiating elements, the isolation is above 20 dB and the isolation is above 15 dB for diagonal radiating elements. The reason is adjacent elements are orthogonal to each other and due to this, one is a horizontally polarized antenna and the other one is a vertically polarized antenna. Diagonal elements are in the same polarization, but are in the opposite direction. Hence, the orthogonal elements have more isolation compared to the diagonal elements.

Figure 7 shows the peak gain and radiation efficiency of the proposed 2 × 2

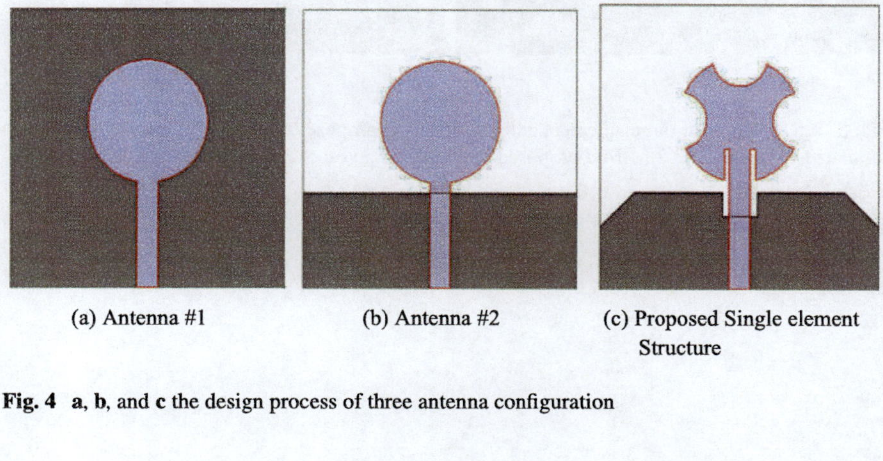

(a) Antenna #1 (b) Antenna #2 (c) Proposed Single element Structure

Fig. 4 **a**, **b**, and **c** the design process of three antenna configuration

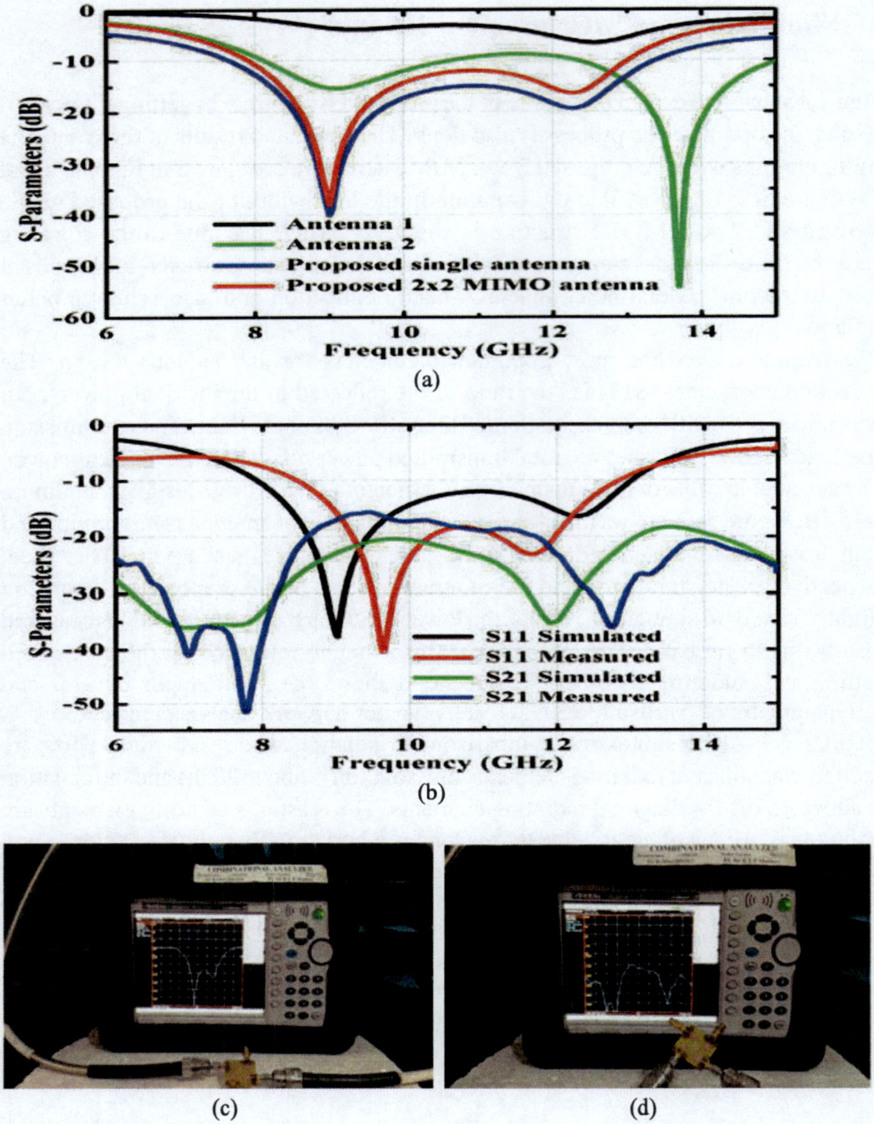

Fig. 5 **a** S11 values of three antenna configuration with proposed 2 × 2 MIMO; **b** measured and simulated values of 2 × 2 MIMO proposed structures; **c** and **d** are photographs of measured S11 and S21 values

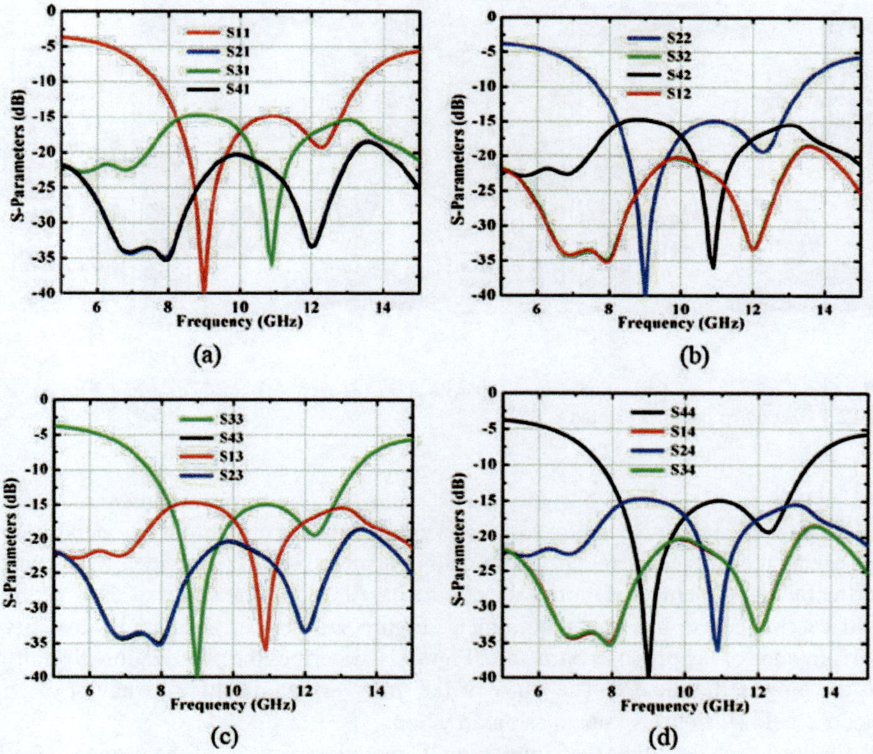

Fig. 6 **a** Simulated S-parameters of proposed structure when port 1 is excited; **b** Simulated S-parameters of proposed structure when port 2 is excited; **c** Simulated S-parameters of proposed structure when port 3 is excited, and **d** Simulated S-parameters of proposed structure when port 4 is excited

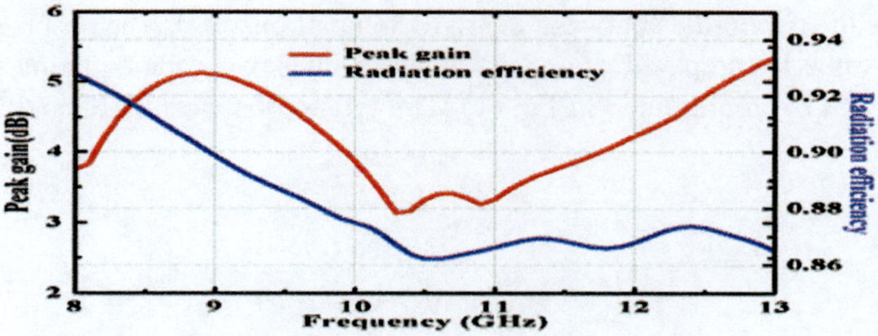

Fig. 7 Simulated peak gain and radiation efficiency results

MIMO structure having a peak gain of 5.09 dB at 9 GHz and 4.50 dB at 12.3 GHz and having more than 86% for the entire band, respectively.

Figure 8a, b shows the surface current distribution of the proposed 2 × 2 MIMO structure at 9 GHz and 12.3 GHz resonant frequencies. It is one of the observation

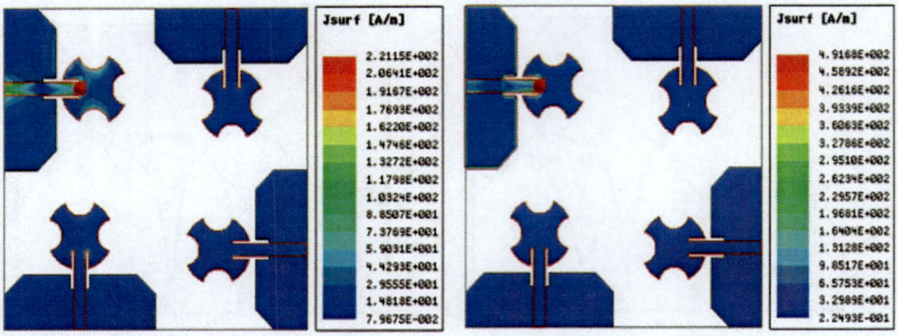

(a) (b)

Fig. 8 **a** Surface current distribution at 9 GHz when port 1 is excited; **b** Surface current distribution at 12.3 GHz when port 1 is excited

parameters in the MIMO design for observing the effect of one antenna on the other antenna and is also useful for observing the flow of current on the metals. By observing the surface current distribution on the metal, it will be decided which part of the radiating element is more excited, and based on that the shape of the antenna will be changed to the user requirement. Figure 9 describes the pattern diversity performance of the proposed structure. Figure 10 describes the polarization diversity of the proposed structure. The study of the time-varying electrical behavior of an electric field (E-field) is known as polarization.

Figure 11a, b describes the simulation 2D radiation patterns of the proposed 2 × 2 MIMO structure when port 1 is excited and the remaining three ports are with 50Ω termination loads.

MIMO Performance

ECC is useful to estimate the diversity performance of the MIMO antenna and also determines the similarities between the radiation patterns [15]. The proposed structure ECC value is computed from S-parameter values using the equation given by Blench et al. [16] represented in Eq. (1). The relation between ECC and DG is in Eq. (2).

$$ECC = \frac{\left| S_{11}^{*}\ S_{12} + S_{21}^{*}\ S_{22} \right|^{2}}{\left(1 - |S_{11}|^{2} - |S_{21}|^{2}\right)\left(1 - |S_{22}|^{2} - |S_{12}|^{2}\right)} \tag{1}$$

$$DG = 10\sqrt{1 - ECC^{2}} \tag{2}$$

The low value of ECC defines that there is less overlapping between the radiating elements. The acceptable value of the MIMO design is below 0.5. Figure 12a describes the simulation values of ECC and DG, which are lower than 0.07 and above 9.985, respectively. In MIMO antennas, S-parameter values of individual elements will not give the overall system performance correctly. Because it considers only

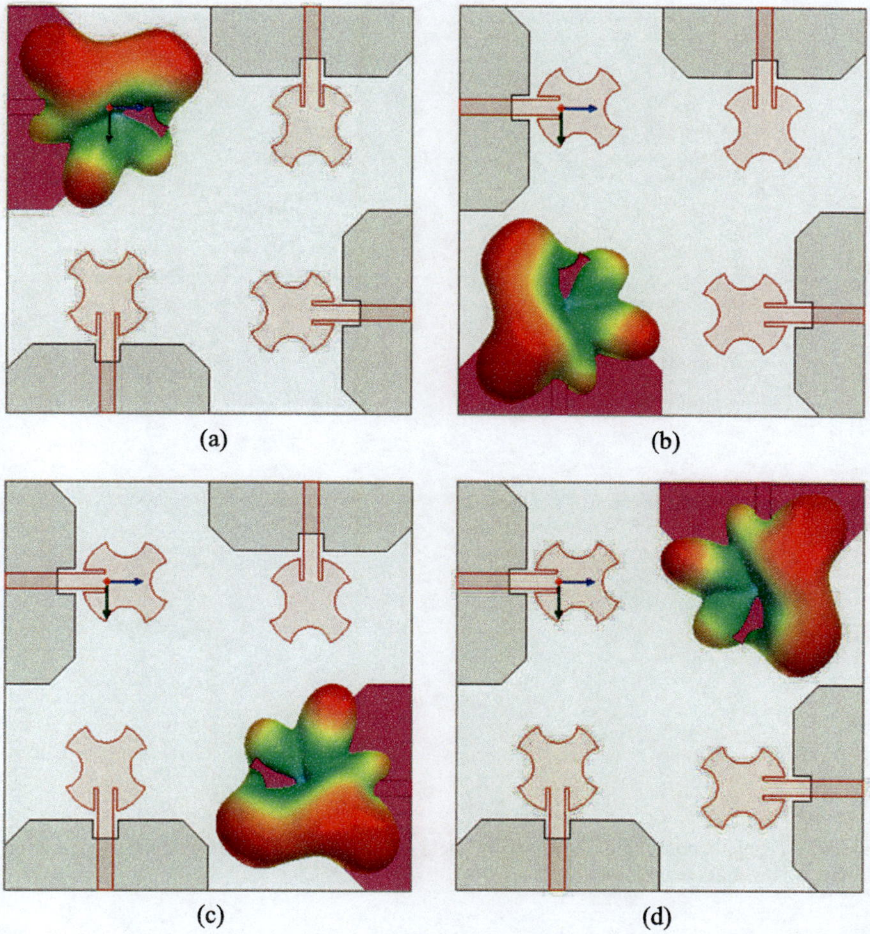

Fig. 9 **a** Radiation pattern at port 1; **b** Radiation pattern at port 2; **c** Radiation pattern at port 3; **d** Radiation pattern at port 4

the characteristic impedance of the individual antenna elements, but not mutual impedances. Changes in the self- and mutual impedances due to the adjacent antenna will not be considered. Hence, the term TARC is required for analyzing system performance in an accurate way. TARC is represented in terms of S-parameters and it is represented in Eq. (3). Figure 12b shows the comparison of S-parameters with TARC values. TARC values are calculated between orthogonal elements in the proposed structure. The acceptable values in the MIMO antenna design are <0 dB.

$$TARC = \sqrt{\frac{(S_{11} + S_{12})^2 + (S_{21} + S_{22})^2}{2}} \quad (3)$$

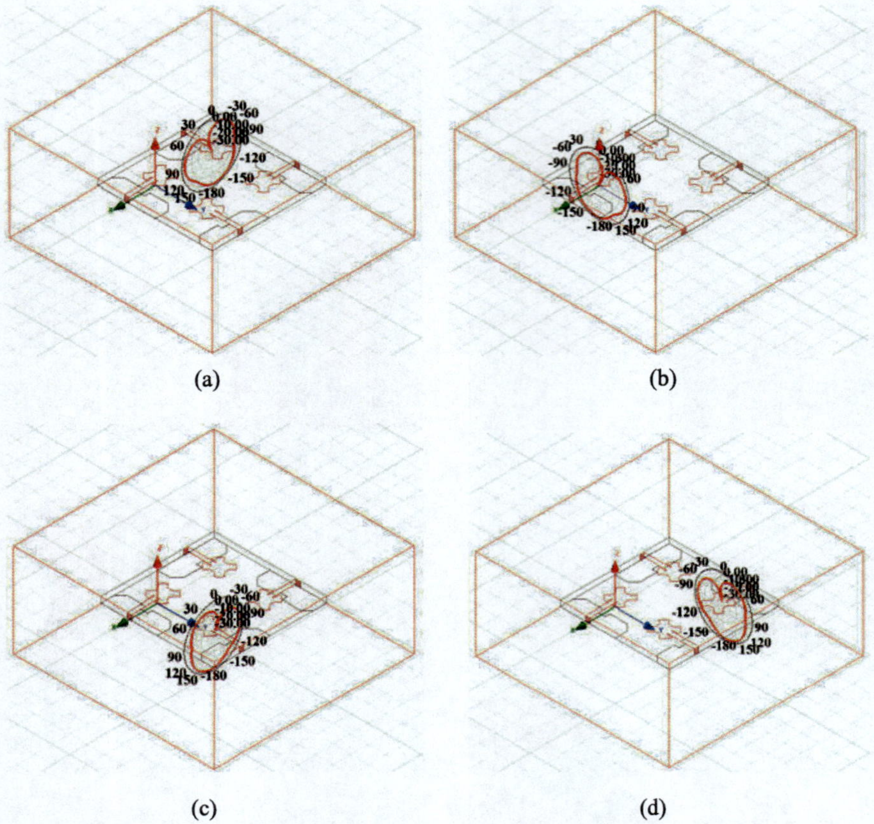

Fig. 10 **a** E-field at port 1 (on XZ-plane); **b** E-field at port 2 (on YZ-plane); **c** E-field at port 3 (on XZ-plane); **d** E-field at port 4 (on YZ-plane)

4 Conclusion

A compact 2 × 2 MIMO diversity antenna is presented in this paper. It is operated in the X-band region, i.e., from 7.9 GHz to 13 GHz and resonates at 9 GHz and 12.3 GHz. The simulation and measurement of S-Parameter values are presented. The isolation between adjacent elements is above 20 dB and between diagonal elements it is above 15 dB. Low ECC, high DG, and acceptable TARC values are achieved. The peak gain and antenna efficiency are also presented for X-band portable device applications.

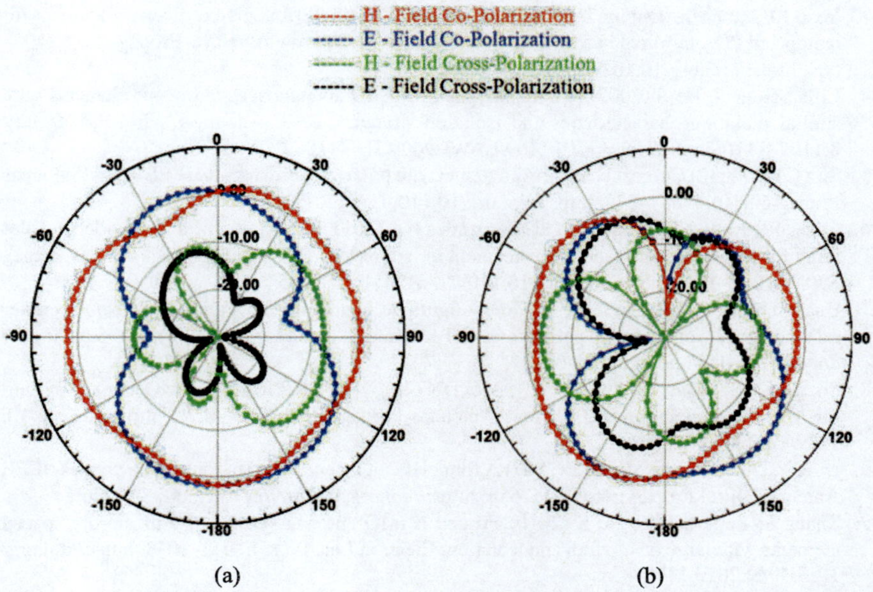

Fig. 11 **a** E (Co and Cross) and H (Co and Cross) fields at 9 GHz when port 1 is excited; **b** E (Co and Cross) and H (Co and Cross) fields at 12.3 GHz when port 1 is excited

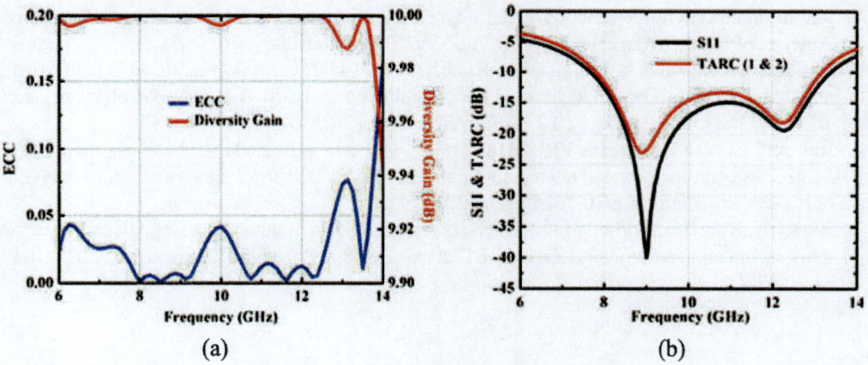

Fig. 12 **a** Simulated ECC and DG results; **b** Simulated S11 and TARC results

References

1. Chen X, Zhang S, Li Q (2018) A review of mutual coupling in MIMO systems. IEEE Access, April 2018. https://doi.org/10.1109/ACCESS.2018.2830653
2. Yuan Q, Chen Q, Sawaya K (2006) Performance of adaptive array antenna with arbitrary geometry in the presence of mutual coupling. IEEE Trans Antennas Propag 54(7):1991–1996, Jul 2006. https://doi.org/10.1109/TAP.2006.877158

3. Yeap SB, Chen X, Dupuy JA, Chiau CC, Parini CG (2005) Integrated diversity antenna for laptop and PDA terminal in a MIMO system. IEE Proc, Microw Antennas Propag 152(6):495–504. https://doi.org/10.1049/ip-map:20045062
4. Li H, Xiong J, He S (2009) A compact planar MIMO antenna system of four elements with similar radiation characteristics and isolation structure. IEEE Antennas Wirel Propag Lett 8:1107–1110. https://doi.org/10.1109/LAWP.2009.2034110
5. Sim CYD (2012) Conical beam array antenna with polarization diversity. IEEE Trans Antennas Propag 60(10):4568–4572. https://doi.org/10.1109/TAP.2012.2207319
6. Costa JR, Lima EB, Medeiros CR, Fernandes CA (2011) Evaluation of a new wideband slot array for MIMO performance enhancement in indoor WLANs. IEEE Trans Antennas Propag 59(84):1200–1206. https://doi.org/10.1109/TAP.2011.2109685
7. Chang DC, Zeng BH, Liu JC (2009) Reconfigurable angular diversity antenna with quad corner reflector arrays for 2.4 GHz applications. IET Microw Antennas Propag 3(37):522–528. https://doi.org/10.1049/iet-map.2008.0119
8. Sharawi MS, Khan MU, Numan AB, Aloi DN (2013) A CSRR loaded MIMO antenna system for ISM band operation. IEEE Trans Antennas Propag 61(8):4265–4274. https://doi.org/10.1109/TAP.2013.2263214
9. Ding Y, Du Z, Gong K, Feng Z (2007) A four element antenna system for mobile phones. IEEE Antennas Wirel Propag Lett 6:655–658. https://doi.org/10.1109/LAWP.2007.913276
10. Zhang S, Zetterberg P, He S (2010) Printed MIMO antenna system of four closely spaced elements with large bandwidth and isolation. Electron Lett 46(15):1052–1053. https://doi.org/10.1049/el.2010.1445
11. Malviya L, Panigrahi RK, Kartikeyan MV (2016) A 2 × 2 dual band MIMO antenna with polarization diversity for wireless applications. Progr Electromagn Res C 61:91–103
12. Xia XX, Chu Q-X, Li JF (2013) Design of a compact wideband MIMO antenna for mobile terminals. Progr Electromagn Res C 41: 163–174, July 2013
13. Ntaikos DK, Yioultsis TV (2013) Compact split ring resonator loaded multiple input multiple output antenna with electrically small elements and reduced mutual coupling. IET Microw Antennas Propag 7(6):421–429. https://doi.org/10.1049/iet-map.2012.0688
14. Moradikorordalivand A, Rahman TA, Khalily M (2014) Common elements wideband MIMO antenna system for WiFi/LTE access point applications. IEEE Antennas Wirel Propag Lett 13:1601–1604. https://doi.org/10.1109/LAWP.2014.2347897
15. Rao Jetti C, Nandanavanam VR (2018) Trident-shape strip loaded dual band-notched UWB-MIMO antenna for portable device applications. ELSEVIER, Int J Electron Commun (AEÜ) 83:11–21. https://doi.org/10.1016/j.aeue.2017.08.021
16. Blanch S, Romeu J, Corbella I (2003) Exact representation of antenna system diversity performance from input parameter description. Electron Lett 39 (9):705–707. https://doi.org/10.1049/e1:200304Y5

Performance Analysis of Activation Functions on Convolutional Neural Networks Using Cloud GPU

Vamshi Krishna Kayala and Prakash Kodali

Abstract One of the tools used for images is Convolutional Neural Networks (CNN) which is a sub-group of the Artificial Neural Network. The activation function is the main characteristic element in CNN. As for any complex application, the starting point is the results of the activation function. The activation function is used to limit the amplitude of the output of a neuron. In order to compute complex functions, non-linearity is introduced by activations to the model. Activation functions of different types can be used with a Convolutional Neural Network in different applications, but better results are given by effective activation functions; they also improve the model performance. The work explores the ability of 10 most used activation functions to evaluate their efficiency in terms of accuracy along with the training time. Sigmoid, Hyperbolic, Tangent (Tanh), Exponential Linear Unit (ELU), Rectified Linear Unit (RLU), Scaled Exponential Linear Unit (SELU), linear, hard sigmoid, softsign, Parametric Rectified Linear Unit (PReLU), and Leaky Rectified Linear Unit (LReLU) activation functions are under consideration. The effects of networks, datasets and processors on training are analyzed.

Keywords CNN · Activation function · Dataset · Training accuracy

1 Introduction

Convolutional Neural Networks are also called as ConvNets; they are exceptionally comparable to the standard neural systems. They are made up of neurons that have learnable weights and predispositions. Each neuron gets an input, performs a dot product an alternatively takes after it with non-linearity. The complete organized network expresses a single differentiable score work from the raw image pixels on one end to the course score at the other. They have the loss function at the final layer which is a completely connected layer. The CNN design makes an explicit suspicion that the inputs are images, hence permits to encode certain properties into

V. K. Kayala · P. Kodali (✉)
Department of ECE, National Institute of Technology, Warangal, India
e-mail: kprakash@nitw.ac.in

T. Laxminidhi et al. (eds.), *Advances in Communications, Signal Processing, and VLSI*,
Lecture Notes in Electrical Engineering 722,
https://doi.org/10.1007/978-981-33-4058-9_4

the design [1] which results in decreased computations compared to the customary neural systems. These make the forward function more effective to execute and immensely diminish the number of parameters utilized within the network. CNNs can be trained using any dataset and on any model. There are several activation functions which can be applied to the network mostly after a convolutional layer and a fully connected layer. These activations play a major part in deciding the exactness of the model. A good model is not effective unless proper activation is chosen suited for the application under consideration.

2 Layers of CNN

2.1 *Convolution Layer*

The convolution layer is the building piece of the CNN that does most of the computational heavy lifting. The convolution layer's parameters comprise a set of learn able filters. Each filter is small spatially (beside width and height) but amplifies to the total depth of the input volume. Degree of the network along the depth axis is continuously broken even with to the depth of the input volume. Three hyper-parameters control the estimate of the yield volume; they are depth, stride and padding.

(1) Depth: It compares to the number of filters used for each learning to extract something different in the input.
(2) Stride: Stride is the esteem with which the filter slides over the picture. When the stride is 1 at that point, the filter is moved by one pixel at a time. Higher esteem of a stride would lead to a smaller output.
(3) Padding: Pad is the additional value given along the boundary of the picture to preserve the input pixel size at the response as well. This padding is mostly done with zeros, to leave the input image unaltered in terms of data.

2.2 *Pooling Layer*

The Pooling Layer is embedded periodically in between progressive convolution layers in a CNN architecture. Its work is to dynamically diminish the spatial size of the representation. This would diminish the parameters and computations within the network and subsequently control the over-fitting. The pooling layer works autonomously on every depth slice of the input and resizes it spatially, utilizing the max or average operation. The max pooling operation pools the maximum pixel value among the pixels.

2.3 Fully Connected Layer

Neurons in a completely connected layer have a full network to all actuations within the past layer. Their enactments can be thus computed with a matrix multiplication taken after by a bias offset. There will be at least one fully connected layer for any model of CNN ideally, and it comes before the classifier like softmax or support vector machine.

2.4 Dropout Layer

The Dropout Layer is used to avoid over-fitting as it can show the negative impact on the performance of the model by learning unwanted features like noise. If the value of the dropout is very less or small, then in most of the cases it leads to a drop in the performance of the model. Under-fitting results in the model not learning the data properly by skipping most of the useful data.

3 Dataset

An information set could be a collection of data. It most commonly compares to the contents of a single database table or a single statistical information matrix, where each column of the table speaks to a specific variable and each row corresponds to a given member of an information set. In the context of machine learning, a dataset is usually any information bundled in a format so the ease of the operation of the data is enhanced. For Convolutional Neural Networks, the dataset is mostly the images in the count of thousands and having some dimensions for a dataset. The image data is usually converted into pixels and represented in the formats like CSV (Comma Separated Values) for ease of operation [2].

3.1 MNIST

MNIST stans for Modified National Institute of Standards and Technology. NIST information set is the base of this dataset, which was given by Yann LeCun. It is full of images of digits which are handwritten and ranging 0–9. The images in this dataset are of dimensions 28×28 and are also black and white. All the images are characterized such that the digits are positioned at the center as much as possible. This dataset is suitable for some of the classification methods like Support Vector Machine, KNN, etc. [3, 4].

3.2 Fashion MNIST

Analysts at Zalando (the e-commerce company) have created a new picture classification information set called Fashion MNIST in trusts on supplanting MNIST. This modern information set contains pictures of different articles of clothing and embellishments such as shirts, bags, shoes and other mold things. The design MNIST training set contains 55,000 pictures and the test set contains 10,000 pictures. Each picture may be a 28 × 28 Gy scale picture (similar to the pictures within the original MNIST), related with a name from 10 classes (t-shirts, pants, pullover, dresses, coats, shoes, sneakers, bags and ankle boots). Fashion MNIST too offers the same train-test-split structure as MNIST, for ease of utilization.

3.3 CIFAR-10

CIFAR-10 could be a collection of pictures that are commonly utilized to train machine learning and computer vision algorithms. It is one of the foremost broadly utilized information sets for machine learning inquiries. CIFAR stands for Canadian Institute for Advanced Research. These images are collected by Alex Krizhevsky, Vinod Nair and Geoffrey Hinton. The CIFAR-10 dataset comprises 60,000 32 × 32 color pictures of 10 classes, with 6000 pictures per class. The classes are categorized as birds, frogs, dogs, cars, horses, trucks, cats, airplanes, deer and ships. Out of 60,000 images, 1000 of them belong to the training set and the rest of them belong to the testing set. The images are of low-resolution; thus, it can be used to experiment on several algorithms in a very short span of time [5, 6].

3.4 CIFAR-100

This information set is similar to CIFAR-10, but it has 100 classes containing 600 pictures each. There are 500 training pictures and 100 testing images per course. The 100 classes within the CIFAR-100 are assembled into 20 super classes. Each picture comes with a "fine" label (the class to which it belongs) and a "coarse" label (the super class to which it belongs) [5].

4 Activation Functions

An artificial neuron comprises two parts, the primary portion which calculates the dot product and the moment portion is the activation function which is utilized to include non-linearity to it. The primary neuron utilizes the step function as its activation. It was based on the basic limit method. There were other activations like tanh, ReLU, ELU [7], sigmoid and also their variants like Parametric ReLU, LeakyReLU and SELU. This work is an extension of the already proposed work [8], where the discussion was on the effectiveness of activation functions sigmoid, ReLU, Tanh and ELU on a CNN with only the MNIST dataset. In this extension of the work, we have taken several other CNN models and the count of the activation function is increased and the experiments have been performed on 4 datasets. Those include CIFAR-10, CIFAR-100, MNIST and Fashion MNIST datasets. All of these are pretrained datasets available as an open source.

4.1 Sigmoid

Sigmoid enactment is additionally known as the logistic function, which is an "S" molded curve that limits the yield between 0 and 1. This work is presently disliked since it has the vanishing angle issue and the output of the sigmoid function isn't centered at zero which makes the optimization go difficult and it has high computing time. Its equation is given by Eq. (1).

$$f(x) = 1/(1 + e^{-x}) \tag{1}$$

4.2 Hyperbolic Tangent (Tanh)

Tanh activation results in the outputs within the extent of -1 to 1. Tanh is zero centered since of the range and is relatively better than sigmoid because it can be optimized exceptionally rapidly. The downside is it is the vanishing gradient. The equation is given by Eq. (2).

$$f(x) = (1 - e^{-2x})/(1 + e^{2x}) \tag{2}$$

4.3 ReLU

ReLU is the widely used activation function; it is preferred for its efficiency and simplicity in application. ReLU has a gradient of 1 for inputs greater than zero as it uses an identity function for positive inputs. ReLU addresses issues of the vanishing gradient commonly observed with other basic activations but it cannot handle the issue of zero gradients for negative inputs. A linear function is obtained for positive inputs and negative inputs get mapped to zero. A neuron is considered to be dead when its value is zero as it doesn't contribute to the learning process. A dead neuron is a liability when the computation is considered [9]. Its equation is given by Eq. (3).

$$f(x) = \begin{cases} 0, x < 0 \\ x, x > 1 \end{cases} \tag{3}$$

4.4 L-ReLU

The LeakyReLU has an advantage over ReLU as it addresses the issue of de-activated neurons. Negative inputs will be mapped to a linear function in this activation, thus, some of the neurons will get reactivated from the dead state. Its equation is given by Eq. (4).

$$f(x) = \begin{cases} \alpha x, x < 0 \\ x, x > 1 \end{cases} \tag{4}$$

4.5 P-ReLU

It stands for Parametric Rectified Linear Unit. In L-ReLU, α is very small and it is a constant value. It is supposed to be a learnable parameter in this variant of ReLU; it is done with the backpropagation method from its data [10].

4.6 ELU

The issue of dying neurons confronted in ReLU is settled by utilizing ELU actuation; this incorporates the thought of exponential operation on the negative input. In this way, ELU performs way better than ReLU. Its equation is given by Eq. (5).

$$f(x) = \begin{cases} \alpha(e^x - 1), & x < 0 \\ x, & x > 1 \end{cases} \tag{5}$$

4.7 Linear

Linear activation is the basic type of activation function which is also called an identity function. It is given by Eq. (6). It is very easy to compute linear functions, but its limitation lies within the fact that it cannot be used to learn complex functions.

$$f(x) = x \tag{6}$$

4.8 Hardsigmoid

It is an approximation of a sigmoid function. The standard sigmoid is slow to compute because it requires computing the exp() function, which is done via a complex code. In many cases, the high precision exp() results aren't needed, and an approximation will suffice. Such is the case in many forms of gradient-descent/optimization neural networks: the exact values aren't as important as approximated values, insofar as the results are comparable with small error. It is implemented as the following equation, Eq. (7).

$$f(x) = \begin{cases} 1, x > 2.5 \\ 0.2x + 0.5, -2.5 < x < 2.5 \\ 0, x < -2.5 \end{cases} \tag{7}$$

4.9 SELU

It is a modified version of ELU. Its equation is given by Eq. (8). Here, λ and α are predefined constants meaning we do not backpropagate through them and they are not hyper-parameters. For scaled standard inputs (mean 0, stddev1), the values are α = 1.6732 and λ = 1.0507.

$$f(x) = \lambda \begin{cases} x, x > 0 \\ \alpha(e^x - 1), x < 0 \end{cases} \tag{8}$$

4.10 *Softsign*

Softsign is an activation function used as an alternative to hyperbolic tangent (tanh) activation. Both tanh and softsign create yield within the extent [−1, 1]. In spite of the fact that tanh and softsign capacities are closely related, tanh merge exponentially though softsign meet polynomial. It is given by Eq. (9).

$$f(x) = 1/(1 + |x|) \tag{9}$$

5 CNN Models

The Convolutional Neural Network model is developed to suit applications like image classifications, data classifications, etc. We can choose custom networks which are combinations of several convolutional layers, pooling layers, max pool layers, fully converged networks and any classifier like Support Vector Machine (SVM) or Softmax. There are several models which are customized to be used for image classification applications used by us in training, and also a standard network called KerasNet is also used to train the datasets.

5.1 *CNN1*

This is a custom-made Convolutional Neural Network for image classification applications referred to as CNNI. This architecture consists of two convolution layers of filter size 5×5, each layer extracting 32 and 64 features, respectively, and two max pool layers of filters 2×2 and a fully converged networked layer extracting 1024 features, and the final layer is a softmax classifier. The layers are stacked upon one after the other as shown in Fig. 1. The input is an image depending on the dataset and the response is the class score of the probability of the image belonging to a particular class.

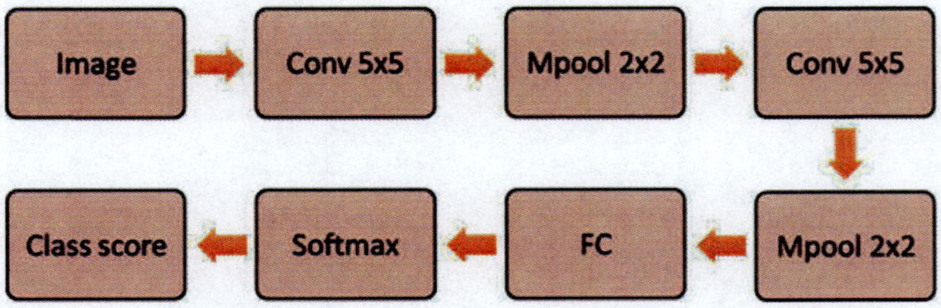

Fig. 1 Layers of CNN1

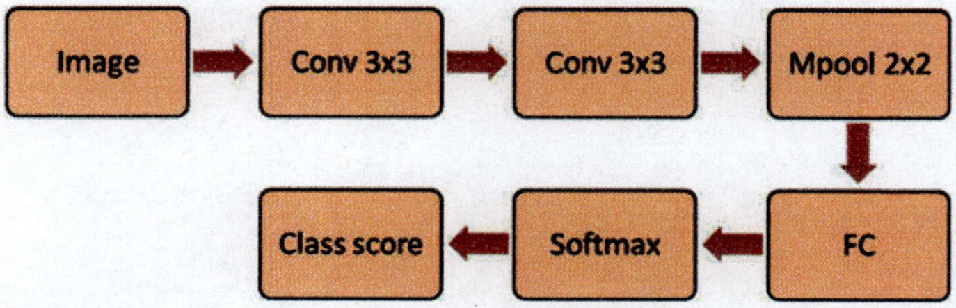

Fig. 2 Layers of CNN2

5.2 CNN2

This is another custom-made CNN with slight modifications compared to CNN1 as shown in Fig. 2. The architecture of CNN2 is made of two convolution 3 × 3 layers stacked upon one another, each extracting 32 and 64 features followed by 2 × 2 max pool layers which are followed by a fully converged network layer extracting 512 features. Those are finally connected to a classifier layer which is softmax in this case; it outputs the final class score of the performance of the model. An activation function is generally used after every convolution.

5.3 KerasNet

KerasNet is a small 8-layer CNN model as shown in Fig. 3. It has 4 Convolution layers each having a filter size of 3 × 3. Out of those four layers, two convolution layers are connected back to back. The first pair of conv. layers extracts 32 features and the second pair of conv. layers extracts 32 features and third pair of conv layers extracts 64 features. It is followed by a max pool layer of filter size 2 × 2. It is then connected by a set of another convolution layer pair same as the previous layer followed by another layer of a 2 × 2 max pool layer. Then a layer of dropout is used, the value of dropout being 0.25. It is then connected to a fully converged network layer then to a dropout layer with a dropout value 0.25, which is finally connected to a classifier layer which is softmax in this case. The activation function layer is

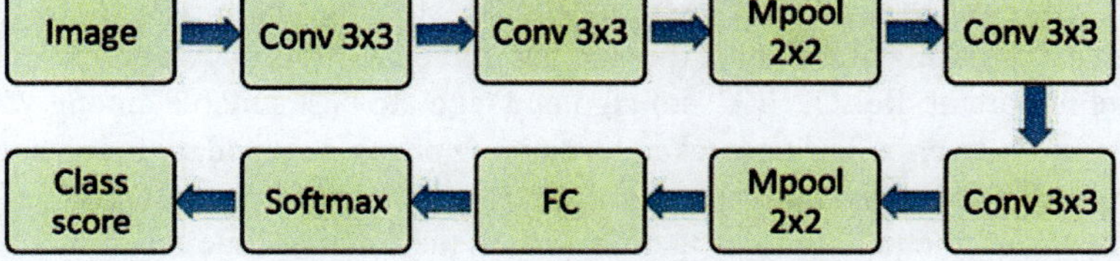

Fig. 3 KerasNet CNN model

connected generally after every conv. layer and fully converged network layer. The fully connected network extracts 512 features.

6 Experimental Procedure

The CNN model is developed on an open-source web application called Google Collaborator. The model is built in Python using an open-source framework like tensor flow, Keras [7], etc. These frameworks help in performing convolution with any image size and any filter size (3 × 3 or 5 × 5 or 7 × 7 or 9 × 9 or 11 × 11), etc. just in a single line of code. In the same way, operations like max pooling, fully connected layer, selecting a classifier, applying an activation function, changing the stride value, etc. can be done in a simple way. All the required layers needed to build a model are stacked up one after the other; the parameters needed for a layer can be given in the corresponding layers itself, and finally the hyper-parameters like optimization method, learning rate, number of learning iterations, batch size, etc. are chosen and then the code is run, after making sure that there are no compilation errors. The time taken, validation accuracy and the test accuracy, validation loss and test loss are displayed for each iteration. Implementation Details: The CNN model is intended to be developed using Tensor Flow and Keras using Python script, trained using the cloud application Google Collaborator. It has Nvidia GPU TeslK80 which has 12 GB graphics memory.

7 Results and Analysis

7.1 Results of Training

Training is done on the tree CNN models mentioned using the four different datasets and all the activation applied on them; this results in a total of 120 training experiments in total shown in Figs. 4 and 5. A consolidated result of all the training is given by adding up the training time and taking the average of training accuracy.

This summary is the combination of all the experiments, i.e. all the datasets on all the networks, which gives ReLU, LeakyReLU, PReLU, ELU, softsign and Tanh in decreasing order as per accuracy and linear, sigmoid, Tanh, ReLU, ELU and softsign in increasing order as per the training timing. As both accuracy and training time are important, ReLU, ELU, softsign and tanh are best suitable among various datasets and networks. The time taken for training phase is secondary for most of the scenarios, so if two activations are giving almost similar performance then a decision could be taken to choose one of them based on the training time taken by each of them.

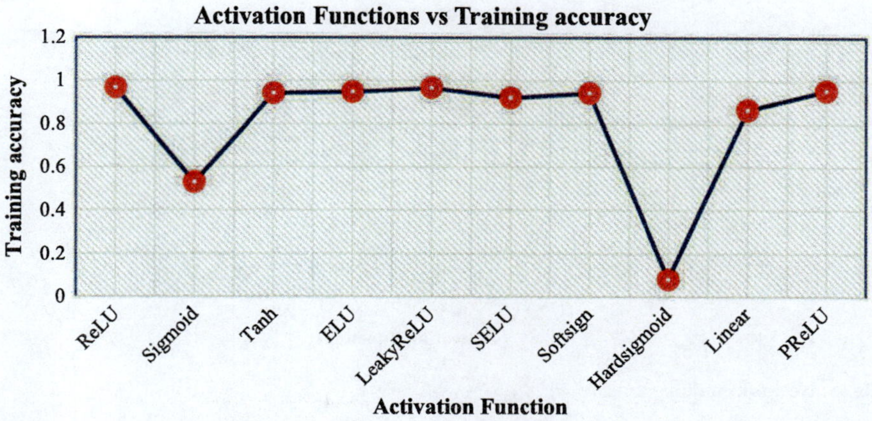

Fig. 4 Consolidated training accuracy

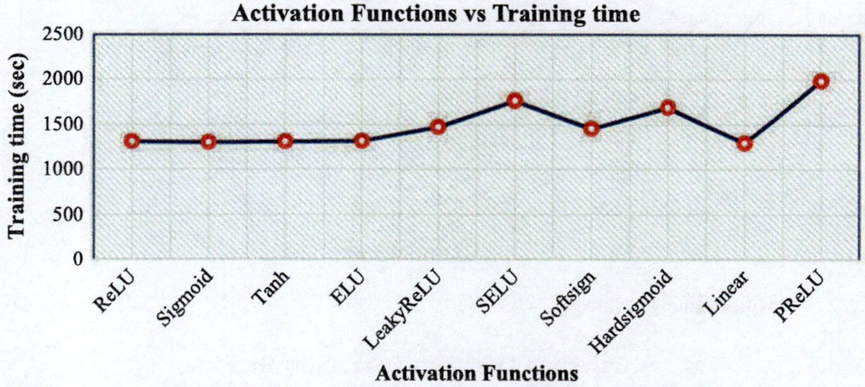

Fig. 5 Consolidated training time

7.2 Comparision of Networks

7.3 Comparision of Datasets

The networks are compared by taking the sum of all the training times and similarly the average of training accuracies for all networks in Figs. 6, 7 and 8. In terms of efficiency, the networks are given as CNN2 > CNN1 > KerasNet. In terms of training time, KerasNet < CNN1 < CNN2. It can be concluded that customized networks are better performing than the standard networks.

The datasets are compared by taking the sum of all the training times and similarly the average training accuracies for all datasets are shown in Fig. 9. In terms of efficiency, the networks are given as MNIST > Fashion-MNIST > CIFAR-100 >

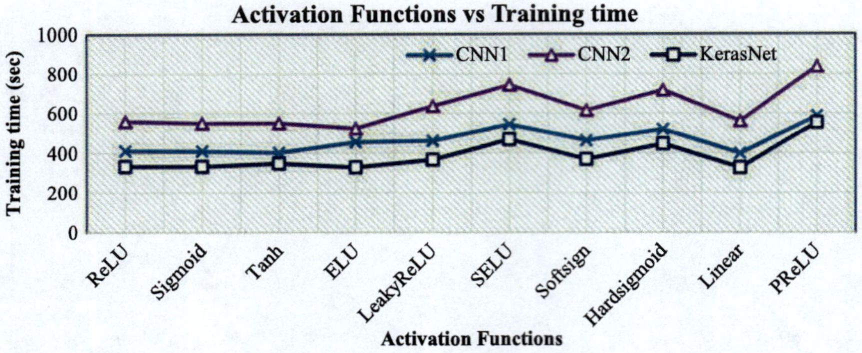

Fig. 6 Networks training time

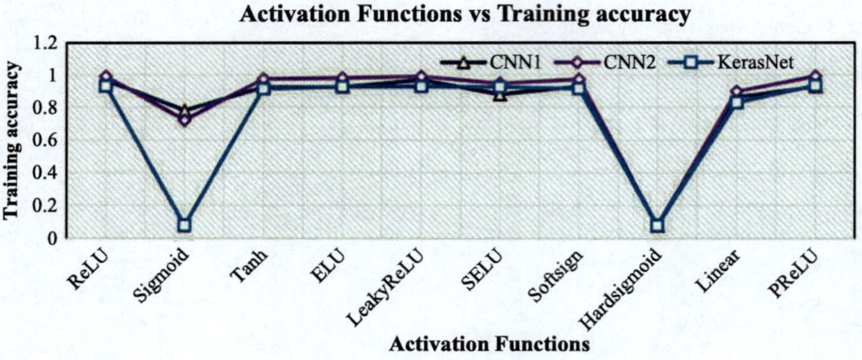

Fig. 7 Networks training accuracy

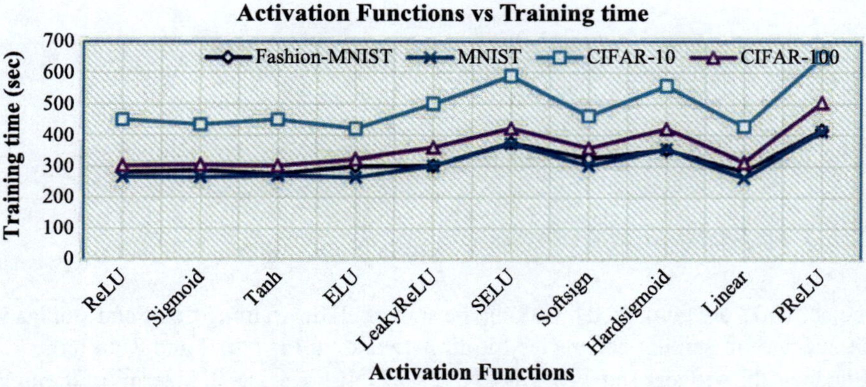

Fig. 8 Dataset Training time

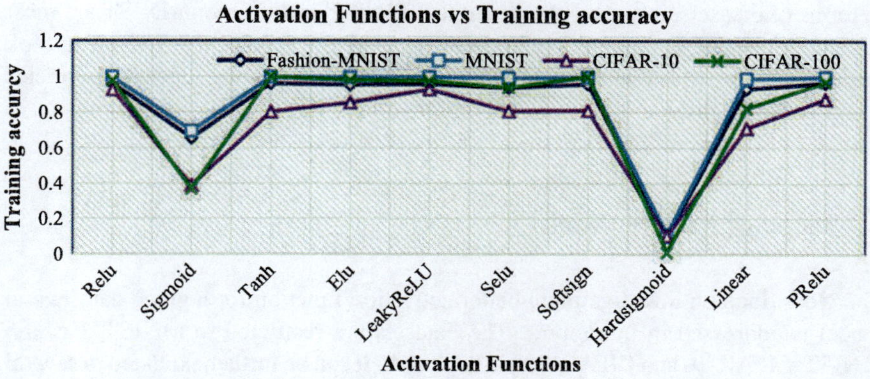

Fig. 9 Dataset training accuracy

Table 1 Comparison of GPU and cloud models

Activation function	MNIST training accuracy (GPU)	MNIST training accuracy (Cloud)
ReLU	0.943433	0.9969
Sigmoid	0.902344	0.9904
Tanh	0.96875	0.9928
ELU	0.984375	0.9924

CIFAR-10. In terms of training time, MNIST < Fashion-MNIST < CIFAR-100 < CIFAR-10. It can be concluded that MNIST and Fashion-MNIST datasets are better performing than others.

7.4 *Training on GPU Versus Cloud*

It can be concluded that cloud GPU gave better accuracy than the hardware GPU, and they are also easy to access for free of cost. It is better to go for cloud-based training to obtain better performance rather than the actual hardware. The numerical data are provided in Table 1 for the accuracy training GPU and Cloud comparison.

8 Conclusion

The final conclusion from the above-performed experiments is that activation functions play a significant effect on the performance of the model. ReLU, LeakyReLU, PReLU, ELU and softsign are the functions with high performance overall. Linear, Sigmoid, Tanh, ReLU, ELU andsoftsign consumed the least time for training overall.

In terms of datasets, it can be concluded that MNIST and Fashion MNIST are most suitable for evaluation in terms of accuracy as well as training time. KerasNet took the least training time overall, but CNN2 and CNN1 were giving the best accuracies compared to KerasNet.

9 Scope of Future Work

The work done on investigating a better activation function for a given dataset and model is addressed in this report. The datasets are restricted to MNIST, Fashion MNIST, CIFAR-10 and CIFAR-100 in this paper. It can be further extended to several other datasets like Imagenet, Caltech101, Caltech-256, etc. which are open-source datasets readily available and they are also widely used for machine learning research works. The models taken for experimentation are also limited to three in our work; it can also be extended to several other standard networks like CaffeNet, GoogLeNet, VGGNet, etc. There is also scope for future work to work on an activation which may be a blend of some of the used activations or any other unused activations which may yield better performance either in terms of time or in terms of accuracy or both compared to the activation functions investigated in this paper.

References

1. https://cs231n.github.io/ CS231n Convolutional Neural Network for Visual Recognition
2. https://en.wikipedia.org/wiki/Data_set
3. LeCun Y, Cortes C, Burges CJC (2010) MNIST handwritten digit database. AT&T Labs. https://yann.lecun.com/exdb/mnist 2
4. https://en.wikipedia.org/wiki/MNIST_database
5. https://www.cs.toronto.edu/~kriz/cifar.html CIFAR10 and CIFAR100 data sets
6. https://en.wikipedia.org/wiki/CIFAR-10
7. Simonyan K, Zisserman A (2014) Very deep convolutional networks for large scale visual recognition. arXiv: 1409.1556
8. Zaheer R, Shaziya H (2018) GPU-based empirical evaluation of activation functions in convolutional neural networks. In: 2018 second international conference on inventive systems and control (ICISC)
9. Duggal R, Gupta A (2017) P-TELU: parametric tan hyperbolic linear unit activation for deep neural networks. In: IEEE international conference on computer vision workshops
10. Veliˇckovi´c P, Wang D, Lane N.D., Li‘o P (2016) Xcnn: Cross-modal convolutional neural networks for sparse datasets. arXiv preprint arXiv:1610.00163
11. Hidenori Ide KK (2017) Improvement of Learning for CNN with ReLU activation by sparse regularization. In: IEEE conference paper, 2017

Pareto Optimal Approach for Contrast Adjustment in Gray Level Images Using IDS Algorithm

Sowjanya Kotte

Abstract Image enhancement/contrast adjustment plays an important role in almost every image processing system. The main aim of the contrast adjustment is to enhance image quality by maximizing the information content in the image. Many researchers implemented heuristic approaches for image enhancement using maximization of objective function based on image quality metrics. And such objective function may fail to yield proper enhancement for a given image (either under enhanced or over enhanced). Because the parameters of the image transformation function are selected based on overall value of the objective function but not individual objectives in the objective function. That means the parameter values so chosen may not satisfy all the objectives of the objective function, and this may lead to improper enhancement of the image. In this context, this paper presents a Pareto optimal approach-based gray level contrast adjustment using Improved Differential Search Algorithm (IDSA) named Multi-Objective Improved Differential Search Algorithm (MOIDSA). Image enhancement is treated as a three-dimensional optimization problem, and MOIDSA is used to solve it. The input image quality is enhanced by simultaneous objectives which is a blend of image performance measures and quality metric as a three-dimensional optimization approach will yield best-compromised solution by satisfying all the individual objectives. This algorithm is tested on some standard test gray level images, and results compared with well existing algorithms have proven its superiority.

Keywords Image contrast adjustment · Parameter estimation · Pareto optimal approach · Improved differential search algorithm · Image quality evaluation

S. Kotte (✉)
Department of ECE, Kakatiya Institute of Technology and Science (A), Warangal 506015, Telengana State, India
e-mail: kotte.soujanya@gmail.com

T. Laxminidhi et al. (eds.), *Advances in Communications, Signal Processing, and VLSI*, Lecture Notes in Electrical Engineering 722,
https://doi.org/10.1007/978-981-33-4058-9_5

1 Introduction

Image processing is a wide and dynamic region of research in processing. It has numerous applications in regular day to day existence undertakings, for example, industrial, transportation and medicine, and so on. Image enhancement is one of the most significant methods, which may be treated as changing an image to other images to enhance the discernment or interpretability of data for human watchers, or to give better contribution to other computerized techniques in contrast adjustment. Genetic Algorithm (GA) for image improvement has been proposed through utilizing a multi-objective (weighted total) work comprising four non-linear mapping functions. It utilizes the hereditary calculation to search for ideal mapping of the dark degrees of the information picture into new dim levels offering better complexity for the image [1]. As of now, some image quality metrics have been presented and utilized for gray level and color image improvement. Contrast adjustment of computerized gray level images by safeguarding the mean intensity of image utilizing PSO has been proposed in [2]. Authors' present Differential Evolution (DE) looking instrument for a worldwide ideal answer for improving the differentiation and subtleties in a dim scale picture. Differentiation improvement of an image is done by gray level change utilizing parameterized transformation function as an objective function [3]. Hybrid bacterial foraging and Particle Swarm Optimization (PSO) have been proposed to solve the image enhancement problem based on entropy and edge information [4]. Improved Particle Swarm Optimization (PSO) has been proposed for image enhancement based on parameterized transformation function as objective function [5]. Particle Swarm Optimization (PSO) and DWT techniques are used to improve the image quality [6]. Parametric sigmoid function-based fuzzy logic technique has been proposed for image enhancement which involves entropy and visual factors as objectives [7]. Contrast information factor-based PSO is proposed to solve image enhancement problem [8]. Blend of number of edge pixels, intensity of pixels, and the image entropy is used as an objective function by various researchers for image enhancement using Differential Evolution (DE) [9], PSO [10–12], Cuckoo Search (CS) [13], Artificial Bee Colony (ABC) [14], Evolutionary algorithm [15], and Harmony Search Algorithm (HSA) [16] as optimization approaches. Authors presented a PSO-based methodology for contrast adjustment using maximization of informative pixel intensity at each pyramid level [17]. Iterative Optimization Process is used to select the optimal parameters in X-ray image enhancement based on entropy maximization [18]. Authors presented a performance analysis of a weighted multi-objective optimization approach for gray level contrast adjustment using State of Matter Search (SMS) algorithm [19]. From the literature study, it has been observed that HE and AHE are the most popular classical methods for gray level image enhancement. These techniques produce poor quality images and found to be exhaustive. Many authors applied global contrast enhancement technique via heuristic algorithms on an objective function given in Appendix. During the process of optimization (one-dimensional optimization), algorithm will select the parameters based on overall objective function value whether it is maximizing or not. So, algorithm does not care which component of the objective function is to maximize. Hence, it may lead to improper enhancement of the given image. Whatever may be enhancement technique the enhanced image quality will be checked with the help of image quality metrics. This idea drives the author to regard PSNR as one objective along with number of edge pixels and image entropy as other objectives. In this present paper, an attempt has been presented for gray level contrast adjustment based on global intensity transformation function using multi-objective IDS algorithm in which all the three objectives will simultaneously search for the optimal possible set of gray levels. That means MOIDSA will yield best-compromised solution (optimal parameters) by satisfying all the three objectives at a time [20]. The suggested approach has been tested on set of standard images which shows promising results. The outcomes are validated using qualitative, quantitative, and statistical analysis.

1.1 Material

The main idea of the image enhancement is to convert the input image into a better quality output image. Expression for local enhancement transformation function is given in [21] (Eq. 1).

$$g(x, y) = \frac{G}{\sigma(x, y)}(f(x, y) - m(x, y)) \tag{1}$$

Mean m(x, y) and standard deviation σ(x, y) are computed in a neighborhood centered at (x, y) having MXN pixels. f(x, y) and g(x, y) are the gray level intensity of the input and output image pixel at location (x, y), and G is the global mean of the input image. Adaptive histogram equalization is also a local enhancement technique which gains the most popularity since its good results are shown in medical image processing [22, 23]. The easiest way to accomplish the task of contrast adjustment is global intensity transformation. Expression for global intensity transformation function is given in Eq. 1 and is applying to each pixel at location (x, y) of the given image is given in Eq. 2.

$$g(x, y) = \frac{k.G}{\sigma(i, j) + b}\left[f(x, y) - c \times m(x, y)\right] + m(x, y)^{a} \tag{2}$$

Global mean (G) of the input image over a n X n window, which are expressed as follows:

$$m(x, y) = \frac{1}{n * n}\sum_{x=0}^{n-1}\sum_{y=0}^{n-1} f(x, y) \tag{3}$$

$$G = \frac{1}{M \times N} \sum_{x=0}^{M-1} \sum_{y=0}^{N-1} f(x, y) \quad (4)$$

$$\sigma(x, y) = \sqrt{\frac{1}{n \times n} \sum_{x=0}^{n} \sum_{y=0}^{n} (f(x, y) - m(x, y))^2} \quad (5)$$

Proper tuning of the parameters a, b, c, and k in Eq. 2 will produce great variations in the processed image by preserving its originality and natural look.

2 Problem Formation

To estimate the quality of output image without human involvement with the proposed approach, we require an objective function that combines important quality measures such as the number of edge pixels, image entropy, and the intensity of the edge pixels [9–15]. Some authors excluded the entropy of the whole image in their objective function [24]. In fact, entropy is one of the key quality measures in the image enhancement. As mentioned in the introduction, the final quality of the enhanced image will be estimated using image quality metrics. So, along with a number of edge pixels and entropy, PSNR is also considered as one of the individual objectives in the multi-objective function.

$$f_1 = \max(n\_edgels(I_e)) \text{or} \min\left(\frac{1}{n\_edgels(I_e)}\right) \quad (6)$$

$$f_2 = \max(H(I_e)) \text{or} \min\left(\frac{1}{H(I_e)}\right) \quad (7)$$

$$f_3 = \max(PSNR(I_e)) \text{or} \min\left(\frac{1}{PSNR(I_e)}\right) \quad (8)$$

n_edgels is the number of edge pixels of the resulting image. *H(Ie)* is the entropy value. PSNR is peak signal to noise ratio of the image enhanced. The main aim of MOIDSA algorithm is to select best-compromised solution (a, b, c, and k) that maximizes the three objectives *f1, f2, and f3* simultaneously. The task of the optimization algorithm is to solve the contrast adjustment problem by tuning the four parameters (a, b, c, and k) to find the perfect combination according to an objective criterion that explains the contrast in the image. In this paper, the limits of these variables are chosen as in [10, 12]; a$\in$ [0, 1.5], b$\in$ [0, 0.5], c$\in$ [0, 1], and k$\in$ [0.5, 1.5]. But they failed to produce good output with the supplied range of b. It has been noticed that small variation in the value of b will have a large effect on intensity stretch. The originality of the image has lost due to normalized intensity values crossed the limit

[0, 255]. To avoid this problem, the limit of b has been modified to [1, G/2], where G is the global mean of the input image [11].

2.1 Multi-Objective Improved Differential Search Algorithm

The IDS algorithm is the latest version of Differential Search Algorithm (DSA) developed by Civicioglu in the year 2013 [25]. DSA is based on the migration of living beings for food during variations of climate. The multi-objective format of IDSA [26], 27] named MOIDSA [20] reduces minimum and increases a maximum of multiple objective functions.

$$\begin{aligned} &Minimize\ \ F(x) = \left\{f_1(x), ...f_{N_{obj}}(x)\right\} \\ &x = [X_1, X_2, ...X_d] \end{aligned} \tag{9}$$

Instead of one solution, optimal solutions are obtained. It is difficult to evaluate all the solutions. So, no global optimum solutions can be fixed. Hence, a group of Pareto optimal solutions comes up, such that these are not surpassed by any other solutions. Here, a decision vector X_1 dominates a counterpart X_2 if X_1. is partially less.

$$\begin{aligned} &\forall i \in \left\{1, 2, ...N_{obj}\right\} : f_i(X_1) \leq f_i(X_2) \\ &\exists_j \in \left\{1, 2, ...N_{obj}\right\} : f_j(X_1) \leq f_j(X_2) \end{aligned} \tag{10}$$

Implementation Strategy for MOIDS algorithm is as follows.

Step 1: Initialization

Population size (*NP*), problem dimension (*D*), and a maximum number of generations (*Gmax*) are initially assigned randomly. Limits of *a, b, c, and k* are also initialized.

Step 2: Organism population

For getting organism population, random generation is adapted

$$X = \begin{bmatrix} x_1^1 & x_2^1 & \cdots & x_{d-1}^1 & x_d^1 \\ x_1^2 & x_2^2 & \cdots & x_{d-1}^2 & x_d^2 \\ \vdots & \vdots & \vdots & \vdots & \vdots \\ x_1^{pop-1} & x_2^{pop-1} & \cdots & x_{d-1}^{pop-1} & x_d^{pop-1} \\ x_1^{pop} & x_2^{pop} & \cdots & x_{d-1}^{pop} & x_d^{pop} \end{bmatrix} \tag{11}$$

$$x_i^j = x_{\min,i} + (x_{\max,i} - x_{\min,i}) * rand() \tag{12}$$

X represents the ensemble of living beings on artificial organisms which is one place in search space is given by a solution containing a, b, c, and k values.

Step 3: Calculate fitness function value.

Evaluate population X using Eq. 9.

Step 4: Locate non-dominated solutions.

Step 5: The non-dominated solution is segregate and stored in the archive with G = 0.

Step 6: Initiate evolution procedure.

Step 7: Search process is performed, and stopover vector is generated using Eqs. 13, 14, and 15.

$$s_i = X_{r1} + scale.(X_{r2} - X_i) \tag{13}$$

$$scale = randg(2 * rand) * (rand - rand) \tag{14}$$

$$s^t_{i,j,G} = \begin{cases} s_{i,j,G} if r_{i,j} = 0 \\ X_{i,j,G} if r_{i,j} = 1 \end{cases} \tag{15}$$

$$X_{i,G+1} = \begin{cases} S_{i,G} \text{ if } f(S^t_{i,G}) \le f(X_{i,G}) \\ X_{i,G} \text{ if } f(S^t_{i,G}) > f(X_{i,G}) \end{cases} \tag{16}$$

Step 8: Evaluate each member of stopover vector

Evaluate stopover vector according to objective functions using Eq. 9 for locating an individual member of stopover vector.

Step 9: Check for dominance stopover vector

Stopover vector stored ideally dominates the artificial organisms vector population. Discard stopover vector if artificial organism dominates. Else, it is added to the population (temp pop).

Step 10: Latest solution vector is added to temp pop

Crowding assignment and non-dominated solution help in selecting next generation. Non-dominated solutions are stored in the repository. When the capacity of the repository is high, the crowded members are selected by crowding assignment operators.

Step 11: Initiate stopping procedure

Generation counter is incremented and termination criteria are checked. If the G has reached the maximum point, the non-dominated solution is printed out, and the process is retarded. Else, step 6 and step 10 are repeated. And the implementation flowchart for MOIDSA is given in Fig. 1.

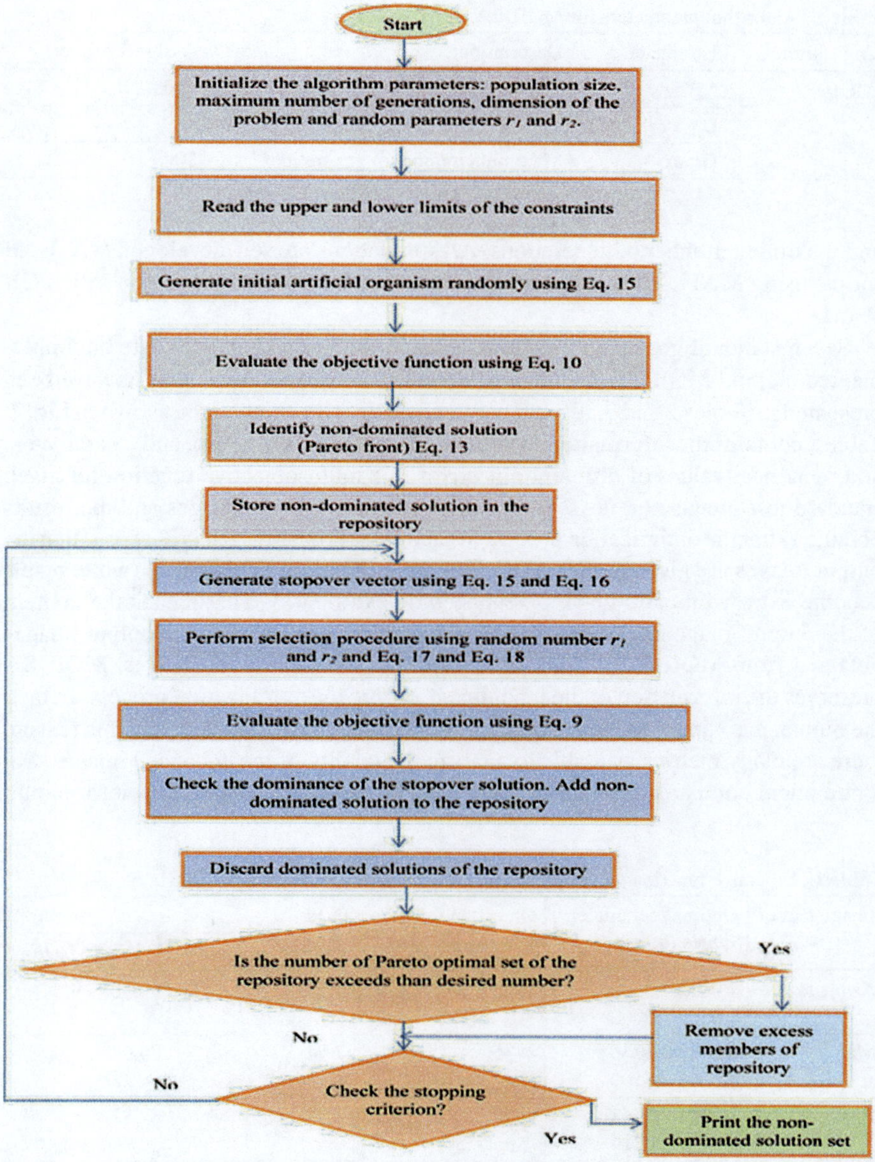

Fig. 1 Implementation flow chart for MOIDSA

3 Results and Discussion

In this section, the MOIDSA is validated by applying it on a set of test images. Control parameters are not involved in MOIDSA hence parameter tuning is not required. Table 1 gives the information about general algorithm parameters like population size

Table 1 Algorithm parameters for MOIDSA

Algorithm	Parameter	Description	Assigned value
MOIDSA	NP	Size of population	100
	D	Dimension of the problem	4
	Gmax	Maximum number of generations	100

and maximum number of generations. All simulations are self-developed MATLAB scripts using MATLAB R2016b on an Intel Core i5 3.10 GHz processor with 8 GB RAM.

Seven standard grayscale images have been used to test and validate the implemented method MOIDSA and obtained results and quantitative analysis have been presented in Tables 2 and 3, respectively, and qualitative comparison shown in Fig. 2. Table 2 contains the information regarding optimal values of a, b, c, and k parameters and numerical values of different objectives in a multi-objective function for given standard test images. Figure 2 shows the instances of input images and the results obtained after the optimization process in each case. From Fig. 2, it is observed that the output images are given by CS, ABC, DE, and PSO are over enhanced (white pixels become extra white and black pixels become extra dark). That means the content of the output image has deviated from the input image, whereas the output images obtained from MOIDSA do not deviate from its original input images. MOIDSA preserves the information of the input image during the optimization process. In fact, the human eye cannot judge the quality of enhanced images always. For this reason, there are many metrics available to analyze the quality of the enhanced images. So, a numerical comparison seems to be necessary here. Table 3 provides information

Table 2 Objective function values of MOIDSA for various gray scale images

Image name	Optimal parameter values (*a*, *b*, *c*, & *k*)	Objective function values		
		$nedgels = 1/OF_1$	$Entropy = 1/OF_2$	$PSNR = 1/OF_3$
Airplane	1.0012 1.0000 1.0000 1.5000	3822	6.71	90.40
Man	1.0083 0.1746 1.0000 0.7500	4435	7.54	85.94
Boat	1.0046 0.2543 1.0000 0.7214	4932	7.18	87.87
Living room	0.9915 0.2390 0.9888 0.8344	5152	7.42	85.67
Breast	1.0028 0.1308 1.0000 1.1102	4414	5.58	85.53
X-Ray	0.9889 0.1174 0.9954 0.9720	2856	6.85	88.90
Lena	0.9722 0.2075 0.9802 0.7875	3440	7.39	88.61

Table 3 Comparison of numerical results of MOIDSA with CS, ABC, DE, and PSO algorithms for standard gray scale images

Image name	Algorithm	No of n__edgels		Entropy		PSNR	RMSE	mSSIM	Time (S)
		Initial	Final	Initial	Final				
	MOIDSA		4435		7.5463	85.9426	0.0129	1.0000	111.5
	CS		4889		7.7377	68.4312	0.1104	0.9986	126.4
Man	ABC	4218	5499	7.5346	7.7506	66.5574	0.1203	0.9983	62.0
	DE		5766		7.7366	66.8014	0,1170	0.9985	62.3
	PSO		5738		7.7558	66.1788	0,1257	0.9982	59.7
	MOIDSA		4932		7.1815	87.8752	0.0103	1.0000	111.4
	CS		5753		7.5202	65.1609	0.1413	0.9977	118.4
Boat	ABC	4535	5325	7.1452	7.4423	63.9177	0.1631	0.9966	62.2
	DE		5944		7.5391	67.4666	0.1084	0.9988	59.6
	PSO		5947		7.5390	67.3955	0.1093	0.9987	59.9
	MOIDSA		5152		7.4218	85.6720	0.0133	1.0000	113.8
	CS		6307		7.7095	65.9731	0,1287	0.9982	123.7
Livingroom	ABC	5068	6205	7.3841	7.6764	62.6154	0.1894	0.9958	62.1
	DE		6514		7.6739	68.1419	0,1003	0.9990	60.0
	PSO		6503		7.6851	67.4183	0,1090	0.9988	60.1
	MOIDSA		4414		5.5800	85.5372	0,0135	1.0000	104.7
	CS		5110		5.6731	70.9635	0,0725	0.9994	121.4
Breast	ABC	3678	5362	5.4212	5.4757	76.0944	0,0401	0.9998	61.3
	DE		5646		5.6861	80.7896	0,0234	1.0000	59.1
	PSO		5687		5.7045	80.8123	0.0233	1.0000	59.9
	MOIDSA		2856		6.8519	88.9046	0,0091	1.0000	102.2
	CS		2743		6.9413	79.7202	0,0264	0.9999	121.9
X-Ray	ABC	2428	2579	6.8201	7,0041	69.0684	0,0901	0.9989	60.2
	DE		4304		5.8791	62.8284	0.1849	0.9953	58.6
	PSO		3012		6.5351	71.6772	0.0667	0.9994	57.9

about the input standard images and outcomes of the MOIDSA for gray level image enhancement. Three image performance measure and three image quality metrics are basically used for the numerical comparison of results. They are (1) value of the objective function, (2) number of edge pixels, (3) entropy, (4) PSNR, (5) Root Mean Square Error (RMSE), and (6) Mean Structural Similarity Index Measure (MSSIM) [28]. From Table 3, it is observed that the value of each image performance measure and quality metric has been increased from its initial value for all images. And it is a known fact, which the image enhancement is not possible without a proper increase in intensity, edge pixels, and image entropy. To test the efficiency of the proposed approach with existing approach/algorithms such as CS, ABC, DE, and PSO were also implemented on standard test images. For this, the population size, maximum

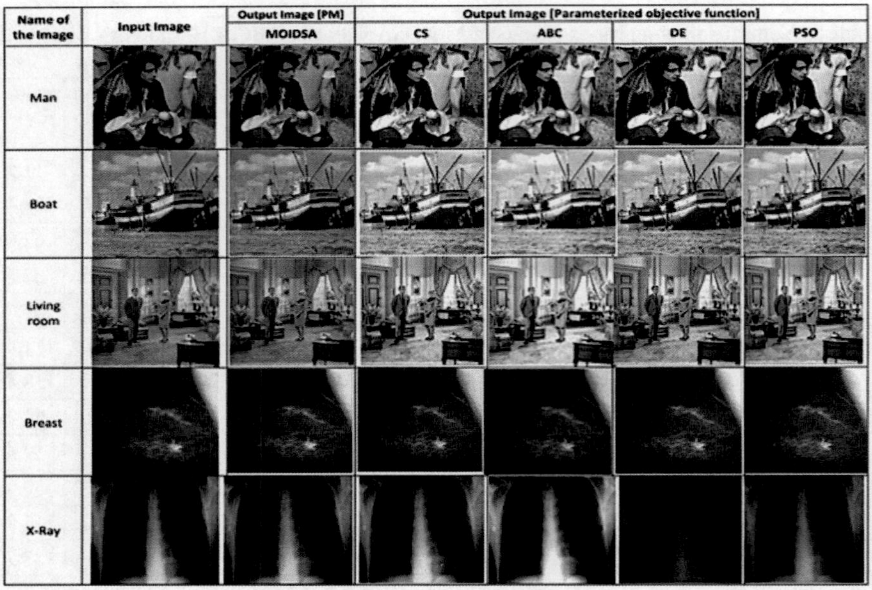

Fig. 2 Qualitative comparisons for enhanced images of MOIDSA, CS, ABC, DE, and PSO algorithms

number of iterations and dimension of the problem have been taken same for all algorithms and the details of the algorithm parameters have been furnished in Table 1.

Performance analysis of the implemented optimization algorithms for image enhancement has been provided in Table 3. It shows the comparison of numerical results of MOIDSA with CS, ABC, DE, and PSO algorithms. It should be noted that MOIDSA is proposed an approach and remaining are self-developed existing algorithms based on a well-known objective approach for gray level image enhancement. The third column of Table 4 gives information about optimal a, b, c, and k parameters for all images with different algorithms. These optimal parameters play a crucial role in the construction of enhanced image using global intensity transformation function. For example, observe MAN image optimal parameters of different optimization algorithms, no such set of parameters are identical. And the interesting point is optimal parameters set given by MOIDSA have the capability of producing best-enhanced image than other algorithms. And it is also observed that performance measures and quality metrics of the enhanced images generated by other algorithms are approximately similar.

The reason is parameter tuning that is purely dependent on objective function during the optimization process. For CS, ABC, DE, and PSO algorithms, the objective is the maximization of the objective function (existing) given in Appendix. But MOIDSA is Pareto optimal IDSA which gives the best-compromised solution (a, b, c, and k parameters) that should have a better tradeoff between all the three objectives which could not happen in the existing objective function. The expression for existing objective function is given in Appendix; it is a combination of three performance measure intensity, edge pixels, and image entropy that has to be maximized to obtain the enhanced image. During the optimization process, optimal parameter selection is based on objective function value. That means performance measure (objectives) in the objective function will not have any priority or balance between objectives. And the duty of any optimization algorithm is to optimize (minimize/maximize) based on objective function. In the present problem, CS, ABC, DE, and PSO algorithms maximized the objective function to its possible extent with corresponding solutions, but it leads to over enhanced images.

From Table 3, it is evident that PSNR and MSSIM values of MOIDSA are higher than other algorithms for all images. But the time consumed for MOIDSA is comparatively higher than other algorithms. The convergence characteristics of MOIDS algorithm for the corresponding images are shown in Figs. 3 and 4 respectively. In

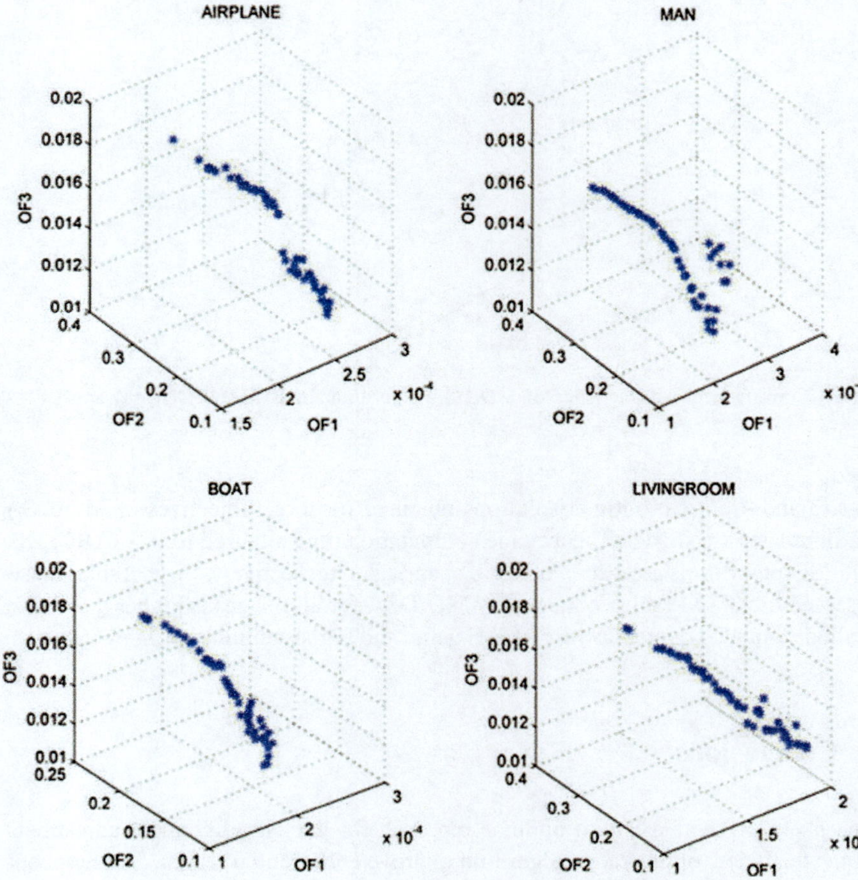

Fig. 3 Convergence characteristics of MOIDSA algorithm for AIRPLANE, MAN, BOAT, and LIVINGROOM images

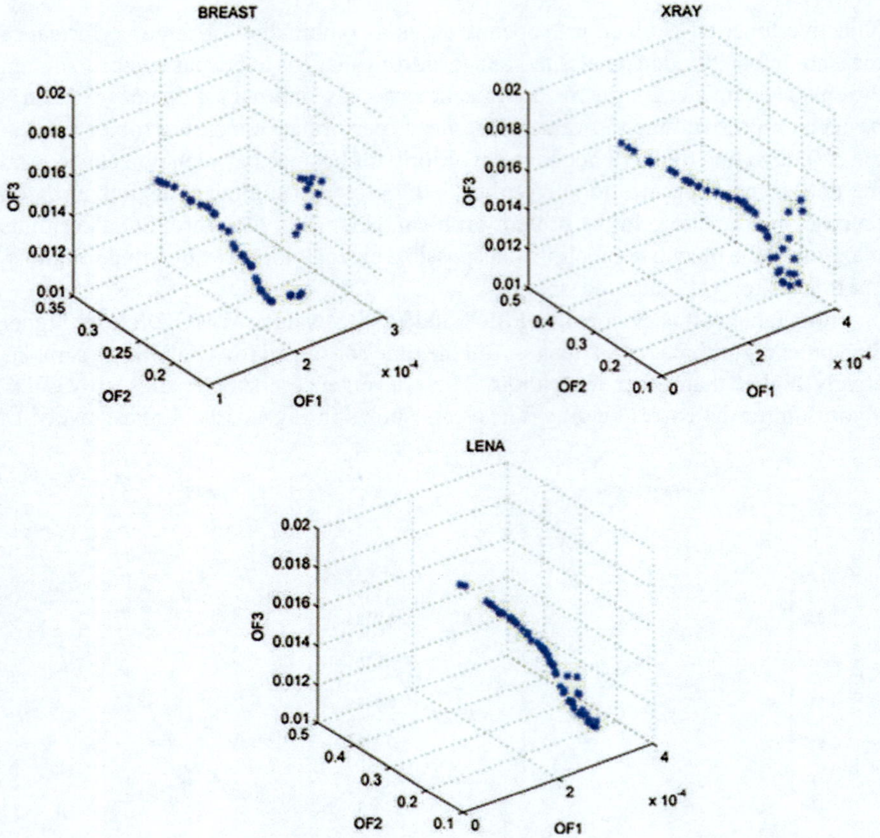

Fig. 4 Convergence characteristics of MOIDSA algorithm for BREAST, X-RAY, and LENA images

Figs. 3 and 4, Pareto optimal solutions obtained for three objectives of MOIDSA for all images are shown. The average computation time required for CS, ABC, DE, and PSO algorithms are 120 s, 60 s, 58 s, and 57 s, respectively. From Table 3, it is observed that PSNR value obtained by MOIDSA for all images is higher than other methods and also quoted lowest RMSE value and highest similarity index value.

4 Conclusions

This paper presents a Pareto optimal approach for gray level contrast adjustment using an efficient optimization algorithm improved differential search. Simultaneous multi-objective optimization approach has been discussed for gray level contrast adjustment. The proposed MOIDSA has been tested on some standard images and proved to be efficient than other existing algorithms such as CS, ABC, DE, and PSO. From the obtained results, it may be concluded that MOIDSA, i.e., three-dimensional optimization approach is better than existing objective-based one-dimensional optimization approach. All quantitative and qualitative results have proved the efficiency of the proposed approach for the gray level contrast adjustment. Hence, it may be concluded that for gray level contrast adjustment MOIDSA outperforms the others.

References

1. Pal SK, Bhandari D, Kundu MK (Mar. 1994) Genetic algorithms for optimal image enhancement. Pattern Recognit Lett 15(3):261–271
2. Kwok NM, Ha QP (2009) Contrast enhancement and intensity preservation for gray-level images using multiobjective particle swarm optimization. IEEE Trans Autom Sci Eng 6(1):145–155
3. Sarangi PP, Mishra BSP, Majhi B,Dehuri S (2014) Gray-level image enhancement using differential evolution optimization algorithm. In: 2014 international conference on signal processing integrated networks, pp 95–100
4. Xuanhua L, Qingping H, Xiaojian K, Tianlin X (2013) Image enhancement using hybrid intelligent optimization. In: 2013 fourth international conference on intelligence system design and engineering applications, pp 341–344
5. Gao Q, Zeng G, Chen D, He K (2011) Image enhancement technique based on improved PSO algorithm. In: Industrial Electronic and Appllication (ICIEA), 2011 6th IEEE conference, pp 234–238
6. Chen H, Tian J (2011) Using particle swarm optimization algorithm for image enhancement. In: 2011 international conference on uncertainty reasoning knowledge engineering, pp 154–157
7. Hanmandlu M, Verma OP, Kumar NK, Kulkarni M (2009) A novel optimal fuzzy system for color image enhancement using bacterial foraging. IEEE Trans Instrum Meas 58(8):2867–2879
8. Hanmadlu M, Arora S, Gupta G, Singh L (2013) A novel optimal fuzzy color image enhancement using particle swarm optimization. In: 2013 sixth international conference on contemporary computing (IC3), vol 8, pp 41–46
9. Coelho LS, Sauer JG, Rudek M (2009) Differential evolution optimization combined with chaotic sequences for image contrast enhancement. Chaos, Solitons and Fractals 42(1):522–529
10. Braik M, Sheta A, Ayesh A (2007) Image enhancement using particle swarm optimization. Proc World Congr Eng I(1):1–6
11. Gorai A, Ghosh A (2009) Gray-level image enhancement by particle swarm optimization. In: 2009 World Congress Nature and Biologically Inspired Computing no. 1, pp 72–77
12. Abunaser A, Doush IA, Mansour N, Alshattnawi S (2015) Underwater image enhancement using particle swarm optimization. J Intell Syst 24(1):99–115
13. Agrawal S, Panda R (2012) An efficient algorithm for gray level image enhancement using cuckoo search. Swarm Evol Memetic Comput 7677:82–89
14. Draa A, Bouaziz A (2014) An artificial bee colony algorithm for image contrast enhancement. Swarm Evol Comput 16:69–84
15. Munteanu C, Rosa A (2004) Gray-scale image enhancement as an automatic process driven by evolution. IEEE Trans Syst Man, Cybern Part B Cybern 34(2): 1292–1298
16. Al-Betar MA, Alyasseri ZAA, Khader AT, Bolaji AL, Awadallah MA (2016) Gray image enhancement using harmony search. Int J Comput Intell Syst 9(5):932–944
17. Nickfarjam AM, Ebrahimpour-Komleh H (2017) Multi-resolution gray-level image enhancement using particle swarm optimization. Appl Intell 47(4):1132–1143
18. Qiu J, Harold Li H, Zhang T,Ma F, Yang D (2017) Automatic x-ray image contrast enhancement based on parameter auto-optimization. J Appl Clin Med Phys 18(6): 218–223

19. Sowjanya K, Kumar PR (2017) Gray level image enhancement using nature inspired optimization algorithm: an objective based approach. World J Model Simul 13(1):66–80
20. Injeti SK (2017) A Pareto optimal approach for allocation of distributed generators in radial distribution systems using improved differential search algorithm. J Electr Syst Inf Technol
21. Gonzales LRC, Woods RE (2010) Steven, Digital image processing, 2nd ed. Tata Mc Graw Hill, New Delhi
22. Saitoh F (1999) Image contrast enhancement using genetic algorithm. In: IEEE SMC'99 conference proceedings. 1999 IEEE international conference on systems, man, and cybernetics (Cat. No.99CH37028), vol 4, pp 899–904
23. Zimmerman JB, Pizer SM, Staab EV, Perry JR, McCartney W, Brenton BC (1988) Evaluation of the effectiveness of adaptive histogram equalization for contrast enhancement. IEEE Trans Med Imaging 7(4):304–312
24. Hashemi S, Kiani S, Noroozi N, Moghaddam ME (2010) An image contrast enhancement method based on genetic algorithm. Pattern Recognit Lett 31(13):1816–1824
25. Civicioglu P (2012) Transforming geocentric cartesian coordinates to geodetic coordinates by using differential search algorithm. Comput Geosci 46:229–247
26. Storn R, Price K (1997) Differential evolution–a simple and efficient heuristic for global optimization over continuous spaces. J Glob Optim 341–359
27. Kotte S, Kumar PR, Kumar S (2016) An efficient approach for optimal multilevel thresholding selection for gray scale images based on improved differential search algorithm. Ain Shams Eng J
28. Wang Z, Bovik aC, Sheikh HR, Simmoncelli EP (2004) Image quality assessment: form error visibility to structural similarity. Image Process IEEE Trans 13 (4): 600–612

Emotion Recognition from Speech Biometric System Using Machine Learning Algorithms

Chevella Anil Kumar and Kancharla Anitha Sheela

Abstract In today's world, emotions play a crucial role in human-to-human and human-to-machine interactions and that is important for both fair and good decisions. The display of emotions gives an important information of the mental state of a human being and this opened up a new research area with a fundamental goal of understanding and recovering desired emotions. In the past years, several biometrics have been used to classify different emotions (facial expressions, speech and physiological signals, etc.) of human beings. Some intrinsic advantages make speech signals became a good source for affective computing. There are several applications in real time for communicating with robots, and also in web-based e-learning and commercial applications to detect emotions from speech, etc. In this work, LPCC, MFCC, and the features from high-amplitude regions of each VOP (vowel onset points)-extracted syllable are selected and used to classify different emotions using the machine learning classifier (SVM). This study uses six emotions: anger, disgust, fear, happy, sad, and surprised.

Keywords SER · LPCC · MFCC · SVM · Consonant region · Vowel region · Transition region · Spectral features · Vowel onset point

1 Introduction

Speech Recognition: Speech is an acoustic signal containing knowledge about an idea which is created in the mind of the speaker and it is a process of identifying words and phrases in speech. In today's advanced technology, speech signals are used for biometric authentication in different fields. Speech characteristics are reasonably slowly varying in a short-term period. The content in the voice signal is expressed

C. Anil Kumar (✉) · K. Anitha Sheela
Department of ECE, JNTUH College of Engineering, Hyderabad, India
e-mail: Chevellaanilkumar@gmail.com

K. Anitha Sheela
e-mail: kanithasheela@gmail.com

T. Laxminidhi et al. (eds.), *Advances in Communications, Signal Processing, and VLSI*, Lecture Notes in Electrical Engineering 722,
https://doi.org/10.1007/978-981-33-4058-9_6

Fig. 1 Emotion recognition system from speech

by the speech waveform's short-term amplitude range. This helps us to distinguish features from expression (phonemes), based on the short-term amplitude. The speech is highly variable due to various speakers, speed, content and acoustic conditions. Hence, its became a very difficult to recognize the emotions from speech.

Speech Emotion Recognition: Recognition of emotion from speech is one of the latest difficulties of word processing. Of addition to human facial expressions, voice has proved to be one of the most effective ways of identifying human emotions automatically. Recognition of emotions from the speaker's speech is extremely difficult due to the following reasons. Due to the different speaking styles and speaking levels, acoustic variation will be introduced which directly affects the characteristics of speech emotions. The same utterance will show different emotions. That emotion can correspond to various parts of the spoken utterance. Therefore, these parts of utterance are very difficult to differentiate.

Another challenge is that emotional expression depends on the speaker and his background and surroundings. The speech style is also changed as the society and climate change, which is another challenge facing in this system. The system for emotional recognition comprises five key steps: input, features extraction, collection of features, classification, and output (Fig. 1).

2 Literature Survey

Starting in the 1930s, several efforts were made to identify affective states from vocal knowledge. For study, some significant voice function vectors were selected. Stevens et al. (1972) studied emotional spectrograms and their acted speech [1]. They found similarities and therefore suggested using data that had been acted upon. Murray and Arnott (1993) presented a qualitative correlation between the characteristic emotion and speech [2, 3]. Petrushin (1998) compared emotion perception in speech between humans and machines, and achieved similar levels for both [4]. Vinay, Shilpi Gupta, and AnuMehra are doing a study on gender-specific emotion recognition by speech signals [5]. Mel-frequency cepstral coefficients were used to manipulate the dynamic variability along with an utterance. For six emotions, Nwe (2001) discussed about MFCC features and hidden Markov models [7]. We have different types of classifiers but SVM classifier provides efficient classification even if less samples are available, and hence it is commonly used to identify speech emotions [9, 9]. Yu et al. (2002) achieved a 73% accuracy using support vector machine. In a call center setting, Lee (2002) tried to differentiate positive and negative emotions using linear discrimination. K-nearest neighbor and SVM combinedly reached a precision rate of 75%.

3 Feature Extraction

Extraction of features from speech requires a lot of attention because recognition output is heavily dependent on this step and it is very important to decide which features are used to differentiate different emotions. Several features (strength, pitch, formant, etc.) are extracted in recent years. In this study, we used feature-level fusion in which LPCC, MFCC, and features from high-amplitude regions of each VOP (vowel onset points)-extracted syllable are selected and used to distinguish different emotions.

3.1 Linear Prediction Coefficents (LPCC)

LPCCs are used to predict a speech signal's basic parameters. In LPCC, current speaking sample is a linear combination of past speaking samples at present time and LPCC algorithm is indicated in Fig. 2.

Input signal is emphasized using high-pass filter since the energy is contained less in high frequencies compared to low frequencies. Hence, to boost energy in high frequency, we have to go for pre-emphasis. For this filter transfer function is expressed as

$$H_p(z) = 1 - a\,z^{-1}$$

The output signal of second step is broken down into frames and discontinuities in signal are reduced by windowing. Due to its smoothness in low-pass [11], the most common window used is the hamming window, and is described as

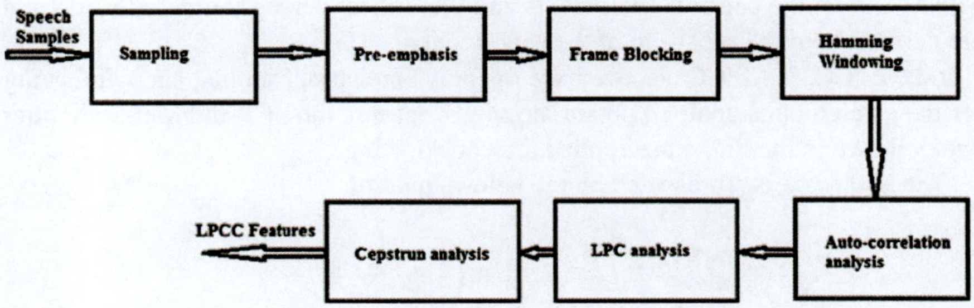

Fig. 2 Linear prediction cepstral coefficient

$$w(n) = 0.54 - 0.46\cos\left(2\Pi\frac{n}{N}\right);\ 0 \le \mathrm{n} \le \mathrm{N})$$

N = length of window. Character of the sound being created by the vocal tract form is assessed by linear predictive analysis. A filter is used to model the vocal tract and transfer function represented as

$$\mathrm{V(z)} = \frac{G}{1 - \sum_{K=1}^{P} a_k z^{-k}}$$

where V(z) is a feature of moving vocal tract. G is filter gain, a_k = linear prediction coefficients, p = order of filter. Autocorrelation method is best for estimating the LPC coefficients and filter gain.

The true cepstrum is defined as the inverse FFT of the speaking magnitude logarithm defined by the following equation:

$$\widehat{s[n]} = \frac{1}{2\pi} \int_{-\pi}^{+\pi} \ln[S(w)] e^{jwn} dw$$

where $S(w)$ and s(n) show the Fourier spectrum pair of a signal [13]. However, using LPCC also we can extract the cepstral coefficients using following equation, where a_k = linear prediction coefficients, p =order of the filter

$$C(n) = \begin{cases} a_m + \sum_{k=1}^{m-1} \frac{k}{m} C_k a_{m-k} & for\ 1 < m < p \\ \sum_{k=m-p}^{m-1} \frac{k}{m} C_k a_{m-k} & for\ m > p \end{cases}$$

3.2 MFCC

MFCC extract the cepstral coefficients and that reflect perception-based audio and are derived from the cepstrum of the Mel frequency (Fig. 3).

Like LPCC, in MFCC also we have to apply emphasis, framing, and windowing for the given input signal. To obtain the Mel spectrum, the FFT and Mel-scale filter banks shown in the Fig. 4 are applied after windowing.

The Mel scale is represented by the below equation:

$$Mel_f = 2595 \ln\left(1 + \frac{f}{700}\right)$$

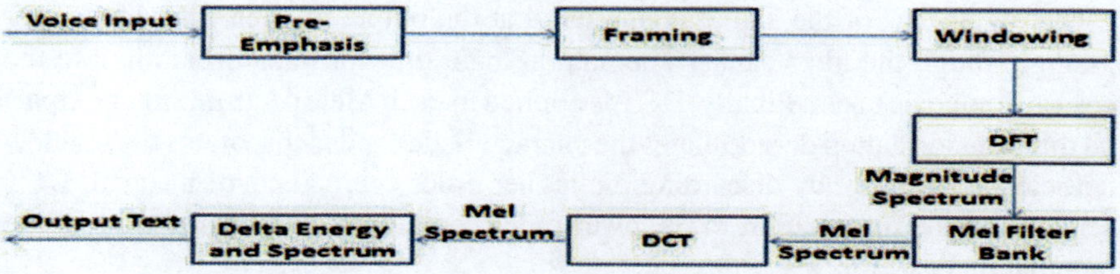

Fig. 3 Mel Frequency cepstal coefficient

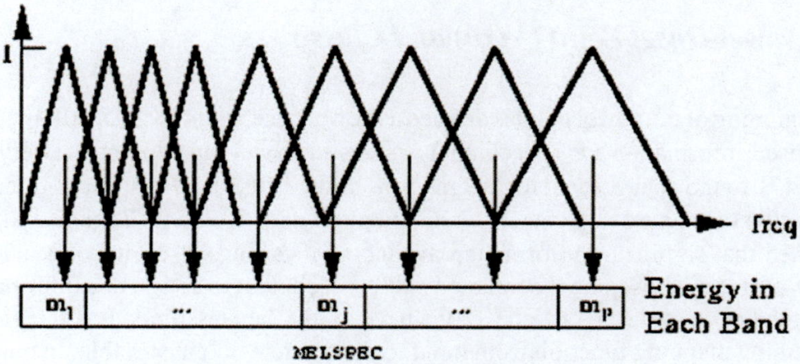

Fig. 4 Mel-scale filter bank

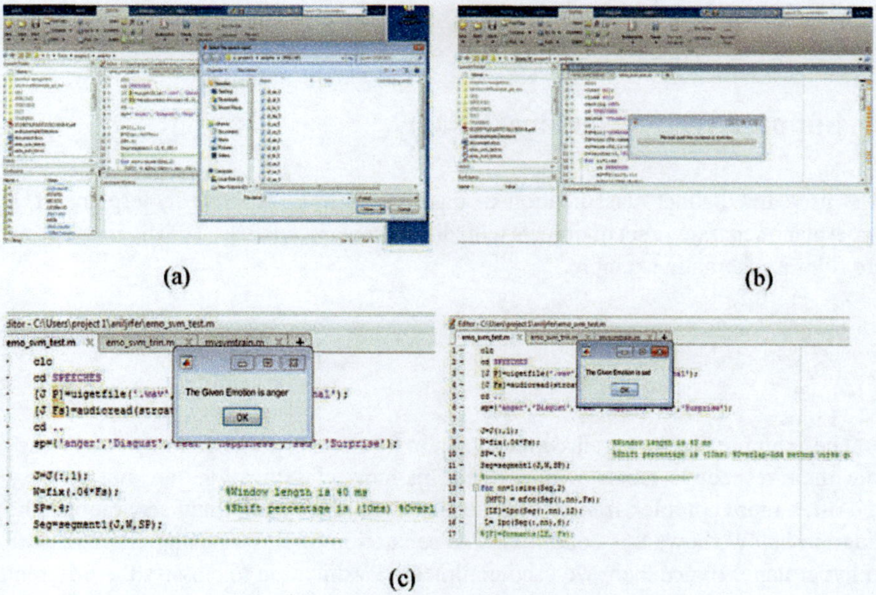

Fig. 5 **a** Dataset **b** Training dataset **c** Testing of different emotions

The log energy of the signal is measured at the output of each filter bank after passing through the filter banks. The natural logarithm for transformation into the cepstral domain is taken. Finally, DCT is applied to each Mel spectrum (filter output) and this transformation de-correlates the characteristics and incorporates the first few coefficients. Because, by discarding the higher order coefficients of a signal, DCT accumulates the information to its lower order coefficients and also it reduces the computation cost.

3.3 *Vowel Onset Point (VOP) and Classifiers*

The beginning of the vowel point is defined as vowel onset point [14, 15, 20]. Essential and biased information for speech analysis lies in vowel onset point. Kewely-Port et al. [17] found that a speaking segment is limited to 20–40 ms in the transition from a stop consonant to a vowel (V) on the place of articulation. Tekieli et al. [18] discussed that significant information available in starting 10–30 ms of V and CV and 40–50 ms of the segment contained sufficient content to assign this information.

To predict classes (targets/labels), we have to use the classifiers. It is the process of obtaining mapping function from input to the discrete output variable. In machine learning, we have various classification algorithms, but in this paper support vector machine (SVM) classifier is suggested, because it achieves good precision even with small test samples.

4 Support Vector Machine (SVM)

It is used for distinct classification of data points by a separate hyperplane. The hyperplane separates a set of objects with different classes/labels. The diagram below provides a schematic example.

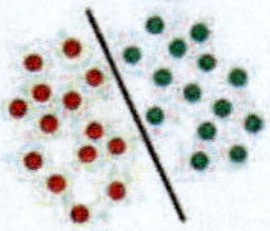

Linear SVM

The example of a linear classifier shown in figure, i.e., SVM dividing two classes into their respective labels with a line. But most classifications are not so easy and often more complex frameworks necessary to make an optimal separation. The diagram below shows this condition. The separation of classes using curve instead of hyperplane. Hence, here we used nonlinear classification to classify the different classes of objects.

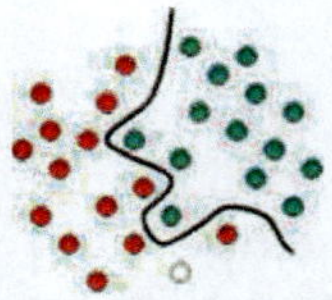

Non-Linear SVM

5 Database

The archive consists of 30 recordings of 5 speakers each delivering six different emotional speeches with duration of 2 s and sampling frequency 44,100 Hz. Since the training samples available for recognition of emotions are not high, we used non-asymptotic mode to prevent data overfitting. In this, we used cross-validation techniques for early training stoppage, where (1−r)N number of input samples are used for research, and rN number of samples are used for validation purposes. The R_{opt} used to divide the training data between assessment and validation is provided by following equation:

$$R_{opt} = 1 - \frac{\sqrt{2w-1}-1}{2(w-1)}$$

where w = total number of free parameters.

6 Results

LPCC, MFCC, and VOP function values are shown below after training and Testing using the SVM classifier and also confusion matrix for six different emotions (Fig. 5 and Tables 1 and 2).

7 Summary and Conclusion

In this study, 9 LPCC, 13 MFCC features, and 5 features from high-amplitude regions (consonant, vowel, and transition regions) of each syllable are extracted from VOP (vowel onset points) and fused as a single set of features, and support vector machine is used to distinguish six emotions from these speech dataset features. The database has 2-s speech recordings with a $f_s = 44{,}100$ Hz. For pre-processing with a frame length of 40 and 10 ms, hamming window is used. Eventually six emotions (anger, disgust, anxiety, joy, sad, and surprised) achieved with approximately 76.6% accuracy of recognition.

8 Future Scope

As the technology grows in today's world, recognition of emotions from speech signals plays an important role in real-time applications and human-to-machine communication. Number of ways exist in which the emotions can be recognized from speech data. The deep convolution neural network (DCNN) structure can be used to more accurately recognize different emotions from speech datasets compared to algorithms for machine learning. Deep convolution neural networks, however, require a vast amount of training and testing data, and GPUs for processing. However, with the introduction of cloud infrastructure, the above difficulty can be solved. In addition, to increase the availability of the database, we have recently created an anechoic chamber with electroglottographic (EGG) equipment in our department for recording ground truth volumes, labeled voice signal database with various emotions.

Table 1 Different output features for speech dataset

	1	2	3	4	5	6	7	8	9	10	11	12	13
13 MFCC features	9.830	9.993	− 3.24	6.993	− 0.87	1.099	0.71	0.352	− 0.54	1.19	0.297	1.161	0.45
9 LPCC features	1	− 3.31	5.181	− 4.84	2.65	− 0.40	− 0.71	0.62	− 0.17	–	–	–	–
5 VOP features	0.043	0.457	1.000	0.457	0.043	–	–	–	–	–	–	–	–

Table 2 Confusion matrix

	Anger	Disgust	Fear	Happy	Sad	Surprise
Anger	4/6	1/6	1/6	–	–	–
Disgust	–	3/6	–	–	1/6	2/6
Fear	–	–	4/6	–	1/6	2/6
Happy	–	–	–	5/6	–	1/6
Sad	1/6	–	1/6	–	4/6	–
Surprise	1/6	–	1/6	–	1/6	3/6

References

1. Williams CE, Stevens KN (1972) Emotions and speech: some acoustical correlates. 52(4B):1238
2. Murray IR, Arnott JL (1993) Towards the simulation of emotion in synthetic speech: a review of the literature of human vocal emotion. J Acoust Soc Am 93(2):1097–1198
3. Gustafson-Capková S (2001) Emotions in speech: tagset and acoustic correlates. Speech Technology, term paper Autumn–01 Sofia Gustafson-Capková
4. Petrushin, VA (1998) How well can people and computers recognize emotions in speech? In: Proceedings of the AAAI Fall Symposium, pp 141–145
5. Vinay, Gupta S, Mehra A (2014) Gender-specific emotion recognition through speech signals. In: 2014 International conference on signal processing and ıntegrated networks (SPIN), 978-1-4799-2866-8/14/$31.00 ©2014, IEEE
6. Abdel-Hamid O et al (2014) Convolutional neural networks for speech recognition. In: IEEE/ACM transactions on audio, speech, and language processing, vol. 22, no. 10, October 2014, 2329–9290 © 2014, IEEE
7. Nwe TL, Wei FS, De Silva LC (2001) Speech based emotion classification. IEEE Catalogue No. 01CH37239 0-7803-7101-1/ 01 / $10.00 © 2001, IEEE
8. Yu F, Chang E, Xu YQ, Shum HY (2001) Emotion detection from speech to enrich multimedia content. In: The Second IEEE Pacific-Rim Conference on Multimedia, October 24–26, 2001, Beijing, China
9. Schmitt AAPATP (2010) Advances in speech recognition. Springer, pp 191–200
10. Koolagudi KSRSG (2012) Emotion recognition from speech using global and local prosodic. Springer
11. Han Y, Wang GY, Yang Y (2008) Speech emotion recognition based on MFCC. J Chongqing Univ Posts Telecommun 20
12. Octavian C, Abdulla W, Zoran S (2005) Performance evaluation of front-end processing for speech recognition systems
13. Rabiner LR, Shafer RW (2009) Digital processing of speech signals, 3rd edn. Pearson education in south Asia
14. Hermes DJ (1990) Vowel-onset detection. J Acoust Soc Am 87:866–873
15. Sekhar CC (1996) Neural network models for recognition of stop consonant-vowel (SCV) segments in continuous speech. Ph.D. thesis, Indian Institute of Technology Madras, Department of Computer Science and Engineering, Chennai, India
16. Wang J-F, Chen S-H (1999) A C/V segmentation algorithm for Mandarin speech signal based on wavelet transforms. In: Proceedings of the IEEE International Conference Acoustics, Speech, Signal Processing, vol 1, pp 1261–1264
17. Kewley-Port D, Pisoni DB (1983) Perception of static and dynamic acoustic cues to a place of articulation in initial stop consonants. J Acoust Soc Am 73(5):1779–1793
18. Teikeli ME, Cullinan WL (1979) The perception of temporally segmented vowels and consonant-vowel syllables. J Speech Hear Res 22:103–121

19. Furui S (1986) On the role of spectral transition for speech perception. J Acoust Soc Am 80(4):1016–1025
20. Yegnanarayana B et al (2005) Detection of vowel onset point events using excitation ınformation. Interspeech 2005, September, 4–8, Lisbon, Portugal

Implementation of Single Image Histogram Equalization and Contrast Enhancement on Zynq FPGA

Prathap Soma, Chundru Sravanthi, Palaparthi Srilakshmi, and Ravi Kumar Jatoth

Abstract In present days, image processing is very much needed in different fields like medicine, acoustics, forensic sciences, agriculture, and industrial applications. These image processing algorithms are most preferably implemented on field-programmable gate arrays (FPGA) as they are reprogrammable to perform required operations. This paper describes the implementation of histogram equalization on Zynq FPGA which consists of ARM cortexA9 processor along with FPGAs. To implement this, Intellectual Property (IP) cores are generated in Vivado high-level synthesis (HLS) tool to figure out and equalize the histogram. These IP cores are brought down to Vivado to create the required hardware. Finally, the design was programmed into Zynq FPGA. The software application is developed using software development kit (SDK). The results are obtained for an image size of 259 × 194. The utilization report shows that implemented design has taken less number of hardware resources as compared with the Kintex KC705 evaluation board. The processing time taken for execution is 9.288 nsec.

Keywords Histogram equalization · Hardware architecture · Zynq 702 · Vivado design suite · Contrast enhancement

P. Soma (✉) · R. K. Jatoth
Department of ECE, National Institute of Technology, Warangal 506004, India
e-mail: prathap.soma@gmail.com

R. K. Jatoth
e-mail: ravikumar@nitw.ac.in

C. Sravanthi · P. Srilakshmi
Department of ECE, National Institute of Technology, Tadepalligudem, Andhra Pradesh, India
e-mail: sharonsravanthi09@gmail.com

P. Srilakshmi
e-mail: srilakshmipalaparthi31@gmail.com

T. Laxminidhi et al. (eds.), *Advances in Communications, Signal Processing, and VLSI*, Lecture Notes in Electrical Engineering 722,
https://doi.org/10.1007/978-981-33-4058-9_7

1 Introduction

Image processing is a process of converting an image into digital form and performs some adequate operations on it to obtain an enhanced image or to extract some useful information from the image. It is among the rapidly growing technologies today, orienting its applications in numerous aspects of the business. Image processing also provides vital research area within modern computer science and engineering disciplines, respectively.

The main objective of image processing is to ensure visualization, image sharpening [1] and restoration [2], measurement of pattern, image recognition [3] to distinguish the objects in an image, and image retrieval (Fig. 1).

Analog and digital image processing are the two types of image processing methods. Analog image processing can be used for hard copies such as photographs and printers. Digital processing involves the manipulation of the digital form of the image by using a computer. Initially, digital signal processing (DSP) algorithms for low-rate applications are implemented using general-purpose programmable digital signal processing chips, and for higher rate applications special-purpose fixed-function digital signal processing chip-sets and application-specific integrated circuits (ASICs) [4] are used. The advancement in technology of field-programmable arrays explored a new path for implementing DSP algorithms. FPGA [5] provides design adaptability and flexibility with optimal utilization of device conserving system power and board space which may not be seen often in DSP chips. FPGAs are also considered to be advantageous over ASICs as they avoid the inability of making modifications in design after manufacturing and high development cost. A wide range of applications such as arithmetic functions for fast Fourier transforms as well as filtering and convolution algorithms can be implemented and reconfigured on the Xilinx FPGA. The further advancement in FPGAs in the objective of integrating a complete ARM Cortex-A9 MP [6] core processor-based system processor drives

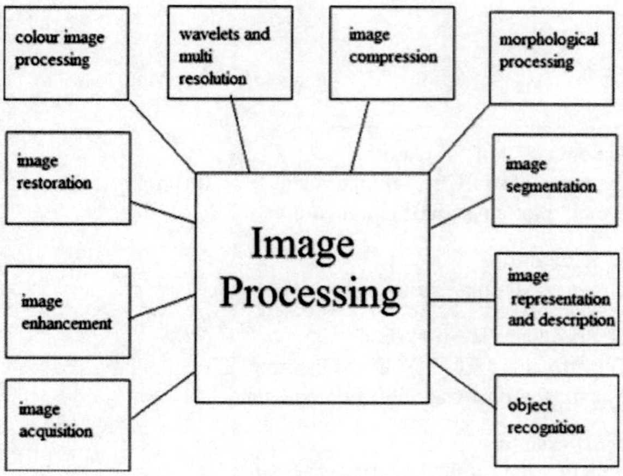

Fig. 1 Block diagram of basic concepts of image processing

to the new invention Zynq. It provides even more ideal and suitable platform for the implementation of flexible SoCs.

This paper is ordered as follows: Section II describes histogram equalization; Section III represents hardware implementation of histogram equalization; in Section IV, results and discussion are provided. Finally, Section V summarizes conclusion.

2 Histogram Equalization

A. Histogram

A histogram [7] precisely represents the frequency of occurrence of numerical data. The image histogram is a graphical representation of the distribution of brightness levels in a digital image. It is a graph between brightness levels and the number of pixels in the image. On seeing the histogram, one can estimate the distribution of brightness levels of an image. On the horizontal axis, brightness levels which are also known as tones are plotted and on the vertical axis, the number of pixels of that particular tonal value in the image is plotted. The values toward the origin in the horizontal axis represent dark tones and the higher brightness levels represent bright and white tones. If most of the data points are closer to the vertical axis, then the image is very dark. If the data points are on the right side of the graph, then the image is very bright.

A color histogram is an N-dimensional histogram where N denotes the number of spectra in an image. For an RGB image, there are three channels, namely, red, green, and blue. So, an RGB image will have a three-dimensional histogram. Plotting histograms for each of the channels separately is a color histogram.

2.1 Contrast Adjustment Using Histogram Equalization

Histogram equalization is an image processing technique in which histogram of an image is stretched so that the contrast of that image is enhanced. By this technique, intensity values of the image can be distributed effectively on the histogram. If the image is over-exposed or under-exposed, histogram equalization is used. Therefore, the technique is used in X-rays, photography, astronomy, etc.

The histogram equalization of an image can be represented as

$$Y = \frac{[(X - X_{\min})]}{[(X_{\max} - X_{\min})]} \cdot Y_{\max} \tag{1}$$

where $X_{\min}$, $X_{\max}$ are the minimum and maximum intensity level original images, $Y_{\max}$ is maximum intensity value of the image ($Y_{\max}$ for a grayscale image

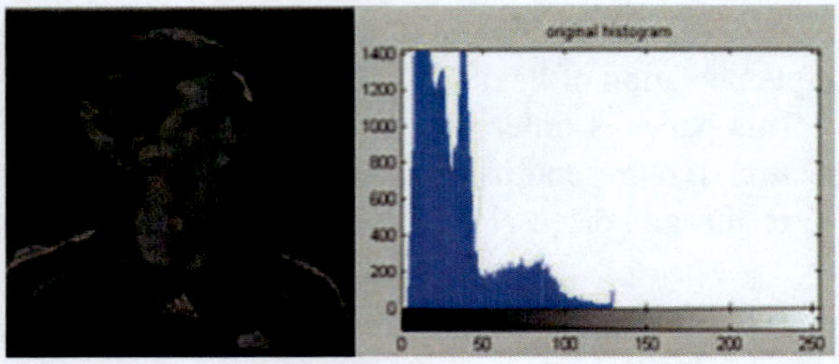

Fig. 2 Original image and its histogram

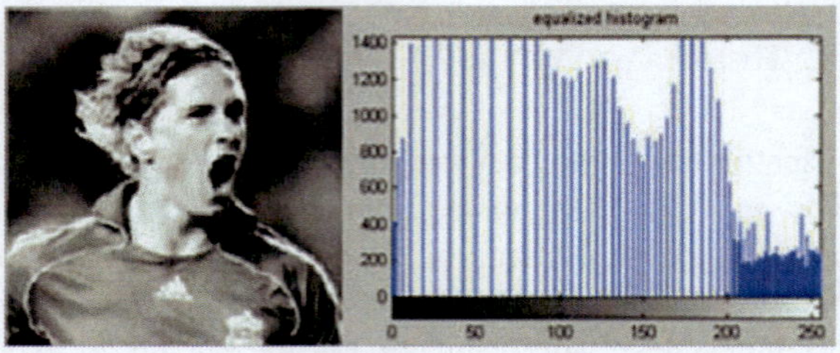

Fig. 3 Equalized histogram and its image

is 255), X is intensity level of original image's histogram, and Y is the modified intensity level after histogram equalization (Figs. 2 and 3).

3 Hardware Implementation

A. Zynq FPGA

Usually, image processing algorithms are implemented on field-programmable gate arrays (FPGAs). These provide the flexibility of reprogramming in the implementation of different logic functions and reconfiguration in computing. FPGAs are made of prefabricated silicon chips whose architecture includes three major components, namely, programmable logic blocks (to implement logic functions), interconnects (for programmable routing), and I/O blocks (which are used to make off-chip connections). The further advancement in FPGAs in the objective of integrating a complete arm Cortex-A9 MP Core processor-based system processor drives to the new invention "Zynq" by Xilinx.

As the metal zinc mixes with variable metals to give alloys of desirable properties, Zynq [8] is very flexible for various applications. Zynq consists of the processing system and a programmable logic. Processing system contains a dual-core ARM Cortex-A9 processor to give a higher performance with a clock speed of 1 GHz. It also consists of peripheral interfaces, memory interfaces, cache memory, and a clock generation circuitry. And the programmable logic composed of FPGAs (Fig. 4).

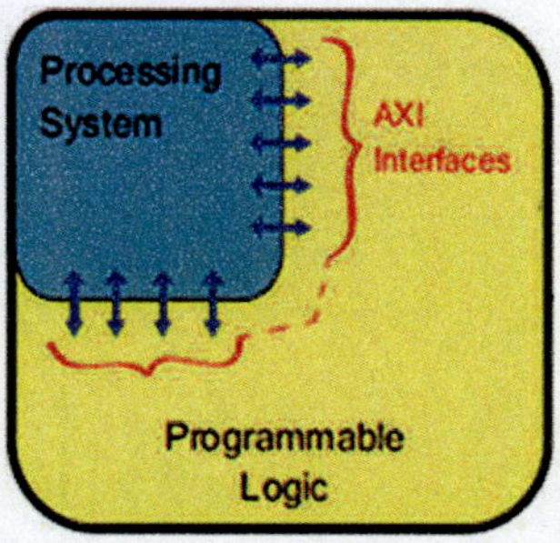

Fig. 4 ZYNQ structural design

B. ***Tools and utilities for Xilinx FPGAs***

There are numerous tools available to develop the code for Xilinx FPGAs. The Vivado design suite [9] is one of the important tools for configuring Xilinx FPGAs. The Vivado is introduced for seven-series FPGA families such as Virtex-7 [10], Kintex [11], Artix [12], and Zynq [8]. Vivado is an advanced version of the Xilinx integrated synthesis environment (ISE) design suite [13], the tool previously provided by Xilinx company for programming FPGAs. Vivado provides intellectual property (IP) cores [14] which are blocks of logic or data that are used in making a field-programmable gate array (FPGA) or application-specific integrated circuit (ASIC) for a product. These IP cores cover almost all the fundamental functions and can help us to considerably reduce our development time while implementing the image processing algorithms in FPGAs. Vivado includes a high-level synthesis (HLS) tool to generate further complex C-based IP cores in a high-level language [15] such as "C, C++ , or System C." The Xilinx software development kit (SDK) [16] is a tool for designing and programming embedded processors inside the FPGA chip.

C. ***Design***

Figure 5 shows the building blocks of histogram equalization. The processor processes the high-level language code written in Vivado SDK. Initially, an image is stored in BRAM [17] as a header form, and then sent to the histogram block and the contrast block simultaneously through a broadcaster. Histogram and contrast blocks are the IP cores generated in HLS software to create a histogram and to equalize the histogram, respectively. Broadcaster is an IP block which transmits the received data to multiple outputs. BRAM stores the pixel values of the histogram generated in the histogram block. Contrasted image is again sent to the BRAM through histogram block.

Figure 6 indicates the hardware architecture of the proposed algorithm.

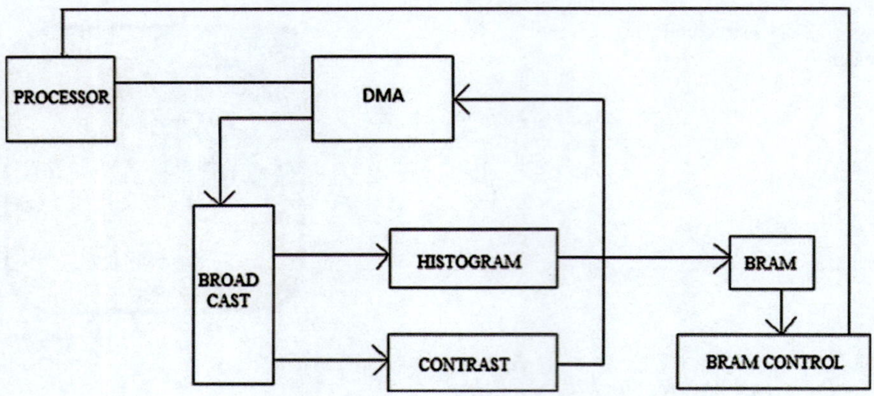

Fig. 5 Basic building blocks for histogram equalization on Zynq FPGA

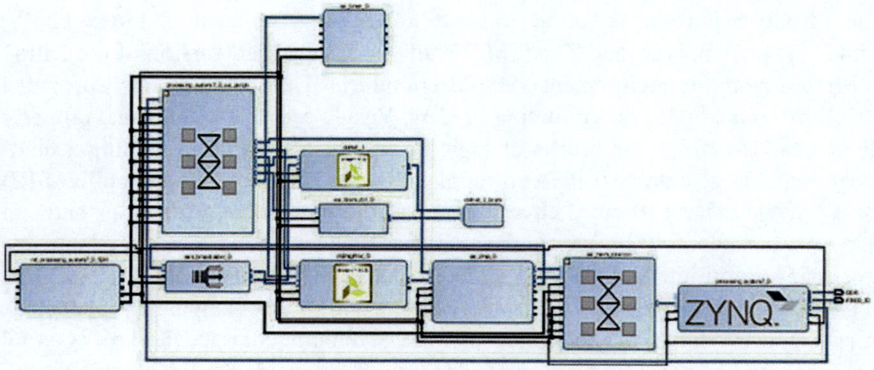

Fig. 6 Hardware architecture of histogram equalization on Zynq FPGA

4 Experimental Results and Discussions

To validate the design, an image of size 259 × 194 was given to the designed blocks through high-level language code in the SDK tool. The output was obtained in the form of a text file, which contains pixel values of the contrasted image. It can be verified using MATLAB.

Tables 1 and 2 show the hardware resources utilized by the Kintex 7 KC705 and Zynq ZC702 evaluation board of histogram equalization algorithm. It is observed that the proposed algorithm on ZC702 scales down the utilized resources (BRAMs) approximately by 12 times.

From Table 3, it is observed that the processing time of the proposed implementation on ZC702 turns down from milliseconds to nanoseconds as compared with the Kintex KC705 (Figs. 7 and 8).

Table 1 Resource usage on Kintex 7 KC705 evaluation board in [18]

S.No	Resources	Available	Utilized	% Resources utilized
1	LUTs	203,800	30,749	15.09
2	FF	407,600	63,361	15.54
3	BRAMs	445	260	58.43
4	LUTRAM	–	–	–
5	BUFG	–	–	–

Table 2 Resource usage on ZYNQ ZC702 evaluation board of proposed algorithm

S.No	Resources	Available	Utilized	% Resources utilized
1	LUTs	53,200	7228	13.59
2	FF	106,400	8340	7.84
3	BRAMs	140	6.5	4.64
4	LUTRAM	17,400	532	3.06
5	BUFG	32	1	3.12

Table 3 Processing time of proposed architecture on different hardware platforms

S.No	Image size	Processing time	
		KC 705 [18]	ZC 702 proposed
1	256 × 256	0.22 ms	9.542 nsec
2	259 × 194	–	9.288 nsec

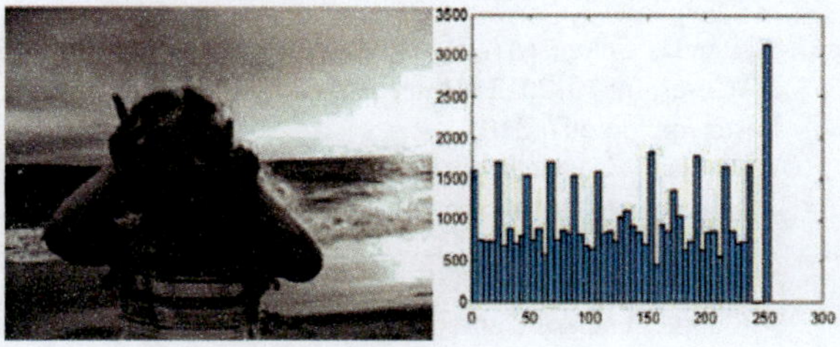

Fig. 7 Image and its histogram of original image

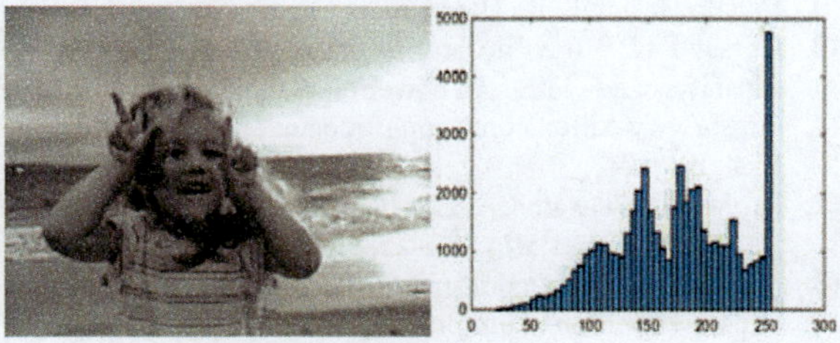

Fig. 8 Image and its histogram after equalization

5 Conclusion

In this paper, we illustrated the implementation of one of the image processing techniques, histogram equalization on the Zynq platform (ZC702 board) of Xilinx, in the hardware block design approach. In this design, various necessary blocks, such as DMA, BRAM, processor, broadcast, etc., were created in the Vivado HLS tool and processed with the help of Vivado and SDK tools of Xilinx.

In future, the same algorithm can be extended to real-time video processing algorithms for contrast enhancement by interfacing the camera module to the FPGA.

References

1. NevriyantoEA, Purnamasari,D (2017) Image enhancement using the image sharpening, contrast enhancement, and standard median filter (Noise Removal) with pixel-based and human visual system-based measurements. In: 2017 International Conference on Electrical Engineering and Computer Science (ICECOS), Palembang, pp 114–119
2. Narmadha J, Ranjithapriya S, Kannaambaal, T (2017)Survey on image processing under image restoration. In: 2017 IEEE International Conference on Electrical, Instrumentation and Communication Engineering (ICEICE), Karur, pp 1–5
3. Zhang X, Li X, Feng F (2019) A survey on free-hand sketch recognition and retrieval.Image Vision Comput
4. Kuon I, Rose J (2007) Measuring the gap between FPGAs and ASICs. IEEE Trans Comput Aided Des Integr Circ Syst 26(2):203–215
5. Teubner J, Woods L (2013) Data processing on FPGAs.In: Data processing on FPGAs, Morgan & Claypool
6. Pleiter, D, Richter, M (2012) Energy efficient high-performance computing using ARM Cortex-A9 Cores. In: 2012 IEEE International Conference on Green Computing and Communications, Besancon, pp 607–610
7. Trahanias PE, Venetsanopoulos AN (1992) Color image enhancement through 3-D histogram equalization. In: Proceedings of the 11th IAPR International Conference on Pattern Recognition. Vol. III. Conference C: Image, Speech and Signal Analysis, IEEE
8. Soma P, JatothRK (2018) Hardware implementation issues on image processing algorithms. In: IEEE International Conference on Computing Communication and Automation 2018 ICCCA 2018, December 14–15, 2018
9. Vivado Design Suite User Guide: Getting Started (UG910)
10. Dorsey P (2010) Xilinx stacked silicon interconnect technology delivers breakthrough FPGA capacity, bandwidth, and power efficiency.Xilinx White Paper: Virtex-7 FPGAs, pp 1–10
11. https://www.xilinx.com/support/documentation/boards_and_kits/kc705/ug810_KC705_Eval_Bd.pdf
12. Parikh PV, Dalwadi BA, Zambare GD (2016) Sobel edge detection using FPGA-Artix®-7. Int J Sci Technol Res 5(6):250–253
13. https://www.xilinx.com/support/documentation/sw_manuals/xilinx11/sim.pdf
14. https://www.xilinx.com/products/intellectual-property/
15. https://www.accellera.org/community/systemc/
16. SDK User Guide: System Performance Analysis (UG1145)
17. Garcia P, Bhowmik D, Stewart R, Michaelson G, Wallace A (2019) A optimized memory allocation and power minimization for FPGA based image processing. J Imaging 5:7
18. Kraft M, Olejniczak M, Fularz M (2017) A flexible, high performance hardware implementation of the simplified histogram of oriented gradients descriptor. Measure AutomMonit 63

Design and Simulation of a Dual-Band Radiometer for Humidity and Temperature Profiling

Akash Tyagi, Bijit Biswas, G. Arun Kumar, and Biplob Mondal

Abstract Remote sensing for weather applications is gaining its importance for accurate measurement of atmospheric parameters, especially in the higher microwave and millimeter wave frequency ranges. In this work, feasibility study on design of a dual-band passive radiometer has been carried out for humidity and temperature profiling of atmosphere. Non-conventional 20–32 GHz frequency band (we mention it as K-band here) and 50–60 GHz frequency band (V-band) have been used for sensing atmospheric humidity and temperature, respectively. Few practical applications of passive radiometer for temperature and humidity sounding have been investigated through literature survey and a design of dual-band radiometer RF front end for atmospheric profiling has been done. System-level simulation has been carried out in SystemVue and results have been produced.

Keywords Atmospheric profiling · Dual band · Humidity sounding · Radiometer · Temperature sounding

1 Introduction

The millimeter wave frequency range includes the absorption band of the atmosphere and is therefore important for atmospheric remote sensing and research; hence,

A. Tyagi · B. Mondal
ECE Department, Tezpur University, Tezpur, Assam, India
e-mail: akashtyagi087@gmail.com

B. Mondal
e-mail: biplob.tezu@gmail.com

B. Biswas
Circuits and Systems Division, SAMEER, Kolkata, WB, India
e-mail: bijit@mmw.sameer.gov.in

G. A. Kumar (✉)
ECE Department, NIT Warangal, Warangal, Telangana, India
e-mail: g.arun@nitw.ac.in

T. Laxminidhi et al. (eds.), *Advances in Communications, Signal Processing, and VLSI*, Lecture Notes in Electrical Engineering 722,
https://doi.org/10.1007/978-981-33-4058-9_8

millimeter wave atmospheric remote sensors, e.g., radiometers, weather radars, are key instruments used for measurements of vertical profiles of atmospheric temperature and humidity, and many other atmospheric parameters in the millimeter wave frequency range. Also, for atmospheric parameter profiling, the primary benefits of remote sensing instruments are in time continuity and low operating costs [1].

Radiometer is a passive remote sensing tool that measures the energy emitted at the frequency range of the millimeter wave. Radiometers are very sensitive receivers designed to measure atmospheric gas-emitted thermal electromagnetic energy using Rayleigh–Jeans law and allow us to derive important atmospheric parameters such as vertical temperature, humidity profile, etc. [2]. They allow deriving meteorological quantities with a high temporal resolution in the order of some seconds under all weather conditions.

Practically deployed radiometers (e.g., RPG-HATPRO) are being used for measurement of humidity, temperature, and liquid water path with accuracy of 0.3 g/m^3, 0.50 K RMS and $\pm$20 g/m^2, respectively, up to 10000 m [3].

Due to its elevated precision and big swath, despite its small spatial resolution, radiometer has now become a popular and effective instrument for earth remote sensing.

Radiometer enables atmospheric and geophysical parameters to be measured remotely by studying and analyzing the received spontaneous electromagnetic emission. Moreover, these atmospheric parameters can only be evaluated at certain specific frequency bands. They allow deriving humidity profile in K-band and temperature profile in V-band, respectively [4, 5]. So usually, radiometers are equipped with multiple receiving channels to derive the atmosphere's characteristic emission spectrum.

2 Design Configuration

A microwave radiometer comprises an antenna system, radio frequency microwave elements (front end), and a back end for intermediate frequency signal processing [6]. The atmospheric signal is very low (approximately -75 dB) which needs to be amplified. Hence, heterodyne methods are often used to transform the signal to lower frequencies allowing commercial amplifiers and signal processing to be used [7]. In order to prevent drifts of the receiver, thermal stabilization is very essential. Block diagram of dual-band (K-band and V-band) radiometer is shown in Fig. 1.

In both K-band and V-band systems of dual-band radiometer, an atmospheric signal is injected through antenna and a directional coupler. An accurate noise signal produced by a noise source is also injected to the receiver through on/off switching and the directional coupler for calibration [8]. This noise signal is used during measurements to determine system nonlinearities and drifts of system noise temperature. The input signal is boosted by a low noise amplifier (LNA) and feeds a mixer with a local oscillator tuned to the specific center frequency. The mixer output is boosted by an ultra-low noise IF amplifier and passed through a band-pass filter

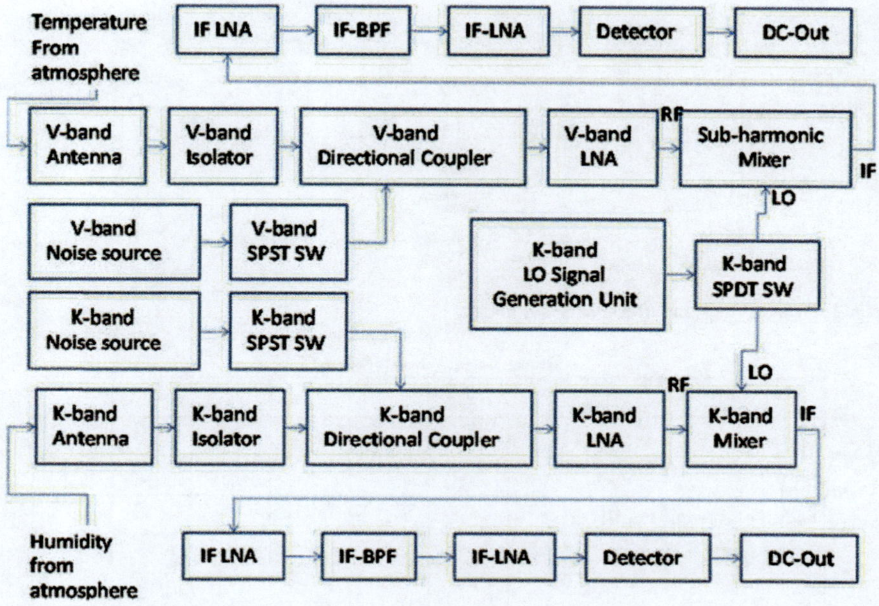

Fig. 1 Block diagram of dual-band radiometer

(BPF) for sensitivity enhancement with certain bandwidth. The BPF is followed by IF boosters and detector.

3 Simulation

In this section, system-level simulations of both K-band and V-band systems of the dual-band radiometer are shown. The results of simulation give an impression on the output voltage for input equivalent power of atmospheric parameters, i.e., humidity and temperature in K-band and V-band, respectively.

3.1 K-band Radiometer System Simulation

SystemVue setup of K-band radiometer is shown in Fig. 2.

Simulation Results. In this section, simulated results/graphs of K-band radiometer are presented in terms of output voltage for input equivalent power of humidity (Figs. 3 and 4).

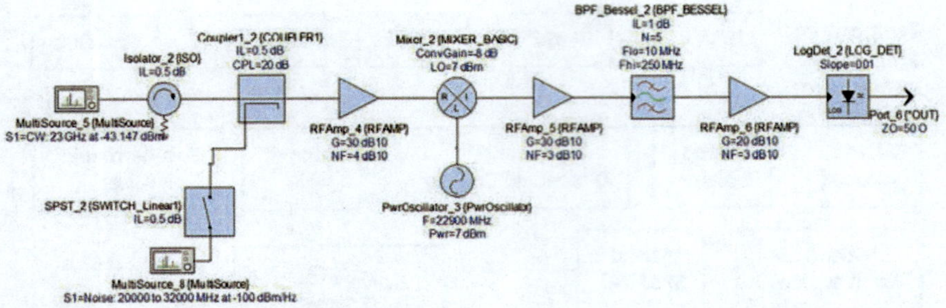

Fig. 2 SystemVue setup of K-band radiometer

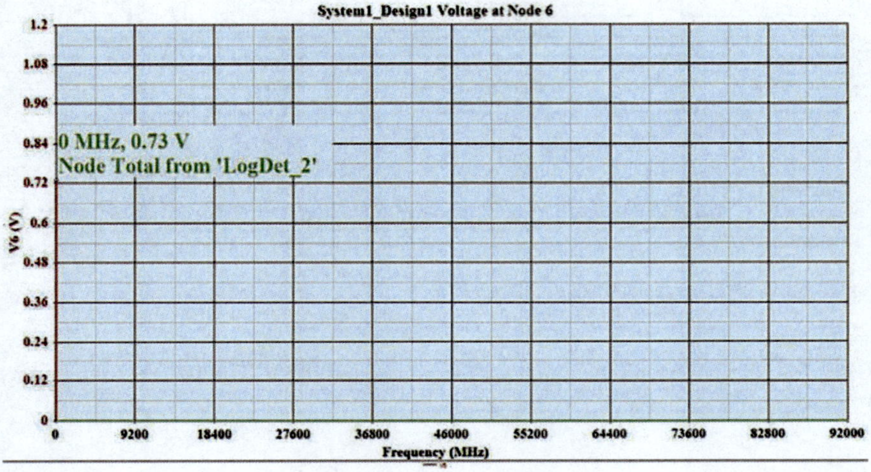

Fig. 3 Output voltage for absolute humidity (7.7282 g/m^3) and relative humidity (30%)

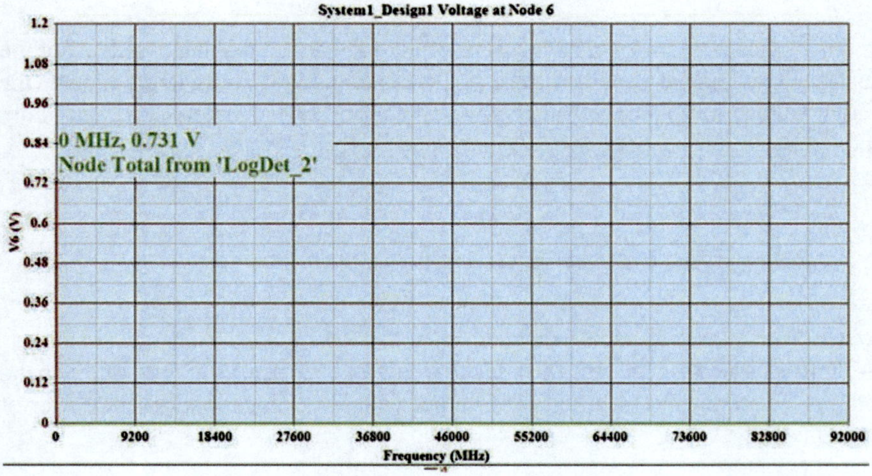

Fig. 4 Output voltage for absolute humidity (13.169 g/m^3) and relative humidity (30%)

Table 1 Parameter value of K-band radiometer simulation

SystemVue block	Parameter
MultiSource_5	Source Type: CW, Center Frequency: 23 GHz
Isolator_2	Isolation: 30 dB
MultiSource_8	Source type: white noise, Start frequency: 20 GHz, Stop frequency: 32 GHz, Power density: -100 dBm/Hz
SPST_2	Isolation: 30 dB, State: 0, Frequency: 23 GHz
Coupler1_2	Default
RFAmp_4	Gain: 30 dB, Noise figure: 4 dB, OP 1 dB: 20 dBm, OPSAT: 23 dBm, OIP3: 30 dBm, Frequency: 23 GHz
Mixer_2	Default
PwrOscillator_3	Frequency: 22.9 GHz
RFAmp_5	Gain: 30 dB, Noise figure: 3 dB, OP 1 dB: 20 dBm, OPSAT: 23 dBm, OIP3: 30 dBm, Frequency: 100 MHz
BPF_Bessel_2	Insertion loss: 1 dB, Flo: 10 MHz, Fhi: 250 MHz, Filter order: 5
RFAmp_6	Gain: 20 dB, Noise figure: 3 dB, OP 1 dB: 20 dBm, OPSAT: 23 dBm, OIP3: 30 dBm, OIP2: 40 dBm, RISO: 50 dB, Frequency: 100 MHz
LogDet_2	Vsat: 10 V

3.2 V-band Radiometer System Simulation

SystemVue setup and parameter values of V-band radiometer are similar to the K-band one as shown in Fig. 2 and Table 1.

Simulation Results. In this section, simulated results/graphs of V-band radiometer are presented in terms of output voltage for input equivalent power of temperature (Figs. 5 and 6, Tables 2 and 3).

4 Conclusion

In this work, an atmospheric remote sensing instrument, radiometer, has been investigated. The origin of the different measurable parameters and their relation to the atmospheric physics have been indicated. In simulation part, we have successfully modeled a passive radiometer RF front end in two frequency bands and completed its simulation in SystemVue environment. This work may be extended toward the realization of the dual-band RF front end of the radiometer for V-band and K-band operation for environmental temperature and humidity sensing, respectively.

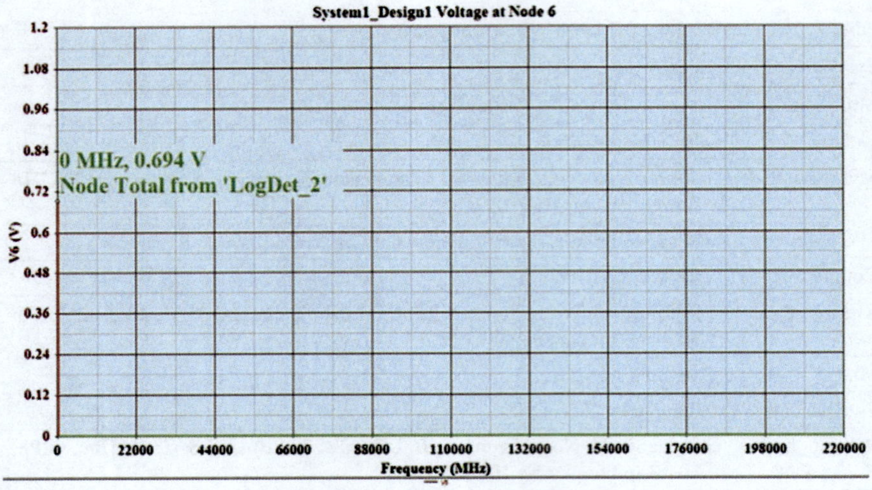

Fig. 5 Output voltage for temperature (77 K)

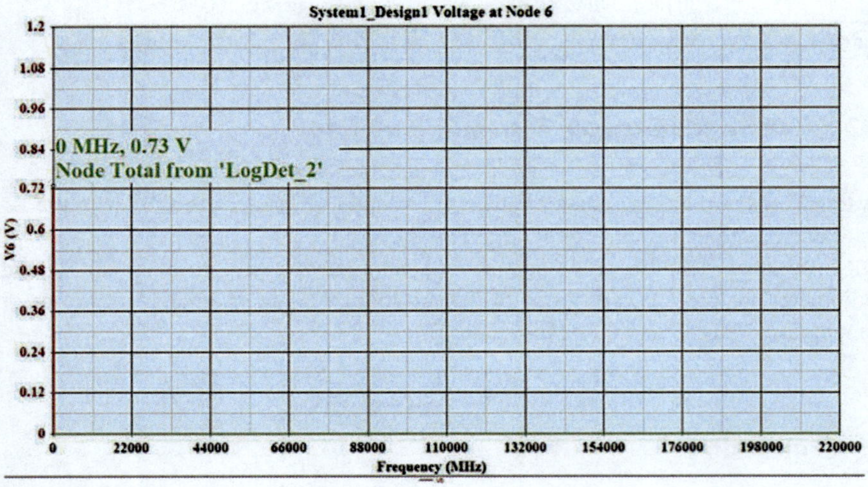

Fig. 6 Output voltage for temperature (300 K)

Table 2 Consolidated result for humidity sensing at 23 GHz

Absolute humidity (g/m^3)	Relative humidity (%)	Equivalent temperature (Kelvin)	Equivalent power (dBm)	Output voltage (Volt)
7.7282	30	293	−43.147	0.73
13.169	30	310	−42.902	0.731

Table 3 Consolidated result for temperature sensing at 55 GHz

Temperature (Kelvin)	Equivalent power (dBm)	Output voltage (Volt)
77	−49.958	0.694
300	−43.052	0.73

Acknowledgements The authors express their gratitude to Mr. Arijit Majumder, Officer-In-Charge, SAMEER Kolkata Centre for allowing this material for publication.

References

1. Kadygrov EN (2006) Operational aspects of different ground-based remote sensing observing techniques for vertical profiling of temperature, wind, humidity and cloud structure: a review. WMO
2. Skou N, Le Vine D (2006) Microwave radiometer systems design and analysis. Artech House, Boston
3. RPG-Radiometer Physics GmbH. https://www.radiometer-physics.de
4. Solheim F et al (1998) Radiometric profiling of temperature, water Vapor and cloud liquid water using various inversion methods. Radio Sci 33(2):393–404
5. Nelson LD et al (1989) Microwave radiometer and methods for sensing atmospheric moisture and temperature. U.S. Patent No. 4,873,481
6. Nelson ME (2016) Implementation and evaluation of a software defined radio based radiometer
7. Küchler N et al (2017) A W-Band radar-radiometer system for accurate and continuous monitoring of clouds and precipitation. J Atmos Ocean Technol 34(11):2375–2392
8. Timofeyev Y (1995) Microwave and millimeter wave temperature and humidity atmospheric sounding. St Petersburg State University, Atmospheric Physics Department

IOT-Based Domestic Aid Using Computer Vision for Specially Abled Persons

G. Sahitya, C. Kaushik, Vasagiri Krishnasree, and Ch. Yajna Mallesh

Abstract Every time the checking of manual working status of the electronic or electrical devices in a room is always a time-consuming process particularly for specially abled people. To avoid this situation, we are implementing a technique of image processing and IOT technology as solution. A device (app) is to be developed which ensures the working status of all devices in room and then acknowledge every update to the responsible person. When the electronic device in the room is turned on, a picture is captured and processed to know the status of the working of a device. The status of working will be updated to the responsible person so that he can take immediate action accordingly. In the further development we can also embed the automatic switching of all the devices in the room according to the presence of the people in the room.

Keywords Electronic devices · IOT technology · Morphological techniques · Python and app

1 Introduction

In our everyday life the remote sensing technology plays a prominent role, where we want every electronic device in our place to be controlled by our hand. If this is done automatically by some means it becomes greater help for specially abled persons. The main constraints in achieving this are the compatibility, accuracy, and power consumption. It is to hand over the control of all the appliances to the machine and makes it do our work. We have developed a system called "Automation with Computer Vision and Image Processing [7]". "Time and power" are the most valuable things. The usage of the new technologies in an effective way will save our time ad also the power consumption. This system will also be working to achieve the same goal. As the controls of the appliances are automatic, we don't have any manual work in controlling them and checking the status of devices like fan and light

G. Sahitya (✉) · C. Kaushik · V. Krishnasree · Ch. Y. Mallesh
Department of ECE, VNR VJIET, Hyderabad, Telangana, India
e-mail: sahitya_g@vnrvjiet.it

T. Laxminidhi et al. (eds.), *Advances in Communications, Signal Processing, and VLSI*, Lecture Notes in Electrical Engineering 722,
https://doi.org/10.1007/978-981-33-4058-9_9

whether properly working or not and get them repaired in advance. There are many models in the market with different concepts. They mainly work on "Remote Sensing Technology" [2]. They are to be controlled with the help of the applications in smartphones [6]. The automation works only when there is some trigger given manually through the application in smartphones. So in this system, all the decisions are to be taken by the human and self-decisions by the automation will be in fewer cases. In our system we are using the process of digital image processing using raspberry PI [1–3, 5–7]. Here we are capturing the pictures in every successive interval and are processed for the presence of human in the room. When case is yes, the appliances which are in the range are going to be switched. For the first time, when the system is being installed the boundary limits for the placement of the lights and fans are going to be given are described. This helps in reducing the processing speed.

2 Materials

Python is a high-level, interpreted, and general-purpose dynamic programming language that focuses on code readability. The syntax in Python helps the programmers to do coding in fewer steps as compared to Java or C++. The language was founded in the year 1991 by the developer Guido Van Rossum. Python has a wide range of application domains. People can do a wide variety of applications like controlling the serial port for device-level communication to complex numerical analysis, web applications, and data visualization.

Python is easy to learn for even a novice developer. Its code is easy to read, can execute a lot of complex functionalities with ease, and supports multiple systems and platforms. It is an Object-Oriented Programming-driven with the introduction of Raspberry Pi, a card-sized microcomputer, and Python has expanded its reach to unprecedented heights. Developers can now build cameras, radios, and games with ease. Python has a plethora of frameworks that make web programming very flexible. Django is the most famous Python framework for web development. It gives rise to quick development by using less code. Even a small team can handle Python effectively. It allows scaling even the most complex applications with ease. Many resources are available for Python. It offers a built-in testing framework to set debugging time and enables fastest workflows. IT giants like Yahoo, Google, IBM, NASA, Nokia, and Disney prefer Python.

3 Method

The block diagram for connecting the appliances through IOT is shown in Fig. 1.

A. **Detection of Lights in a Room**

The room will be designed in such a way it has lights with a border around edges which can be recognized using color detection in Open-CV and Image Processing techniques. In this case, we have arranged red color border around lights and then calculated the intensity of color at that contours. First, our required Python packages are imported. To start detecting red color border coated around lights in an image, we first need to load the input RGB image from disk and converted it into HSV image. The HSV image properties are closer to the perception of humans. With RGB, a pixel is represented by three parameters, namely, blue, green, and red. Each parameter usually scales from 0 to 255 with 8-bit resolution.

With HSV, a pixel is represented by three parameters, namely, Hue, Saturation, and Value. Unlike BGR, HSV does not use primary color to represent a pixel. It uses hue for shade of color, saturation for color intensity, and value for how bright or dark the color is. The mask having lower range and upper range for the hue to search red color is defined. The lower range is the minimum shade of red and Upper range is the maximum shade of red that will be detected. The red color has hue value ranging from 0 to 10 as well as from 170 to 180 and mask is created for the input image. A mask is simply specific part of an image. Here, we are checking through HSV image and then checking for colors that are between the lower range and upper range. The areas that are matched set to white and other are black.

The image is blurred (i.e., smoothing) to reduce high-frequency noise. To reveal the brightest regions, we need to apply threshold for setting the regions to white and black. Due to this a small amount of noise is induced in image. To remove noise, series of erosions and dilations are performed on the thresholded images. After performing these operations, we can observe our image is much cleaner than previous image. Thus, the detection of lights in a room will be detected. The algorithm for detection of lights in a room is given in Fig. 2.

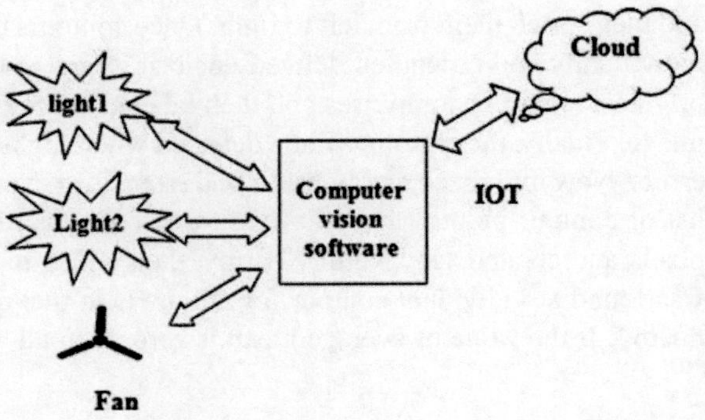

Fig. 1 Block diagram of connecting appliances through IOT

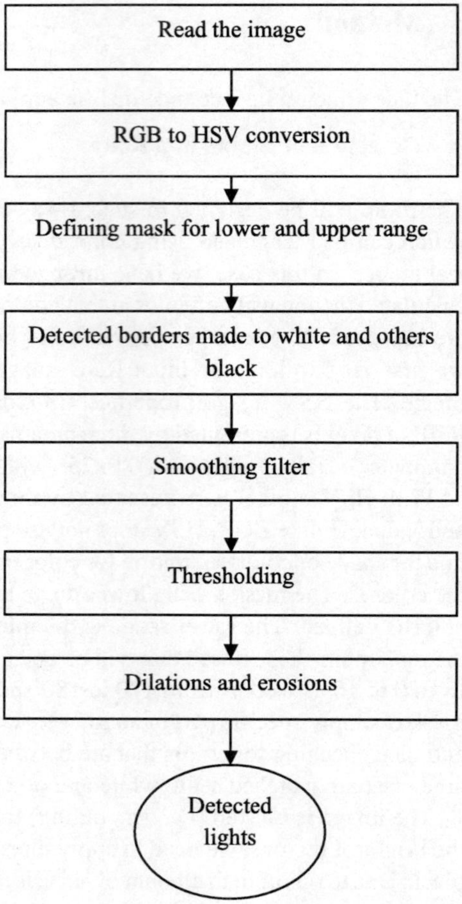

Fig. 2 Algorithm for detection lights in a room

B. Status of Lights in a Room (ON/OFF)

After detection of lights in the room, to find out the status of the lights, the labeled blobs are needed to be drawn on input image. To do that, first the contours are drawn and then sorted them from left to right. Once contours have been sorted looped them individually and computed defined enclosing circle which represents the area that brightest region encompasses and then labeled the region and drawn them on input image. Finally, the regions will be detected where lights are in the room without any errors. Now initialized empty list to find intensities in each contour and then for each list of contour points, created a mask that contains filled in and access the image pixels and created 1D "NumPy" array, then added to list. Now, the average mean was found at individual contour for all pixels in that region where D-type must be float64. If the value of average mean is zero, then all the lights are OFF and if it is maximum, then the lights are ON. Thus, the status of lights in a room will be stated using Open-CV. Next, checked for the other appliances available in the room for all combinations and then the status of fans was found in a room. The algorithm for status of lights in a room is given in Fig. 3.

Fig. 3 Algorithm for status of lights in a room

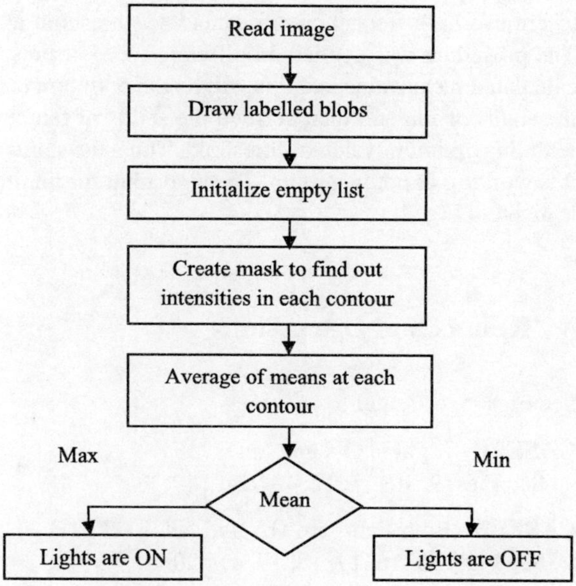

Fig. 4 The algorithm for finding the status of fans in a room

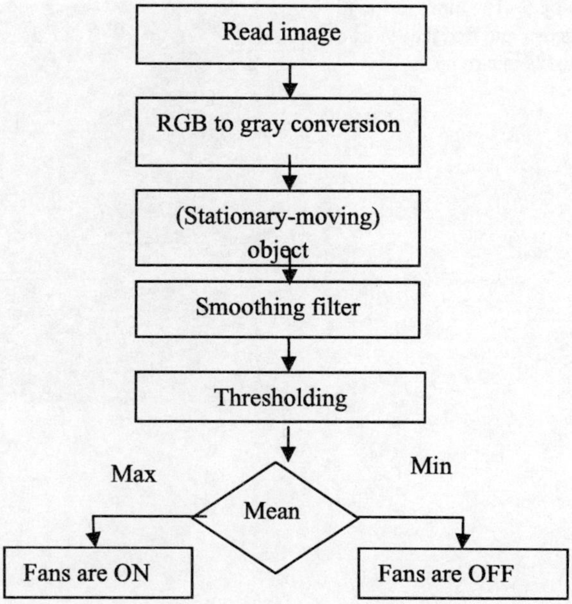

C. Status of Fan in a Room

The status of fan will be stated whether it is working or not using mean value of bright region when the image is subtracted from image of room where fan is stationary. First, we need to load our images and convert them into gray scale and then subtract moving object image from stationary image and then performed smoothing of image to remove high-frequency noises and set threshold to get dark and bright regions. The procedure was applied on different types of fans and for all combinations and calculated mean value and this will be the optimum threshold value for determining the status of fan and then verified the status of fan currently available in the room with this optimum value of threshold. Thus, the status of fan will be stated whether it is working or not in a room. The algorithm for finding the status of fans in a room is given in Fig. 4.

4 Results and Discussions

See Figs. 5, 6, 7, and 8.

CASE: 1—Light is **ON** and fan is **OFF**
See Figs. 9, 10, 11, 12, 13, and 14.

CASE: 2—Both lights are **ON** and fan is **OFF**
See Figs. 15, 16, 17, 18, 19, and 20.

Fig. 5 The input image to detect and find the status of lights in a room

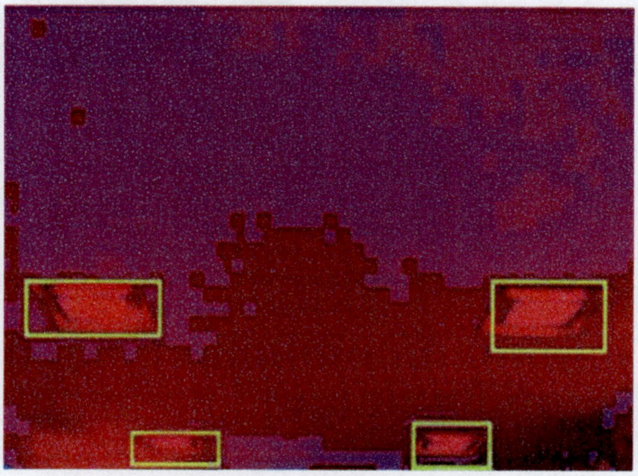

Fig. 6 Input image converted to HSV

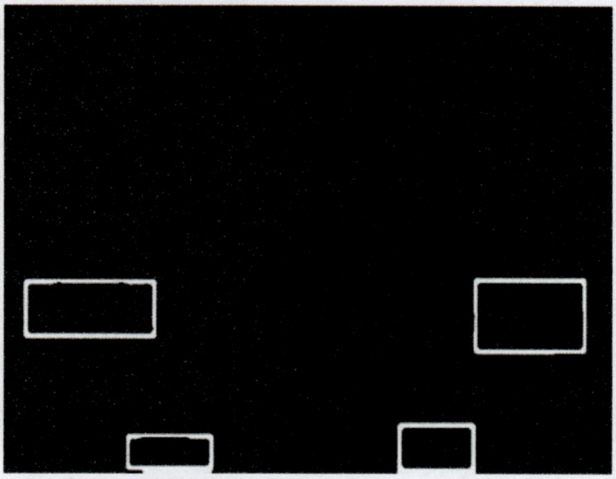

Fig. 7 Blurred and applying threshold to reveal brighter regions of the image

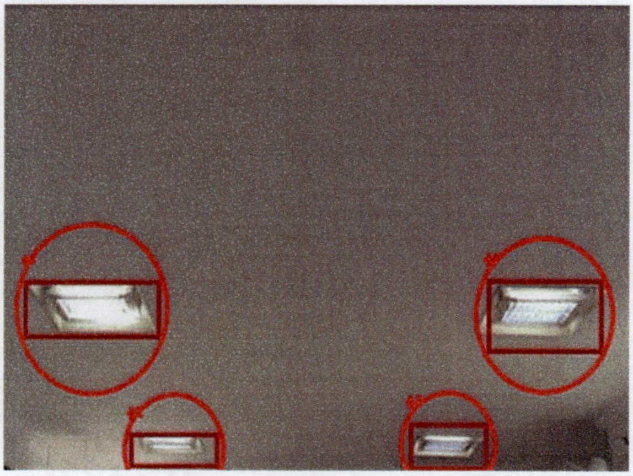

Fig. 8 Detecting the regions of lights location

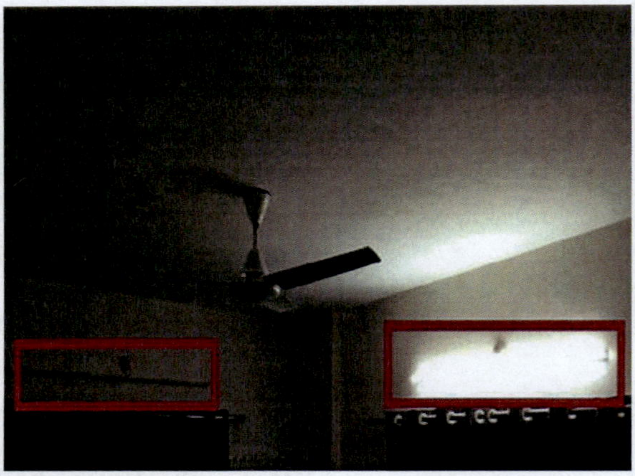

Fig. 9 Input image with light with ON state and a light with OFF state and fan with OFF state

Fig. 10 Source image converted into HSV image

Fig. 11 Blurred image

Fig. 12 Threshold image

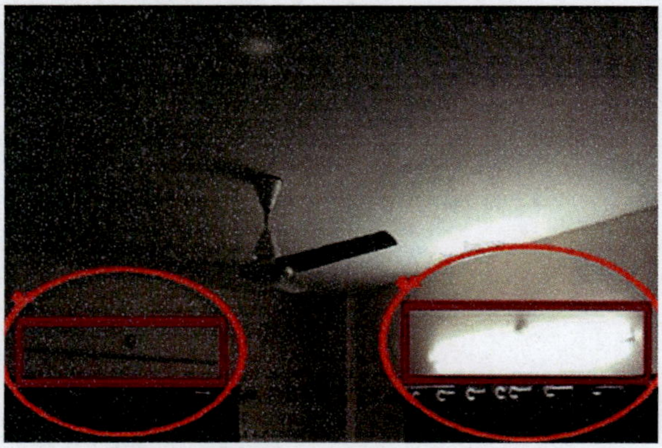

Fig. 13 Detected the position of the lights

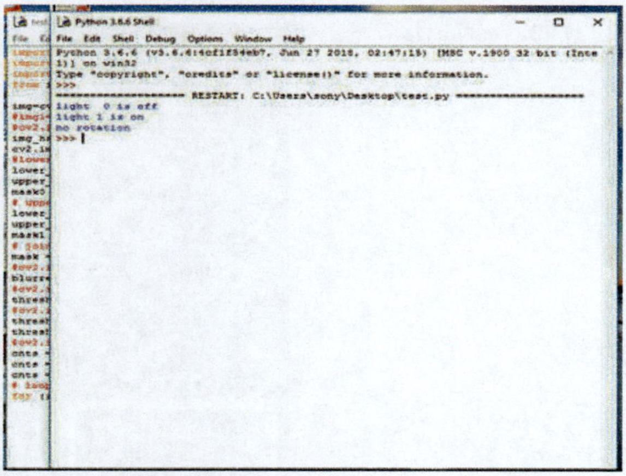

Fig. 14 Output obtained when image is processed

Fig. 15 Input images with both the lights in ON state and the fan in OFF state

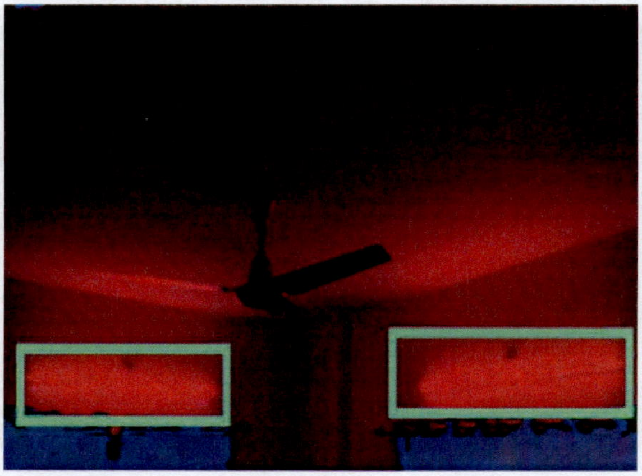

Fig. 16 Source image converted into HSV image

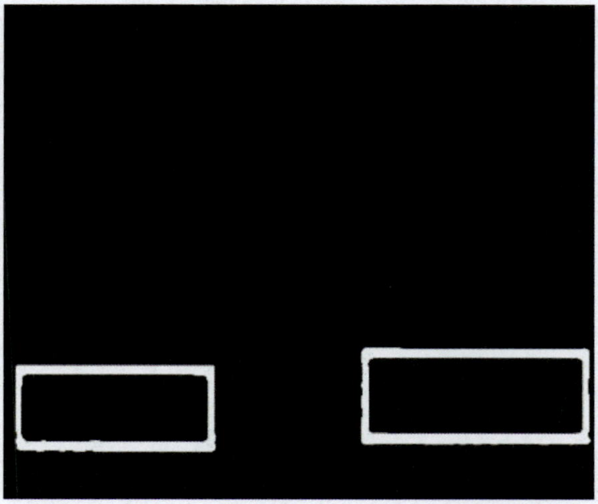

Fig. 17 Blurred image

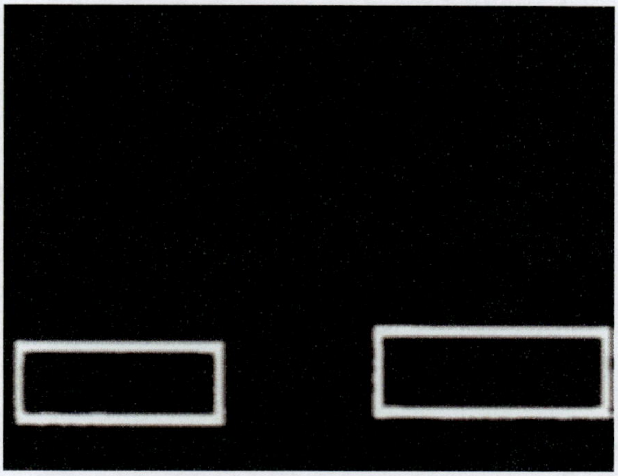

Fig. 18 Image after applying "thresholding"

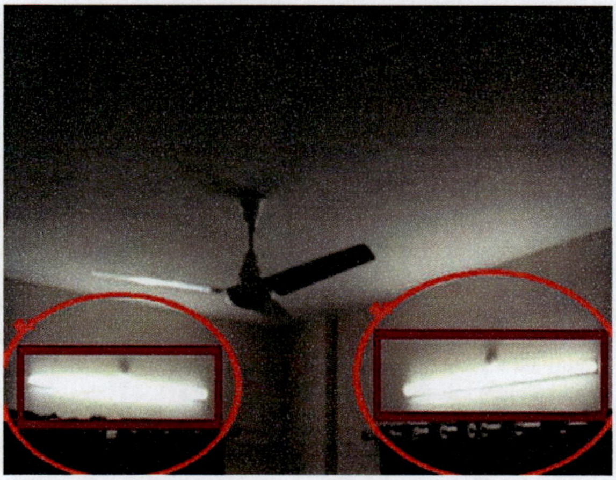

Fig. 19 Detection of lights

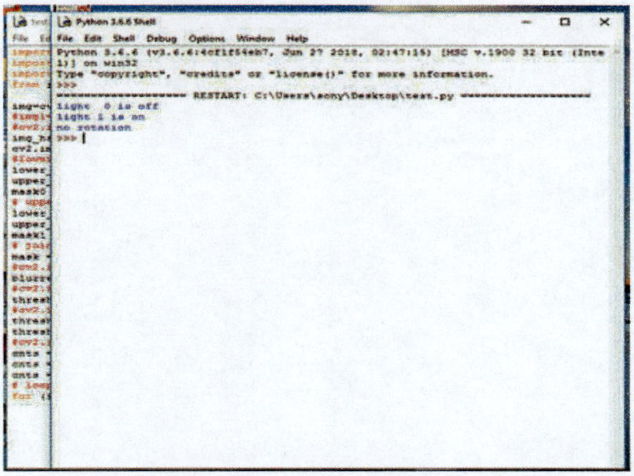

Fig. 20 Output obtained when the image is processed

5 Conclusions

The checking and maintenance of the status of the household appliances like fans and tube lights is a tedious job for specially disabled people. Hence the app developed in this algorithm is going to replace the manual task with automatic status finding and intimating to a responsible person, it is going to be a very helpful aid for special people. The algorithm can either give an audio or visual indication at home directly so that it can serve as an indicator for either visually disabled or physically disabled get benefited. In addition to this by passing the information to the responsible person through cloud, the updating of status of home appliances can be done regularly and maintenance can be carried out in time.

References

1. Shroff N, Kauuthale P, Dhanapune A, Patil SN (2017) IOT based home automation system using Raspberry pi-3. Int J Eng Technol 4(5):2824–2826
2. Patchava VK, Kandala H, Babu PR (2015) A smart home automation technique with Raspberry Pi using IOT. In: Proceedings of International Conference on Smart Sensors and Systems, Bangalore
3. Severance C (2013) Eben upton: Raspberry pi. 46:14–16
4. Pandey A, Mishra S, Tiwari P, Mishra S (2018) IOT based home automation system using Rspberry pi kit. Int J Adv Eng Res Dev 5(4):307–309
5. Saxena S, Gupta R, Varshney S, Singh SP, Kumari, S (2016) Internet of things based home automation using raspberry PI. Int J Eng Sci Comput 6(4):3849–3851
6. Gamba M, Gonella A, Palazzi CE (2015) Design issues and solutions in a modern home automation system. In: Proceedings of International Conference on Computing Networking and Communications (ICNC 2015), pp 1111–1115
7. Gonzalez and woods, Digital image processing, 2nd edn. Prentice Hall

Aeroload Simulation of Aerospace Vehicle Using Fin Load Simulator

MVKS Prasad, Sreehari Rao Patri, and Jagannath Nayak

Abstract An Aerospace vehicle is designed to meet its objective of either reaching a particular target point or intercepting an incoming hostile target. The vehicle is designed to have some subsystems to achieve this objective. The subsystems include On-board computer, Rate gyros and accelerometers, Actuation system, Seekers, Radio Proximity Fuse, etc. Each subsystem has its own functionality. Along with its own functionality it has to work in integration with other subsystems to meet the objective. The rate gyros and accelerometers sense the rates and accelerations and provide them as inputs to navigation, control, and guidance. The navigation system generates the Aerospace vehicle positions, velocities, and accelerations which are in turn inputs to the guidance. The guidance generates the error between the current location of the Aerospace vehicle and the expected position calculated from the intended trajectory. This error is corrected by control by generating the required rates and hence the deflection commands to the Actuation system. In ideal situation the control surfaces deflect as much as the given commands. But in space due to the Aerodynamic load coming on the control surfaces, the control surfaces deflect either more or less depending upon whether the load is aiding (helping) the control surfaces movement or opposing. Aeroload simulation is to be carried out on ground to check the robustness of the Aerospace vehicle to meet the mission objectives. This can be done by applying flight load in terms of torque to control surfaces on ground and the feedback can be measured and compared against the commands. The mission objectives are to reach the target point within acceptable error even with slight difference between control system command and feedback due to Aerodynamic load along with

M. Prasad (✉)
Research Centre Imarat (RCI), DRDO, Hyderabad 500069, Telangana, India
e-mail: malapakaprasad@gmail.com

S. R. Patri
National Institute of Technology Warangal, Warangal 506004, Telangana, India
e-mail: patri@nitw.ac.in

J. Nayak
Centre for High Energy Systems and Sciences, DRDO, Hyderabad 500069, Telangana, India
e-mail: nayak_jagannath@rediffmail.com

T. Laxminidhi et al. (eds.), *Advances in Communications, Signal Processing, and VLSI*,
Lecture Notes in Electrical Engineering 722,
https://doi.org/10.1007/978-981-33-4058-9_10

other inaccuracies. In this paper Aeroload simulation of an Aerospace vehicle using a hardware Fin Load Simulator is presented.

Keywords Aerospace vehicle · On-board computer · Rate gyros and accelerometers · Actuation system · Seekers · Radio proximity fuse · Aeroload simulation · Fin load simulator

1 Introduction

Before carrying out any Aerospace vehicle mission simulation is carried out to validate the design as well as hardware systems of the vehicle. The design includes Navigation, Guidance, and Control. Navigation algorithm calculates positions, velocities, and accelerations of the vehicle from time to time taking rates and accelerations as inputs as sensed by rate gyros and accelerometers. The target point or destination point is available as either a point or in terms of trajectory in Guidance. So the guidance takes the vehicle information from navigation and uses destination information and calculates the error between the two. This can be nullified by accelerating the vehicle in lateral axes. Using the control algorithm these accelerations a_y and a_z are converted into rate commands and these rates are achieved by commanding control system.

The Navigation, Guidance [1], and Control algorithms are validated by running the software simulation in one PC with Aerospace vehicle model developed as Plant model to generate rates and accelerations in the same way as generated by Rate gyros and accelerometers [2] in flight. These results are basis for validating hardware.

Hardware-In-Loop Simulation (HILS) is carried out to validate Aerospace vehicle subsystems such as On-Board Computer (Navigation, Guidance, and Control programs are embedded), Rate Gyros and Accelerometers package, Actuation System, Seeker [3] (Used for terminal Guidance), Radio Proximity Fuse (used for final few milliseconds of mission). In hardware Actuator-In-Loop simulation, the control commands generated from control (residing in On-Board Computer) are given to hardware actuators and feedbacks are given to the vehicle plant model developed in RT Linux [4] running in Real Time in PC [5]. In Actuator simulation with no load applied on Actuators, the feedbacks match with commands and mission will be normal. But when load is applied on Actuators as it happens in Aerospace vehicle mission, the control feedbacks no longer match with control commands. Then we have to see whether mission objectives will be met or not. So hardware Actuator-In-Loop with load is to be carried out. An equipment known as Fin Load Simulator is used in HILS for this purpose.

Before carrying out load simulation, first simulation is carried out with actuator model, then with hardware actuator with no load. When the results of above two configurations are as expected, Aeroload simulation [6] is carried out.

2 Actuator Model Development and Implementation in Aerospace Vehicle Simulation

First HILS is carried out with Actuator model of an Electro-Mechanical Actuation system [7]. In this case an actuator model is developed as a second-order model. The main parameters for a second-order model are natural frequency of operation of actuator ω_n ($\omega_{n=}2.0*pi*f_n$) and damping factor ξ. The frequency of operation is considered as $f_n = 18$ Hz and the damping factor is taken as 0.7 ($\xi = 0.7$) [8]. The HILS setup for actuator model in loop simulation is shown in Fig. 1.

The control surface deflections generated by the OBC are given to the second-order actuator model residing in Aerospace vehicle model. Any particle in space can be completely defined with six parameters and they are three translational accelerations (a_x, a_y, a_z) and three rotational accelerations ($\dot{p}$, $\dot{q}$, $\dot{r}$). So the Aerospace vehicle model which is also known as 6DOF (6 Degrees of Freedom) model [9] is generated as six equations with the characteristics of that vehicle as inputs. The generalized equations of translational and rotational accelerations for an Aerospace vehicle are given below (Eqs. 1 to 6).

$$a_x = \frac{\mathrm{Th} - \mathrm{drag}}{m} \tag{1}$$

$$a_y = \frac{y_\beta \cdot \beta}{m} + \frac{y_\delta \cdot \delta y}{m} + (X_{CG} - X_{NS})\dot{r} \tag{2}$$

$$a_z = \frac{z_\alpha \cdot \alpha}{m} + \frac{z_\delta \cdot \delta_p}{m} - (X_{CG} - X_{NS})\dot{q} \tag{3}$$

$$\dot{p} = \frac{L_p \cdot p}{I_{xx}} + \frac{L_{\delta r} \cdot \delta_r}{I_{xx}} + \frac{C_{L \cdot Q \cdot S \cdot d}}{I_{xx}} \tag{4}$$

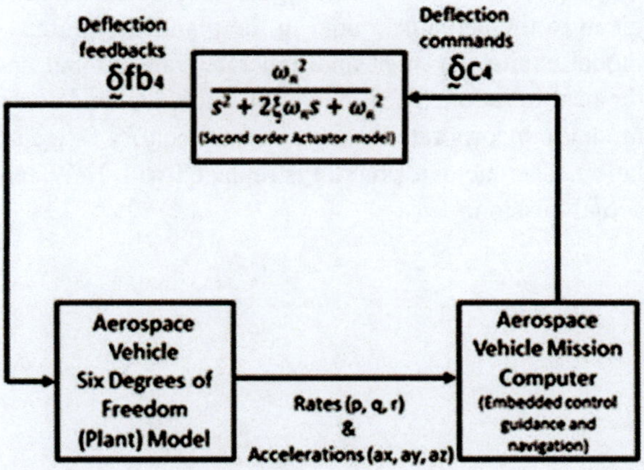

Fig. 1 HILS setup for actuator model in loop simulation

$$\dot{q} = \frac{M_\alpha \cdot \alpha}{I_{yy}} + \frac{M_\delta \cdot \delta_p}{I_{yy}} \quad (5)$$

$$\dot{r} = \frac{N_\beta \cdot \beta}{I_{zz}} + \frac{N_\delta \cdot \delta_y}{I_{zz}} \quad (6)$$

The forward acceleration a_x is calculated as drag subtracted from thrust (Th) and the value is divided by mass (m). Acceleration in y-axis, i.e., a_y is calculated as Aerospace vehicle body force y_β due to side slip angle β multiplied by β added with vehicle control force y_δ due to control deflection multiplied by effective control deflection δ_y in y-axis. The sum force is divided by mass to get acceleration. This acceleration is then added to the contribution of rotational acceleration in yaw, i.e., $\dot{r}$ (multiplied by the difference between center of gravity and sensor location ($X_{CG} - X_{NS}$)) to linear acceleration in y-axis. Acceleration in z-axis, i.e., a_z is calculated as Aerospace vehicle body force z_α multiplied with α (angle of attack) and added with force due to control deflection (Z_δ) multiplied by effective control deflection δ_p in Z-axis. The sum of these two terms is divided by mass to get acceleration. Then this sum is subtracted with contribution of rotational acceleration in pitch, i.e.,$\dot{q}$ (multiplied by the difference between center of gravity and sensor location ($X_{CG} - X_{NS}$)) to linear acceleration in z-axis.

The rotational acceleration $\dot{q}$ is calculated as sum of moment of vehicle due to α (M_α) multiplied by α and moment due to δ (M_δ) multiplied by effective deflection in Pitch δ_p divided by moment of inertia about y-axis (I_{yy}). The rotational acceleration $\dot{r}$ is calculated as sum of moment due to β (N_β) multiplied by β and moment due to δ (N_δ) multiplied by effective deflection in Yaw δ_y divided by moment of inertia about z-axis (I_{zz}).

Similarly the roll acceleration $\dot{p}$ is calculated from rolling momemt due to damping L_p, the effect due to control, i.e., due to effective roll deflection, due to rolling moment with coefficient C_L. Then the sum is divided by I_{xx} moment of inertia with respect to x-axis to get roll acceleration.

These rotational accelerations are converted into rates. These rates along with translational accelerations are given to OBC where navigation, guidance, and control are executed. The deflections generated by control in OBC to correct the errors are given to the actuator model in the plant. The output deflections from the actuator model excite the plant and generate translational and rotational accelerations for the next iteration. The navigation, guidance, and control are validated with actuator model to incorporate the lag and nonlinearity for gain and phase margin [10] calculation. Then actuator model is replaced with H/W actuation system to simulate the actual mission.

3 Hardware (H/W) Actuation System Simulation with Aeroload

The setup for carrying out simulation with hardware actuation system [11] is shown in Fig. 2. With respect to Fig. 1 here the Actuator model is replaced with H/W Actuation system. So the deflection commands from the mission computer are given to hardware actuation system instead of actuator model and deflection feedbacks are given to plant model. The simulation results, i.e., rates, accelerations, deflection commands, and feedbacks are compared with results with actuator model in order to validate the H/W actuation system.

In this setup simulation is carried out with no load on the actuation system. But in actual mission load comes on the actuation system either opposing or aiding known as Aeroload.

A setup is established in HILS to carry out load simulation of Aerospace vehicle. The setup for carrying out load simulation is shown in Fig. 3. Here the control deflection commands generated by OBC based on Aerospace vehicle rates and accelerations sent from 6DOF are given to the actuators mounted on Fin Load Simulator. The Aerodynamic load which is also known as hinge moment (Hm[]) is calculated in 6DOF as given in Eqs. 7 to 10.

$$\mathrm{Hm}[0] = \mathrm{Talpha_h} * \mathrm{alp13} + \mathrm{Tdelta_h} * \mathrm{findefop}[0][0] * \mathrm{r2d} * \mathrm{Qsd}; \tag{7}$$

$$\mathrm{Hm}[1] = \mathrm{Talpha_h} * \mathrm{alp24} + \mathrm{Tdelta_h} * \mathrm{findefop}[1][0] * \mathrm{r2d} * \mathrm{Qsd}; \tag{8}$$

$$\mathrm{Hm}[2] = \mathrm{Talpha_h} * \mathrm{alp13} - \mathrm{Tdelta_h} * \mathrm{findefop}[2][0] * \mathrm{r2d} * \mathrm{Qsd}; \tag{9}$$

$$\mathrm{Hm}[3] = \mathrm{Talpha_h} * \mathrm{alp24} + \mathrm{Tdelta_h} * \mathrm{findefop}[3][0] * \mathrm{r2d} * \mathrm{Qsd}; \tag{10}$$

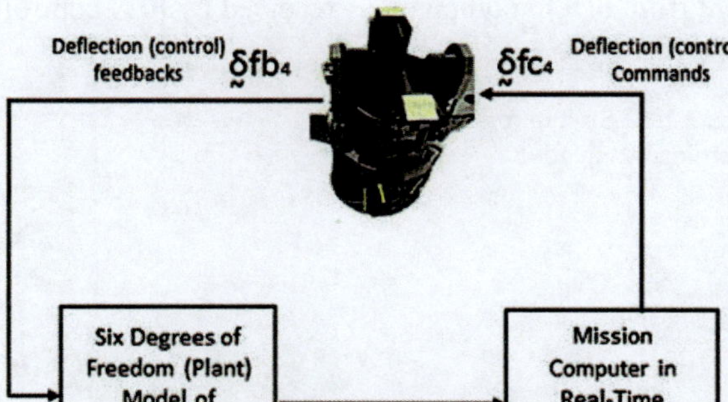

Fig. 2 HILS setup for H/W actuator-in-loop simulation

where alp13 and alp24 are given by

$$\text{alp13} = \text{Alpha}_t * \cos(\text{phir}); \tag{11}$$

$$\text{alp24} = \text{Alpha}_t * \sin(\text{phir}); \tag{12}$$

And Alpha_t is calculated as

$$\text{Alpha}_t = \sqrt{(\alpha_{\text{Body}})^2 + (\beta_{\text{Body}})^2} \tag{13}$$

Talpha_h & T_deltah are calculated as one-dimensional interpolation of alpha and delta with Mach number. Findefop[0][0], Findefop[0][1], Findefop[0][2], and Findefop[0][3] are deflection commands of fins 1, 2, 3, and 4, respectively. Phir is Aerodynamic rotation angle of Aerospace vehicle. Q is dynamic pressure (0.5*rho*v*v), s is surface area, d is diameter of Aerospace vehicle, and v is velocity of the Aerospace vehicle. Radian to degree conversion is carried out with r2d term. Hm[0], Hm[1], Hm[2], and Hm[3] represent the Aeroload for Fin1, Fin2, Fin3, and Fin4, respectively. This hinge moment expected for each actuator is individually sent through Digital to Analog converter of 6DOF computer to Fin Load Simulator controller as shown in Fig. 3.

The Fin Load Simulator controller will apply this load (torque) to individual actuators through load cell of Fin Load Simulator. The load from 6DOF computer and deflection commands from OBC are synchronized with liftoff. The torque is thus applied throughout the mission and corresponding control deflection feedbacks are recorded. These deflection feedbacks are compared with deflection commands. Mission performance is checked through rates, accelerations, and other parameters.

The Fin Load Simulator with one actuator mounted on it is shown in Fig. 4. As a first exercise Fin Load Simulation is carried out for one actuator and remaining three are run with no load with actuators kept on nozzle positioned on ground. Actuator closed-loop HILS is carried out with actuator1 load profile of mission and torque as sent from 6DOF computer, as received by FLS controller, torque feedback received

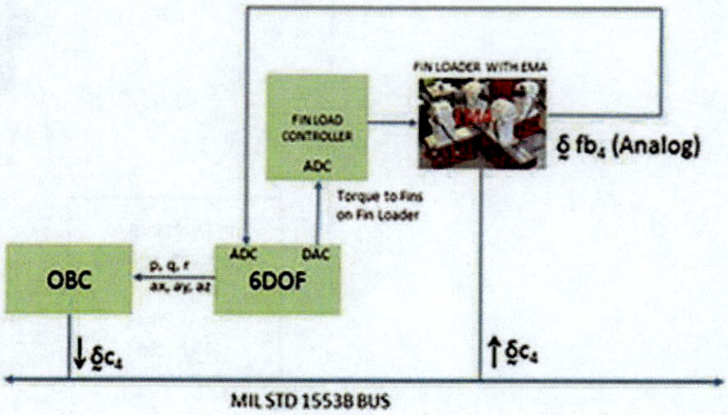

Fig. 3 HILS setup for Aeroload simulation

Fig. 4 Fin load simulator with one actuator

from FLS, Actuator commands and feedbacks are recorded. In the next section the results are presented with details.

4 Actuator-In-Loop Results with Fin Load

The flight deflection command and feedback for fin1 called delta1 is shown in Fig. 5 along with its 1 to 6 s expanded plot in Fig. 6. On average a difference of 0.2° is seen between command and feedback (feedback is lower due to Aeroload in mission).

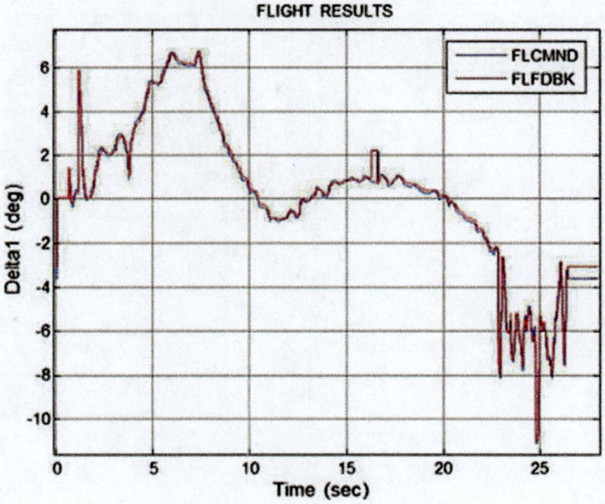

Fig. 5 Flight command and feedback of delta1

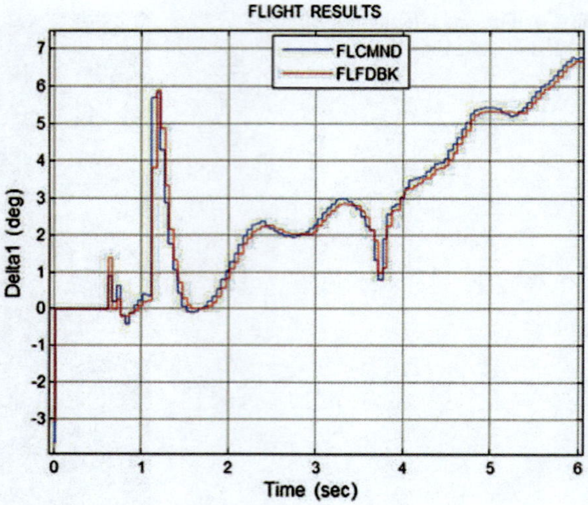

Fig. 6 Expanded plot of flight command and feedback of delta1

Actuator closed-loop HILS results for delta 1 are shown in Fig. 7 and expanded plot in Fig. 8. This is not showing any difference in value between command and feedback as HILS run for this case is carried out without load. The flight load which is the same load applied to fin1 is shown in Fig. 9. This shows a peak load of 230 Nm at 19 secs. The peak Aeroload comes based on the altitude, Aerospace vehicle velocity, dynamic pressure, and Aerodynamics working on the system.

Actuator closed-loop HILS is carried out by applying flight delta1 load to fin1 and no load to remaining three actuators. Figure 10 shows delta1 command and feedback and Fig. 11 shows expanded plot from 1 to 6 s. Figure 11 shows a difference of 0.2 degrees between command and feedback on average (feedback is lesser in value). This matches with flight delta1 shown in Fig. 6.

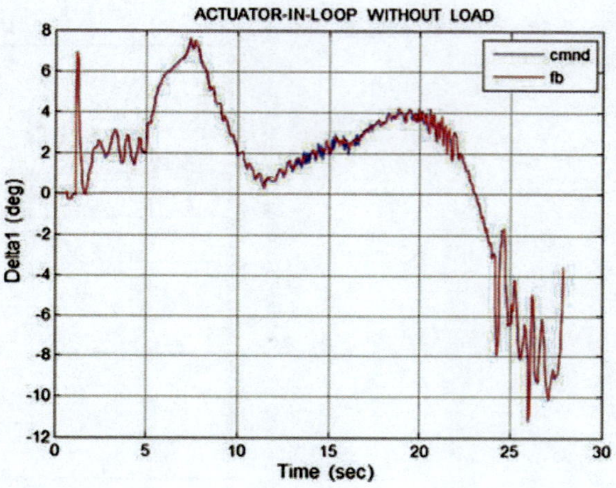

Fig. 7 Delta1 command and feedback of actuator-In-Loop without load

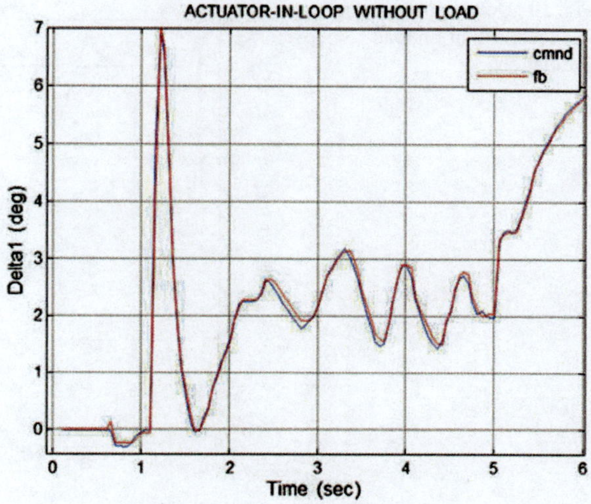

Fig. 8 Expanded plot of delta1 command and feedback of actuator-In-Loop without load

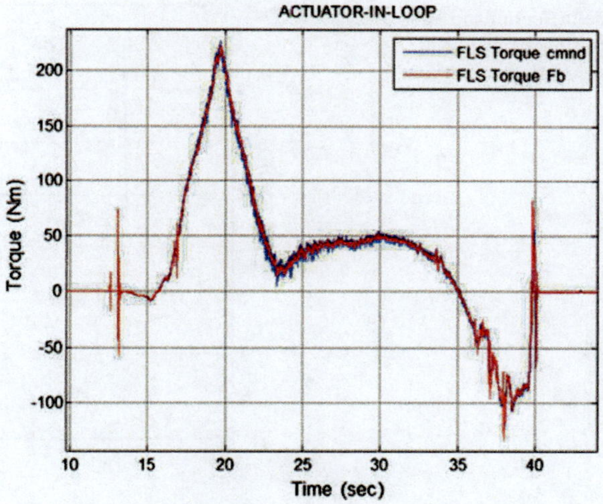

Fig. 9 Flight Aeroload of delta1

5 Conclusion

From flight delta1 command and feedback, from HILS delta1 command and feedback with no load and from HILS delta1 command and feedback with load it is very clear that the reduction in feedback compared to command is result of Aerodynamic load. The reduction in value of feedback is also matching as it is 0.2 degrees. From HILS results without load it is very evident that delta1 command and feedback match in value. So the effect of load on control feedback observed in flight is reproduced in HILS by carrying out HILS with load. In future Aeroload simulation will be extended by applying load to all four actuators instead of one actuator. This helps in accurate design of control and guidance of Aerospace vehicle to meet mission objectives.

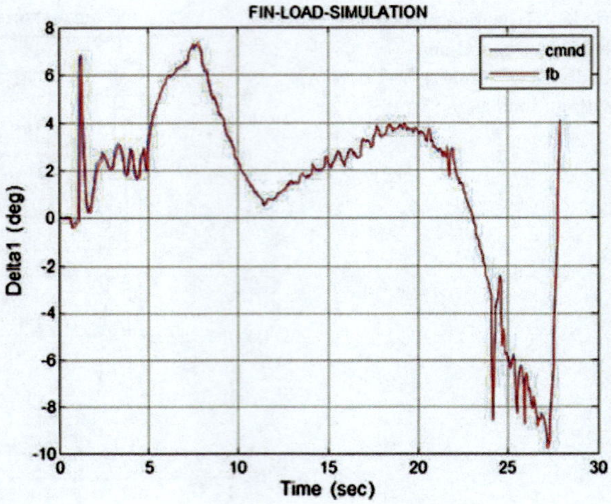

Fig. 10 Delta1 command and feedback of fin load simulation

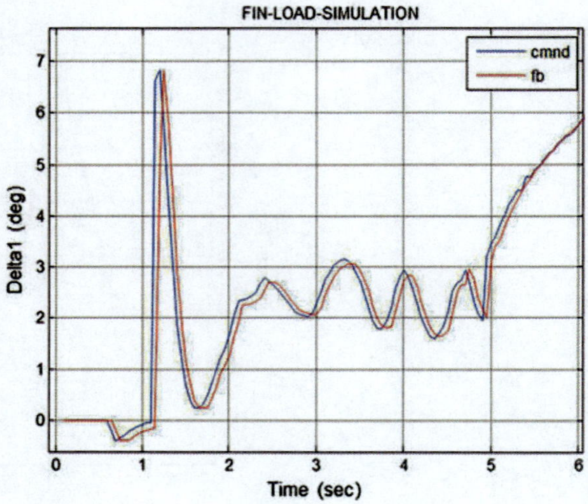

Fig. 11 Expanded plot of delta1 command and feedback of fin load simulation

Acknowledgements The authors express their sincere gratitude to Director RCI for his excellent support during this activity of Aeroload simulation. The authors thank Mr. L. Sobhan Kumar Technology Director DHILS for his encouragement during this activity. The authors are thankful to Mrs. Amrita Singh and Mr. M. Srinivasa Rao Scientists Directorate of Hardware-In-Loop Simulation (DHILS) for their excellent support in carrying out this activity. The authors are also indebted to Programme Area Defence members and employees of DHILS for their excellent support.

References

1. Gangadhar M, Singh A, Prasad MVKS (2017) Hard real time distributed aerospace system, jointly published by Dr. APJ Abdul Kalam missile complex. DRDO and Institute of Defence Scientists and Technologists in 2017, Bengaluru
2. Prasad MVKS, Gangadhar M, Singh A (2018) Rate gyroscope sensor simulation and validation in Hardware in Loop Simulation (HILS) for an Aerospace vehicle. ICITE 2018, Osmania University, Hyderabad
3. Srinivasa Rao B, Satnami S, Prasad MVKS (2016) Modeling and simulation of IIR seeker for an aerospace vehicle, INDICON-2016, Bengaluru
4. The RTLinux® embedded enterprise versus the Linux "RT" patches: FSM Labs White Paper, 2001
5. Krishna CM, Shin KG (1997) Real time systems. McGraw-Hill
6. Chun T-Y, Chol K-J, Woo H-W, Hur G-Y, Kang D-S, Kim J-C (2007) Design of the Aeroload simulator for the test of a small sized electromechanical actuator. In: 2007 International Conference on Control, Automation and Systems, South Korea
7. Schneiders MGE, Makarovic J, van de Molengraft MJG, Steinbuch M (2005) Design considerations for electro-mechanical actuation in precision motion systems. Elsevier IFAC Publications
8. Chin H (2009) Feedback control system design. MITMECHE
9. Singh A, Gangadhar M, Prasad MVKS (2014) Distributed real time modelling of an Aerospace vehicle. In: International conference on computing and communication Technologies (ICCCT-2014), Hyderabad
10. Burns RS (2001) Advanced control engineering. Butterworth-Heinemann
11. Prasad MVKS, Sreehari Rao P, Nayak J (2019) Actuation system simulation and validation in Hardware in Loop Simulation (HILS) for an Aerospace vehicle. ICITE 2019, Osmania University, Hyderabad

Design and Analysis of Wearable Step-Shaped Sierpinski Fractal Antenna for WBAN Applications

Sandhya Mallavarapu and Anjaneyulu Lokam

Abstract A novel multiband fractal wearable antenna is proposed and designed for various wireless wearable applications. Employing fractal shaped geometrical structures has been shown to progress several antenna features to varying extents. Yet a direct relationship between antenna characteristics and geometrical properties of fundamental fractals has been missing. This research is in-tended as a step forward to plug this gap. The major focus of this paper is on the design of the textile antenna, where Sierpinski carpet fractal geometry is used to generate multiband behavior and producing small antennas that make it suitable to integrate into clothing. The characteristics of the proposed antenna are verified with two substrates, one is a jeans fabric which is a flexible and low-cost material and the second one is an FR-4 which is a commercially used PCB substrate. The radiating element and ground plane are made of conductive copper sheets made to attach to the substrate. The antenna has been optimized to operate in multiple bands between 1 and 6 GHz. The performance of the antenna such as re-turn loss, impedance, Directivity, and gain is comprehensively investigated and carried out. All results have been verified using simulation with the help of a CST MWS 2018.

Keywords Wearable antenna · Sierpinski fractal antenna · MRR · CST · Wi-fi · Wi-MAX

1 Introduction

To endure the requirements of modern wireless communication standards antennas should be designed to give higher gain, higher performance, multiband support, and wider bandwidth. Also, they must be of low cost and conventionally smaller designs which don't spoil the wearer's comfort. To fulfill these necessities Fractal Antennas are discovered. The fractal term was coined in 1975 by Benoit B. Mandelbrot who is a French mathematician. Fractal formed antennas reveal some exciting features

S. Mallavarapu (✉) · A. Lokam
National Institute of Technology, Warangal, India
e-mail: sandhyamallavarapu@student.nitw.ac.in

T. Laxminidhi et al. (eds.), *Advances in Communications, Signal Processing, and VLSI*, Lecture Notes in Electrical Engineering 722,
https://doi.org/10.1007/978-981-33-4058-9_11

that come from their geometrical properties. There are various fractal structures available in the literature some of which are Koch curves, Koch snowflakes, Sierpinski gasket, and Sierpinski carpet. The self-similarity and space-filling of certain fractal geometries result in a multiband behavior and miniaturization of self-similar fractal antennas [1–5]. The Sierpinski carpet fractal antenna based on fractal structure is low volume antenna having moderate gain and can be operated at multiband of frequencies leads to a multi-functional structure [6, 7]. There are many advantages of applying fractals to build up various antenna configurations. The combination of structural complexity and self-similarity makes it possible to design antennas with wideband and multiband performances. By applying fractals to antenna structures [8].

- The size of the antenna reduced.
- Multiband resonant frequencies achieved.
- Optimized for gain.
- Attain wideband band.

2 Design Methodology of Antenna

To start with the reference geometry of the Sierpinski fractal was built from a Microstrip square patch [9] and recursively undergoes few iterations to generate multiband characteristics. This fractal wearable antenna is an iterative representation of step shape from which a square patch is removed. It was built on a 54 × 54 mm^2 square patch. The substrate materials used for the design of the antenna were FR-4 and Jeans with dielectric constants of 4.4 and 1.7, respectively. The dielectric properties of the flexible jeans fabric were verified by the method of microstrip ring resonator (MRR) [9–11]. The ground plane and the patch formed with copper (annealed). A simple edge feeding is used with a microstrip line of dimension 3 × 14 mm^2.

The first iteration of the Sierpinski fractal can be obtained by dividing the original square patch into 4 equal squares of 27× 27 mm^2 each. Then remove the upper left square portion to get the first iteration. For the second iteration, again divide each of the remaining three squares from iteration1 into 12 squares of size 13.5 × 13.5 mm^2 and then remove the upper left squares from each of the squares as shown in Fig. 1. Due to this iterative procedure the size of the antenna was greatly reduced. For the reference antenna, the area A(0) = 54× 54 = 2916 sq. units.

Whereas for iteration 1, A(1) = 27 × 27 = 729, i.e., the size reduction was calculated as 25%. For iteration 2, further three squares of areas A(2) = 13.5 × 13.5 = 182.25 were removed, i.e., the size reduction was calculated as 50%. If this process is continually repetitive the size has greatly reduced.

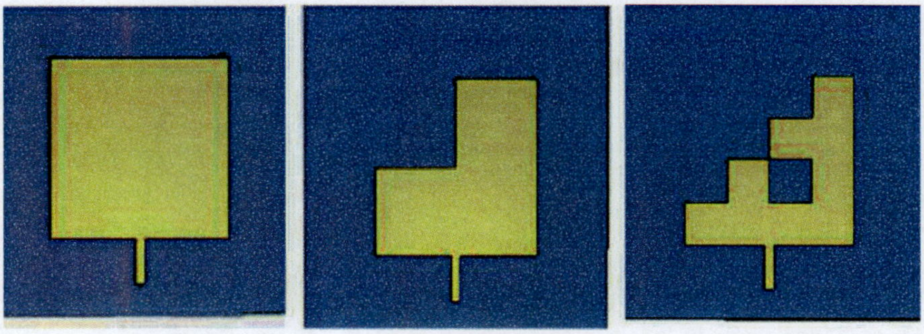

Fig. 1 The geometry of simulated Sierpinski step-shaped fractal reference square patch for two iterations

3 Simulation Results

Initially, the square patch was designed and simulated. Then Sierpinski carpet fractal geometry is applied for two iterations. All the simulations were carried out using CST MWS software. For optimization purposes, simulations have been carried out at various feed positions. The better impedance matching was found at the center of the patch. Figure 2 shows the simulated results for the return loss of the reference and first iteration on flexible jeans substrate. The second iteration follows the same multiband characteristics shown in the subsequent sections along with other performance parameters.

3.1 Current Densities

The current density determines the dispersal of current over the patch of the antenna. The dispersal of Current for the proposed antenna on jeans substrate at frequencies 1.652, 4.07, and 5.82 GHz for first iteration are shown in Fig. 3. And the distribution of current for the proposed antenna on jeans substrate at frequencies 3.88 and 4.98 GHz for second iteration are shown in Fig. 4. The magnitude of current density indicates coupling to the antenna. The maximum current density represents the highest coupling whereas the minimum current density represents the lowest coupling to the antenna.

From the distributions shown above, it is observed that the current distribution is nearly homogeneous at all central frequencies of first iteration and current density has the utmost value at slots and edges.

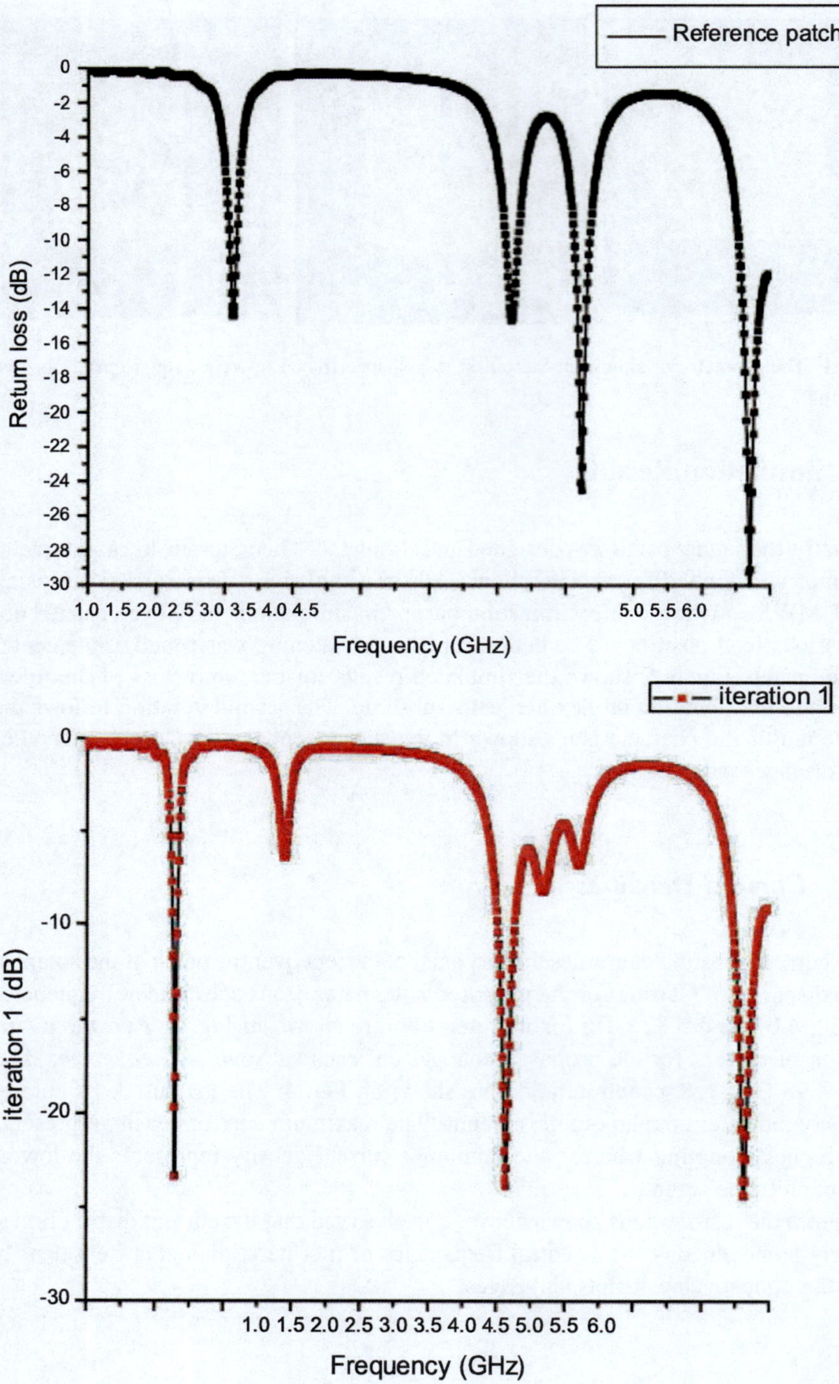

Fig. 2 The simulated Sierpinski fractal showing return loss for two iterations

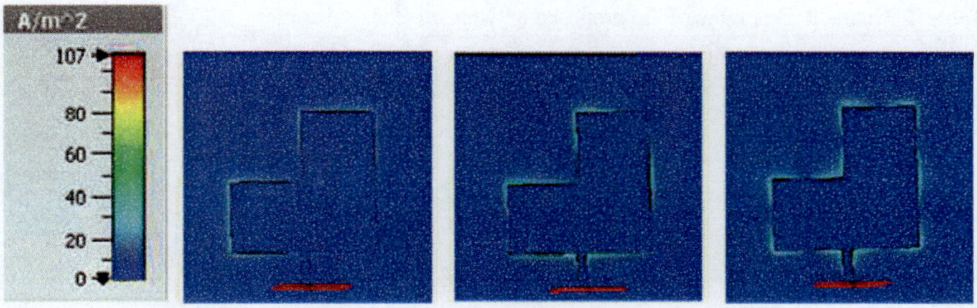

Fig. 3 Current densities of proposed antenna for iteration 1 at 1.652, 4.07, and 5.82

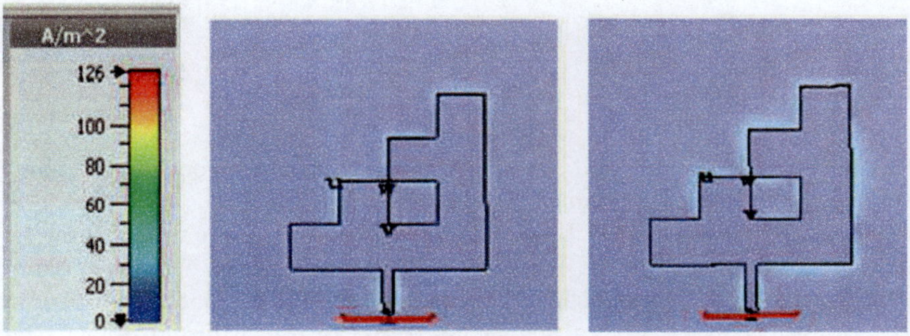

Fig. 4 Current densities of proposed antenna for iteration 2 at 3.88 and 4.98

3.2 Gain and Directivity

The peak gain and directivity of the proposed antennas along with other parameters is shown in the characteristic Tables 1 and 2 for both the substrates, respectively, up to two iterations. From the table, it is observed that a peak gain and directivity was increased as the iteration number increased particularly for the jeans fabric. It gives multiple operational frequency bands to fractal geometries with directive patterns. This type of performance is obtained with a simple feeding mechanism [12]. Therefore, fractal boundary wearable patch antennas are motivating replacement in the multiband antenna.

Table 1 Characteristic table of the proposed antenna on the FR-4 substrate

Iteration	Frequency	Return loss	Gain	Directivity
	2.61	−16.53	1.05	3.44
Zero		−14.64	0.68	4.75
	3.85	−27.29	2.23	11.59
	4.78	−21.64	0.31	4.84

Table 2 Characteristic table of the proposed antenna on jeans substrate

Iteration	Frequency	Return loss	Gain	Directivity
	2.26	−14.21	2.34	7.02
Zero		−14.53	2.845	5.36
	4.62	−23.96	2.76	6.72
	5.86	−26.33	3.18	7.21
	1.65	−16.28	1.15	5.59
First	4.07	−22.88	3.02	5.91
	5.82	−23.43	3.69	8.58
Second		−15.24	1.27	4.69
	4.98	−33.41	4.95	9.69

4 Conclusion

This research has demonstrated the partially textile Sierpinski carpet fractal antenna for wireless wearable applications. The jeans and FR-4 materials have been used as substrates, where a rigid PCB substrate is replaced with fabric materials for the wearable applications. Both antennas showed good performance in terms of return loss characteristics, Omni-directional pattern, gain, and efficiency along with reduced size. The major characteristics of this antenna are its ease in construction using fractal geometry, operable at multi-band frequencies. The simulated result shows that the antenna is suitable for multiband applications. As the fractal iteration increases, the perimeter of patch increases and the effective area of antenna decreases. The Sierpinski carpet with step-shaped iteration reduces the overall size of the antenna by 50% for the second iteration. Proposed antennas can find applications in wearable communications where a simple device is integrated into the clothing and in commercial applications like Wi-Max and Wi-Fi. They are also simple to fabricate and smaller than conventional antennas of similar performance.

References

1. Sran SS, Sivia JS (2016) Design of C shape modified Sierpinski carpet fractal antenna for wireless applications. Int Conf Electr Electron Optim Tech ICEEOT 2016:821–824
2. Puente-Baliarda C, Romeu J, Pous R, Cardama A (1998) On the behavior of the sierpinski multiband fractal antenna. IEEE Trans Antennas Propag 46(4):517–524
3. Ahmad S, Saidin NS, Isa CMC (2012) Development of embroidered Sierpinski carpet antenna. In: 2012 IEEE Asia-Pacific Conference on Applied Electromagnetics APACE 2012—Proceedings, pp 123–127
4. Chen WL, Wang GM (2008) Small size edge-fed Sierpinski carpet microstrip patch antennas. Prog Electromagn Res C 3:195–202

5. Chaouche YB, Messaoudene I, Nedil M,Bouttout F (2018) CPW-fed hexagonal modified Sierpinski carpet fractal antenna for UWB applications. In: 2018 IEEE international symposium on antennas propagation & SNC/URSI national radio science meeting APSURSI 2018—Proceedings, pp 1045–1046
6. Sierpinski G, Multiband F (2001) Mod mod Mod 49(8):1237–1239
7. Abdullah N, Shire AM, Ali MA, Mohd E (2015) Design and analysis of Sierpinski carpet fractal antenna. ARPN J Eng Appl Sci 10(19):8736–8739
8. Werner DH, Ganguly S (2003) An overview of fractal antenna engineeringresearch. IEEE Antennas Propag Mag 45(1):38–57
9. Balanis CA (1996) Antenna Theory_ Analysis and Design.pdf
10. Ahmed MI, Ahmed MF, Shaalan AA (2017) Investigation and comparison of 2.4 GHz wearable antennas on three textile substrates and its performance characteristics. Open J Antennas Propag 05(03):110–120
11. Rashidian A, Aligodarz MT, Klymyshyn DM (2012) Dielectric characterization of materials using a modified microstrip ring resonator technique. IEEE Trans Dielectr Electr Insul 19(4):1392–1399
12. Sankaralingam S, Gupta B (2010) Determination of dielectric constant of fabric materials and their use as substrates for design and development of antennas for wearable applications. IEEE Trans Instrum Meas 59(12):3122–3130
13. Azari A (2011) A new super wideband fractal microstrip antenna. IEEE Trans Antennas Propag 59(5):1724–1727

Evaluation of Real-Time Embedded Systems in HILS and Delay Issues

L. A. V. Sanhith Rao, Sreehari Rao Patri, I. M. Chhabra, and L. Sobhan Kumar

Abstract The present generation war game requires precision engagement with agility. Embedded systems of Aerospace vehicle like Seekers/Sensors are used to provide the necessary Guidance in the terminal phase. Seeker-based guidance can shape the latex demand within the capability of the Aerospace vehicle for a precision impact in the terminal phase. In addition, the autonomous guidance of Passive Imaging Infrared (IIR) seeker is less susceptible to external counter measures. Thorough performance evaluation of IIR Seekers and Guidance schemes is very essential for the effectiveness of the mission. For terminal engagement, Aerospace vehicle dynamics and accuracy are the prime factors which can be met by appropriate homing guidance design. Latest advances in seeker/sensor technology for locating target need to be integrated with the guidance system for steering and stabilizing the guided vehicle. Hardware-in-loop Simulation (HILS) of IIR Seekers integrated with the Real-Time Six Degrees of Freedom plant & target model helps in the evaluation of Aerospace vehicle embedded system design. Establishing the HILS test-bed with seeker and target along with Dynamic Motion Simulators and other subsystems is a major challenge. Various tests and the detailed procedures adopted to evaluate the Embedded systems of Aerospace vehicle in HILS are explained in this paper. The delay issues associated with the HILS runs were also discussed at the end.

Keywords Seeker · HILS · Embedded systems · Real time · Delay

1 Introduction

Seekers locate and track the target to provide in-flight guidance for the flight vehicle and increase the probability of kill based on received energy from the

L. A. V. S. Rao (✉) · I. M. Chhabra · L. S. Kumar
DRDO, RCI, Hyderabad, India
e-mail: sanhith139@rediffmail.com

S. R. Patri
National Institute of Technology Warangal, Warangal 506004, Telangana, India

T. Laxminidhi et al. (eds.), *Advances in Communications, Signal Processing, and VLSI*, Lecture Notes in Electrical Engineering 722,
https://doi.org/10.1007/978-981-33-4058-9_12

target. Present generation seekers based on the terminal engagement requirement can be RF/IIR/MMW/Laser or even electro-optical and IR. The data collection from these seeker-based sensors (multimode, multispectrum, etc.) with the interfaces is an important element for guidance system engineering in achieving precision strike with agility under all weather conditions. The seekers for the guidance are mostly stabilized. However, conformal arrays strapped on the body are also being attempted.

Presently, a variable range flight vehicle using homing guidance system based on Charged Coupled Device (CCD)/Imaging Infrared (IIR) stabilized seeker with Lock on before Launch (LOBL) configuration has been integrated for guiding the vehicle from liftoff to terminal engagement. The seeker with its stiff stabilization loop (Bandwidth > 15 Hz) and accurate ($\approx$1 milli radian) tracking loop (with agility > 2 Hz bandwidth) needs to be tested with the 5-axis motion simulator to represent real flight combat scenario. Innovative ideas were implemented for evolving various semi-natural (near natural) configurations to validate the embedded guidance and control software with number of flight hardware including the sophisticated seeker systems. The 6 DOF real-time vehicle model was validated during control flight with simulated guidance for flight vehicles. This was extended to a full-scale HILS along with target dynamics for guided flights. A full-scale dynamic tests [1] have been evolved for seeker characterization to fine-tune the performance of seeker before HILS.

A seeker model before dynamic tests is helpful in specifying the requirements of dynamic tables. A target motion system dynamics has to be much higher than that of the guidance loop and preferably more than seeker track loop. Further, Aerospace vehicle autopilot bandwidth requirement demands much higher dynamics of flight motion simulator (FMS). The isolation ratio of seeker can be tested with a higher dynamic FMS. A typical HILS test-bed has been established within the present limitation for validating the guidance and control system. At present, HILS configuration has been upgraded depending on the available state-of-the-art vehicle and target motion simulation systems.

Attempts are being made to introduce IR target growth, atmospheric attenuation, and background Scene generation in HILS. This will avoid the uncertainties arising due to simplifications in mathematical models for validating the image processing algorithms under dynamic conditions. Facilities are geared up to connect entire hardware, actual/simulated ground computers with sophisticated and flexible input–output interfaces to bring connectivity in real-time simulation environment.

In this paper the homing guidance requirements and techniques along with the seeker are brought at the beginning. Simulation techniques, modelling features are followed by dynamic tests with results. Finally, HILS with an IIR seeker of anti-tank aerospace vehicle has been highlighted with the results. Relevant conclusions and references are given at the end.

2 Homing Guidance Requirements and Techniques

For guided vehicle terminal engagement, vehicle dynamics and accuracy are the prime factors which can be met by appropriate homing guidance design. Latest advances in sensor/seeker technology for locating target and vehicle need to be integrated with the guidance system for steering and stabilizing the guided vehicle. The functional block diagram of any Flight vehicle can be shown as given in Fig. 1.

In this context, a typical anti-tank weapon using homing guidance system based on IIR stabilized seeker is considered. This uses lock on before launch (LOBL) for guiding the weapon. The seeker system as well as entire guidance and control system needs to be evaluated independently before integration. A typical dynamic test plan for characterizing the seeker was evolved which helped in evaluating the performance of the seeker for meeting the need of guidance system design. Additionally, total dynamic target scene generation for IIR needs to be evolved for validating the image processing algorithm required for homing guidance systems. Presently the image processing for finding the target signatures by IIR seeker is carried out by captive flight trials using helicopter.

The flight vehicle homing guidance requirements with stabilized seekers change with miniaturization as well as state-of-the-art algorithm design. The Precision Guided Munition (PGM) also have similar requirements with stabilized seeker where the lock on has to be carried out after ejection with appropriate automatic target recognition techniques. In this case integration with mother vehicle avionics including

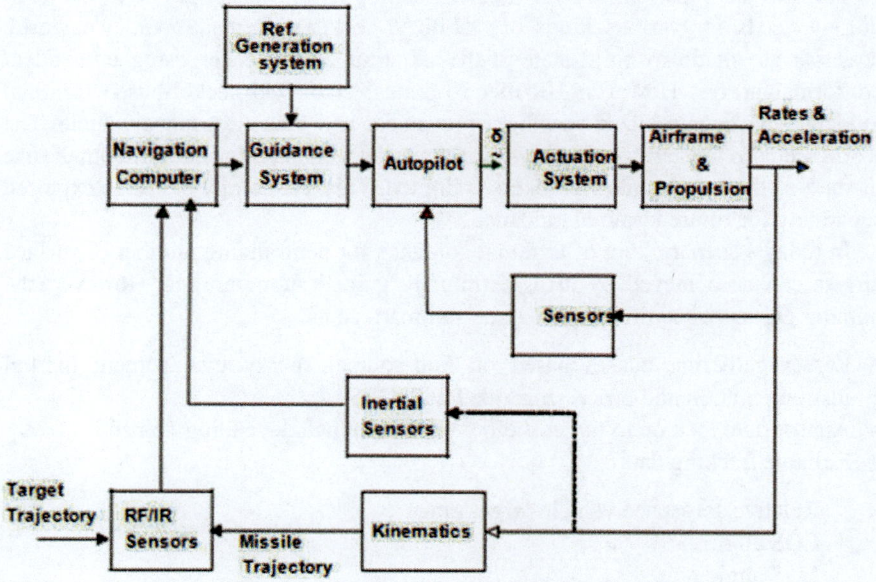

Fig. 1 Functional block diagram of aerospace vehicle

transfer of navigation data after appropriate sensor data fusion using GPS/INS techniques is also needed. Further mother vehicle strapdown GPS/INS guidance scheme may be enhanced with the aid of IIR stabilized seeker with reasonable range (<10 Km). The ABMs also need IIR-based stabilized seekers with a range around 30 Km for endo- and exo-atmospheric engagement against high-velocity air targets (with relative velocity 3–6 Km/s).

Classical guidance schemes based on homing using pursuit path, constant bearing path, and proportional navigation [2] have been used traditionally. Augmented proportional navigation law with addition of acceleration command to account for target manoeuvring has also been pursued. Today's modern guidance schemes for meeting the homing guidance requirements may be summarized as below:

- Variants of PN (e.g., APN) with estimation techniques and other image processing techniques
- Enhanced guidance law (e.g., Zero Sliding Guidance law etc.) with more number of state estimations and prediction of target recursively
- Use of optimal control based on linear quadratic techniques
- Use of neutral nets in a hybrid fashion.
- Parallel structures with distributed storage and processing for faster numerical computations.
- Learning ability for adjusting weight and biases for nonlinear dynamics.
- Adaptability in the changing environment.

A typical seeker-based terminal guidance scheme can be described in the Fig. 2.

The stiff stabilization loop for precise tracking by the seeker is one of the technology needs for various kinds of stabilized seeker systems. In today's world, attempts are made to build state-of-the-art strap down seeker using embedded/ conformal arrays. However, IIR/Electro-optical stabilized seeker-based terminal guidance has been realized and flight-proven for anti-tank Aerospace vehicle. The world scenario for stabilized seekers is progressing well but information about the use of state-of-the-art strap down seekers is limited. This technology has to be explored separately for future terminal guidance.

In today's scenario aim of terminal guidance for neutralizing number of surface targets can be achieved by using terminally guided submunitions. However, the homing guidance requirements may be summarized as

- Lesser gathering basket based on mid-course, energy management inertial instrumentation and processing aided with GPS.
- Appropriate lock on to target (before or after launch depending on need)
- Reliable tracking data

    Relative aerospace vehicle-target range
    LOS angle
    LOS angle rate
    Bore sight error angle

- Appropriate guidance law leading to minimum miss distance in the presence of

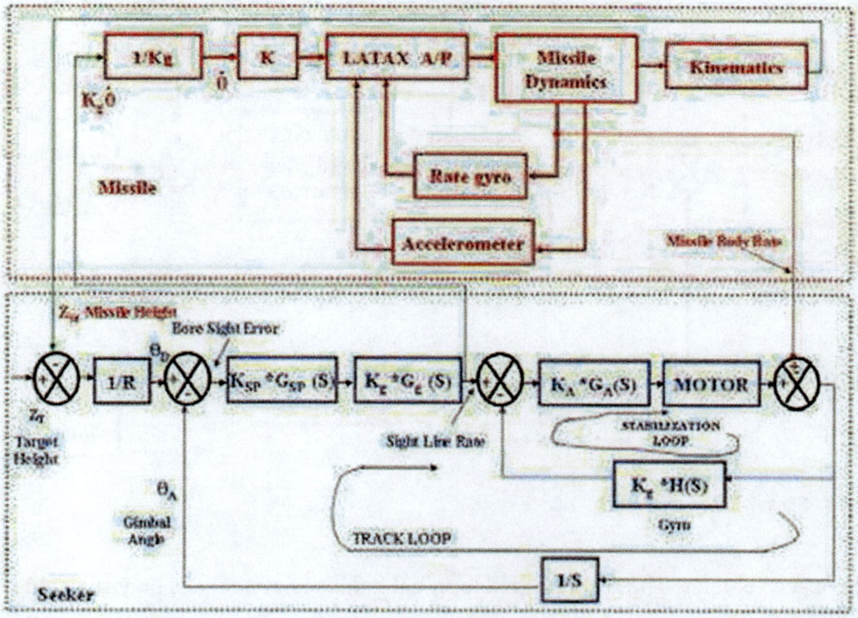

Fig. 2 Typical seeker-based terminal guidance scheme

Target manoeuvres
External and internal disturbances

- Effective flight control system

Steering capability for the guidance law
Required Latex generation
Stabilization of bare airframe
Reduction of sensitivity to disturbance inputs
Use of three-loop autopilot with synthetic stability loop

Dynamic tests as well as state-of-the-art HILS techniques are to be pursued for freezing terminal guidance system design.

3 Seeker Modelling and Dynamic Tests

The use of modelling and simulation in the development of military weapon system began to expand several years ago as the cost of flight testing began to rise. Since then, the role of modelling and simulation has expanded to include hardware in loop simulation (HILS), which helps for system design and development. It also plays an important role in the development of seeker-based terminal guidance for guided

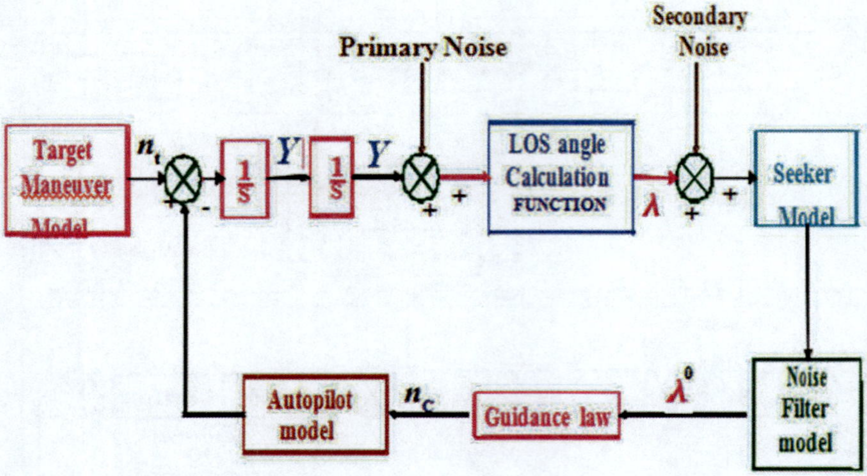

Fig. 3 Seeker-based terminal guidance system

aerospace vehicles. The seeker-based terminal guidance system can be visualized as shown in Fig. 3.

The mathematical modelling and its match with H/W characterization with appropriate test-bed leading to total HILS can be summarized as below:

A. *Mathematical Model*

- Seeker dynamic system model
- Seeker system front end model
- Seeker model integration with aerospace vehicle model
- Aerospace vehicle + seeker integrated model in mission scenario.

B. *Test-bed Preparation*

- Simulink/MatrixX Software and generated/developed S/W in Non Real Time (NRT)/Real time (RT) environment for seeker model.
- Use distributed processing environment with Guidance laws, Navigation and Control modules for RT Rapid Prototype environment using PCs and state-of-the-art RT simulation computers

C. *Seeker H/W Characterization*

- Dynamic characterization of H/W seeker
- IIR system characterization including dome

D. *Test-bed Preparation*

- Use the RT test-bed with high-speed data link for control and visualization.
- Independent test-bed for seeker system characterization.

This needs to be followed by Hardware in Loop Simulation (HILS) which has been described in next section. During the development of CCD-based seeker of ATM, it was felt necessary to evolve a full-scale dynamic test-bed for characterizing the seeker. The homing guidance loop with its inner tracking and stabilization has been described in Fig. 2. The independent characterization of the stabilization loop as well as track loop is necessary for meeting the performance requirements of PN guidance law as observed during the course of seeker-based guidance system design. The test-bed was geared up with a high fidelity (±30°/s body rate@40 Hz). Single axis rate table (SART). Further target motion system (TMS) with dynamic response much higher than the track loop (>1.5 Hz @ 10°/s SLR) was also introduced.

The following tests have been performed for characterizing the seeker dynamically.

- Isolation Ratio
- Decoupling Ratio
- Bore sight shifting test (Step body rate)
- Track loop bandwidth
- Bore sight step response
- Step target motion (Bore sight Impulse response)
- Sight Line rate calibration

The ratio of Aerospace vehicle body rate to gimbal angle rate and sight line rate is defined as Isolation and decoupling ratios, respectively. Isolation ratio defines the degree of Isolation between Aerospace vehicle body and Seeker gimbal, i.e., irrespective of Aerospace vehicle body disturbances seeker gimbal will continue to stare at the target. Decoupling ratio dictates the tracking of target irrespective of disturbances in body rate. A picture of test setup is given in Fig. 4. However detailed setup of both the tests is shown in Fig. 5.

The bore sight shifting against a sudden body jerk was tested in the dynamic test bench as shown in Fig. 6.

It was observed that LOS error was less than 0.36° (<1/3 FOV) even for 100°/s step body rate experienced. Bore sight step response test (test-bed similar to bore sight shifting test) has been designed to test shift in bore sight at the start of track

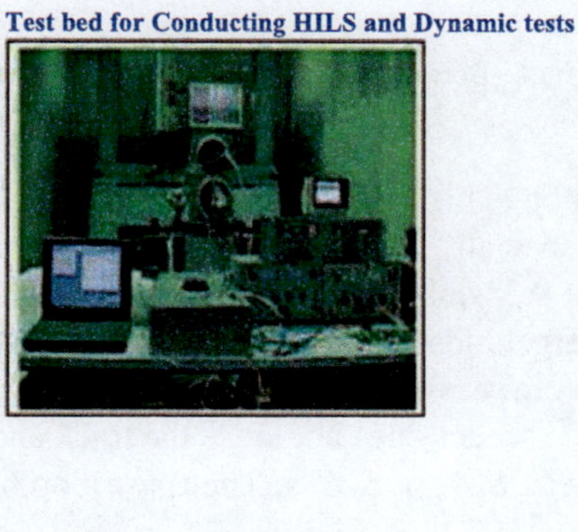

Fig. 4 Setup for seeker HILS and dynamic test

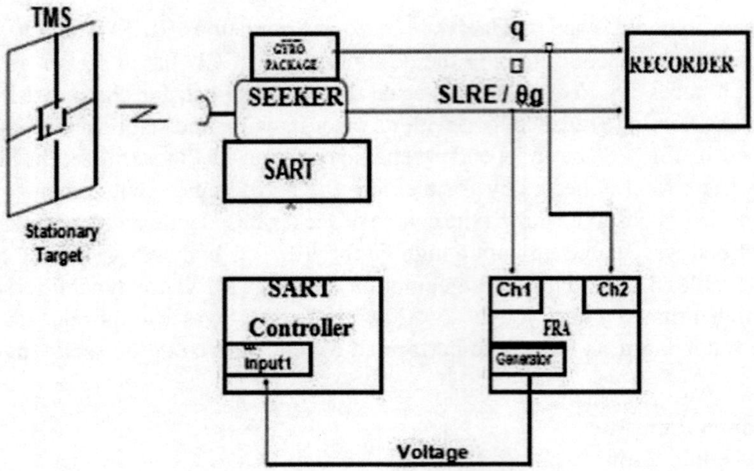

Fig. 5 Isolation and decoupling test

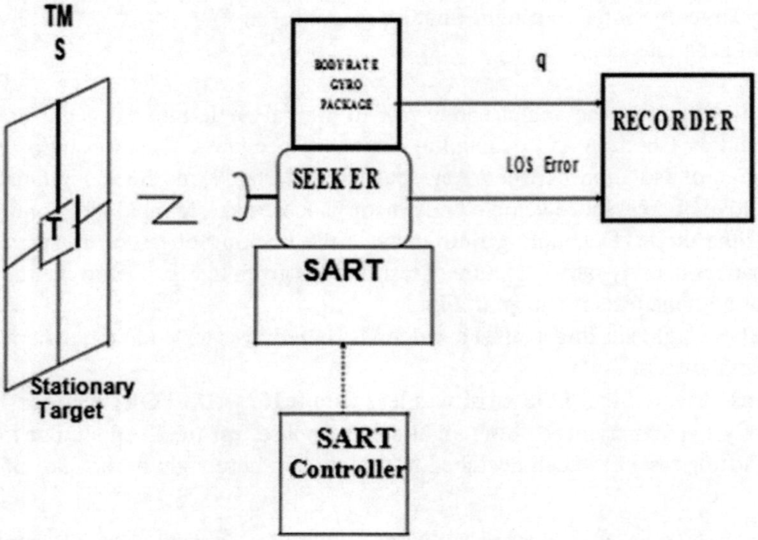

Fig. 6 Bore sight shifting test

loop. This requirement is very typical for lock on after launch (LOAL) situation especially for PGMs. A typical bore sight step requirement of 1° gives an overshoot of 0.3° only (<FOV). To determine the impulse response of bore sight against a step target motion the test-bed used is given in Fig. 7.

In case of IIR seeker of an ATM, even for a step input of 0.5° to the target (1/3 FOV) the seeker does not loose the track and bore sight error settles within 800 ms with a peak SLR of 5–6°/s. The track loop bandwidth was tested with physical sinusoidal

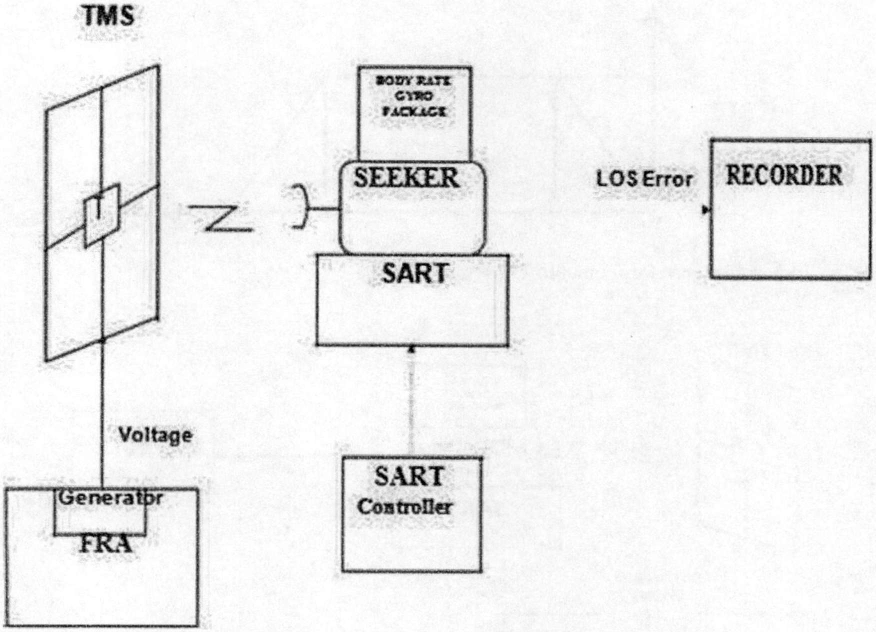

Fig. 7 Step target motion test

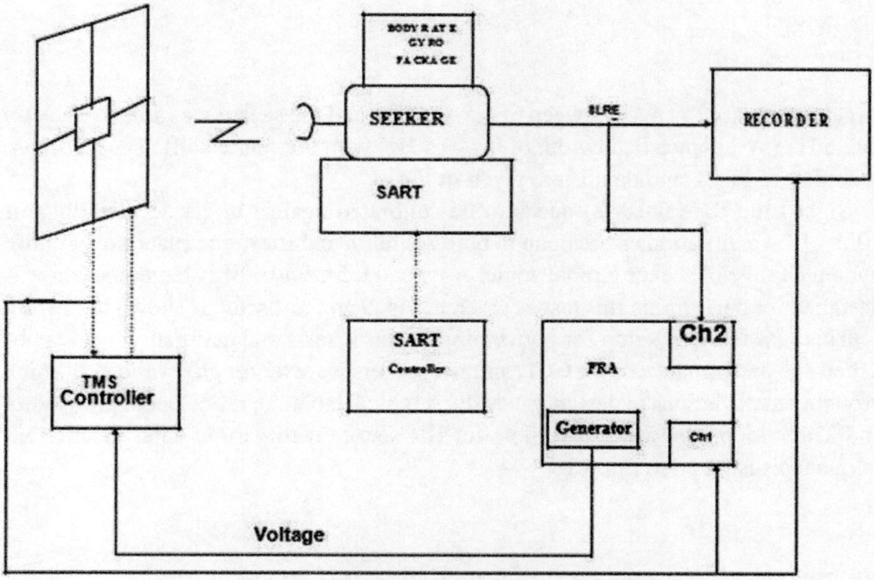

Fig. 8 Track loop bandwidth

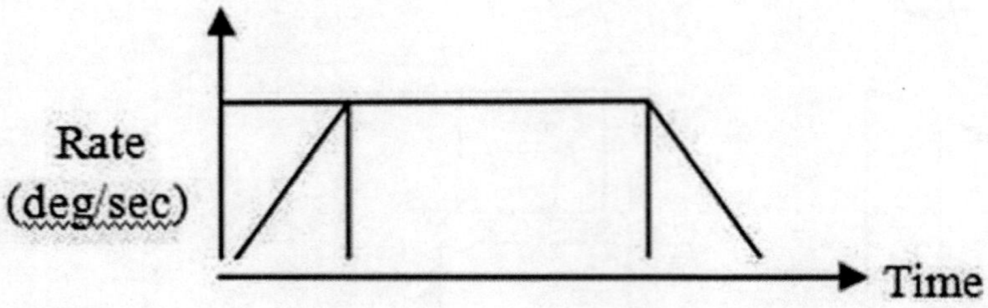

Fig. 9 Trapezoidal waveform input to TMS

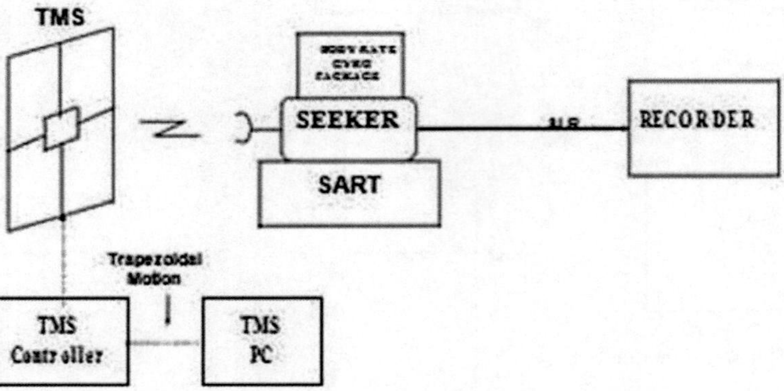

Fig. 10 SLR calibration

target motion (±2°/s over a sweep of 0.5–3.0 Hz) and the seeker was able to track up to 2.5 Hz (>Guidance bandwidth of 0.5–0.6 Hz) with 90° phase shift. The test setup is similar to Fig. 4 and details are given in Fig. 8.

Sight Line Rate (<10°/s) needs to be calibrated against target accelerations of 10°/s^2. The calibrations were done in both Azimuth and Elevation planes over entire dynamic range of seeker gimbal angle. A typical trapezoidal SLR target motion was designed for performing this test as given in Fig. 9 and test setup is shown in Fig. 10.

This new test-bed setup for performing dynamic tests and calibration was established and performance of the CCD and IIR seekers were thoroughly validated which brought out deviations in design precisely. A typical Isolation ratio, Decoupling ratio, and Track loop bandwidth test plots for IIR seeker with various rates at different frequencies is given in Fig. 11.

4 Hardware in Loop Simulation and Results

The dynamic tests have also helped to make the seeker ready for integrated HILS [3–10]. During HILS tests, a Six DOF rigid body model was integrated with the

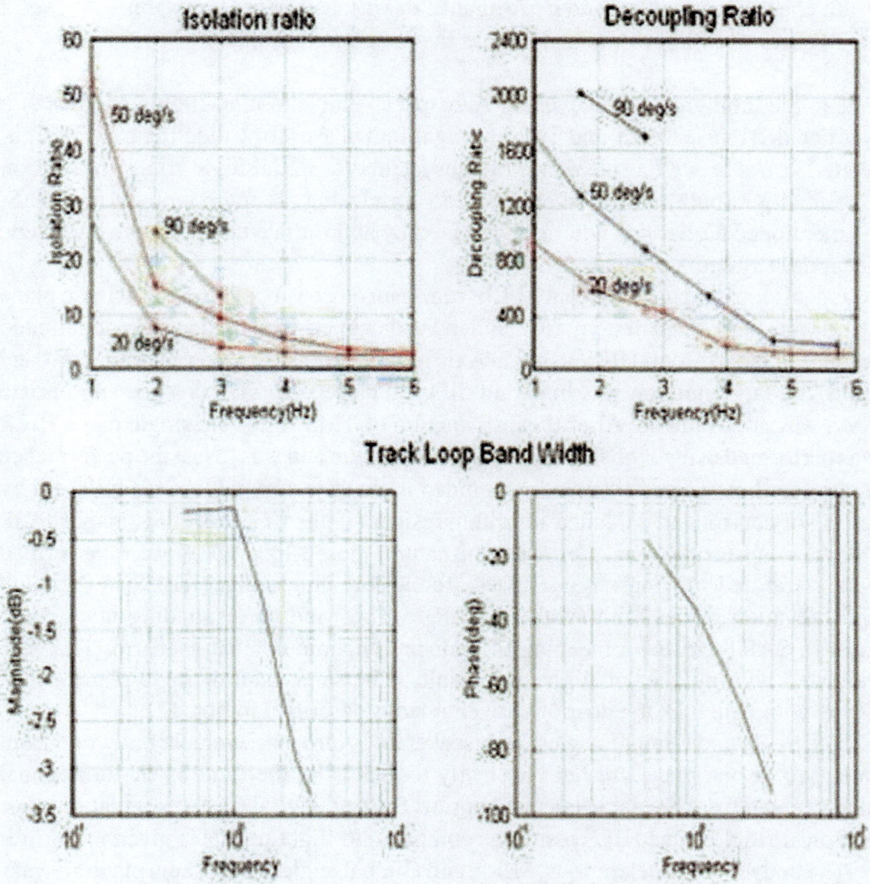

Fig. 11 Typical dynamic test results

Hardware seeker and on board computer. At the beginning a simplified seeker model was introduced which was later replaced by the real hardware. A typical seeker dynamic model has been already described in Fig. 2. This needs to be augmented with Front end CCD/IIR processing details. IR target growth simulator will be introduced in HILS for simulating the growth of target along with atmospheric attenuation during the course of the flight. This will be enhanced subsequently by target scene generation system. However, helicopter-based captive flight trials were performed for validating the image processing algorithms independently.

A test plan was worked out for performing HILS with CCD/IIR seeker. The main objectives of seeker in loop HILS are

- Validation of Control and Guidance system.
- To study seeker dynamic performance with Aerospace vehicle and trajectory dynamics.

- To check functionality and performance of various Flight Hardwares
- To check the Flight H/W and S/W in the integrated manner.

Due to the limitation of the existing Aerospace vehicle motion simulator (MMS) it was decided to use SART and TMS during initial phases for validating basic design-related issues as well as clearance of subsystems for initial flight trials. In addition a point target (bulb) was used on the TMS for closing the track loop during HILS. As mentioned earlier this will be augmented by appropriate target growth and scene generation systems in the future.

Open-loop guidance, guidance FLIP run, semi closed-loop HILS, and single plane HILS were performed in stepwise manner. Further open-loop guidance was continued leading to closed-loop HILS using 5-Axis motion simulator (after upgradation) and TMS. No target motion was imparted till semi closed-loop HILS since appropriate TMS was not available. After the introduction of TMS full-scale single plane HILS was performed using high fidelity SART. Initially the aim was to test the performance of the stabilization and track loop in guided flight trajectory which was followed by testing of control and guidance algorithm resident in the CGC during guidance FLIP (Flight Input Profile) run. Certain semi-natural closed-loop HILS were performed with static and moving targets. It was found that single plane HILS in Azimuth and Elevation planes with synthetic SLR to TMS and corresponding plane body rates to SART creates a meaningful realistic dynamic combat scenario. This was validated with number of flight trial results. The semi-natural single plane HILS, where complete 6 DOF equations are enabled is described in Fig. 12.

The SLRs and Gimbal angles from seeker and Aerospace vehicle body rate from rate gyro in one plane are fed physically to CGC and the same in the orthogonal plane are fed from the data generated apriori from all digital 6 DOF simulation runs. A typical single plane HILS results as compared to flight results is given in Fig. 13.

The body rates, accelerations, SLRs, and Gimbal angles in the same plane are validated with actual flight results. Final closed-loop HILS with 5-Axis motion simulator was also performed within the limitation of the facility. This setup is shown in Fig. 14.

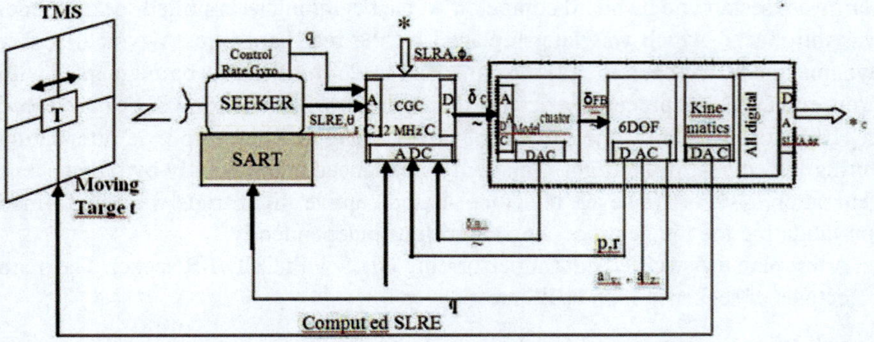

Fig. 12 Single plane HILS

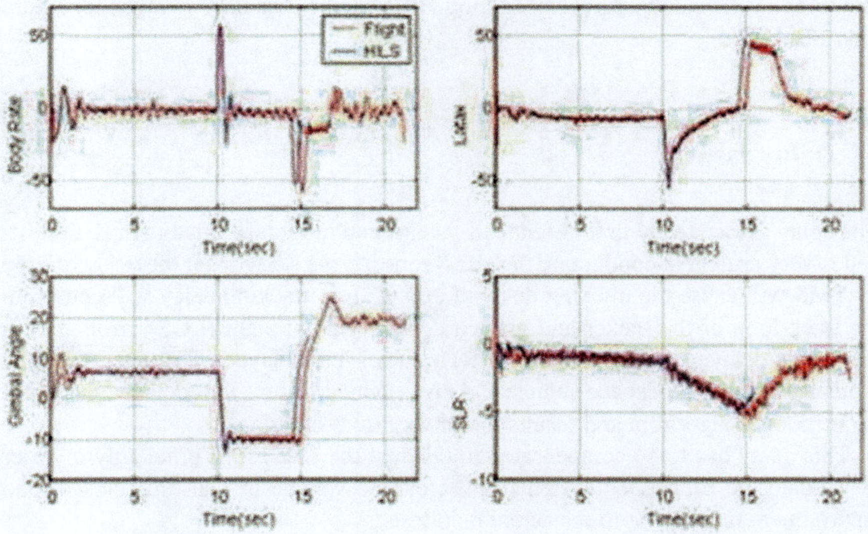

Fig. 13 Typical seeker HILS results

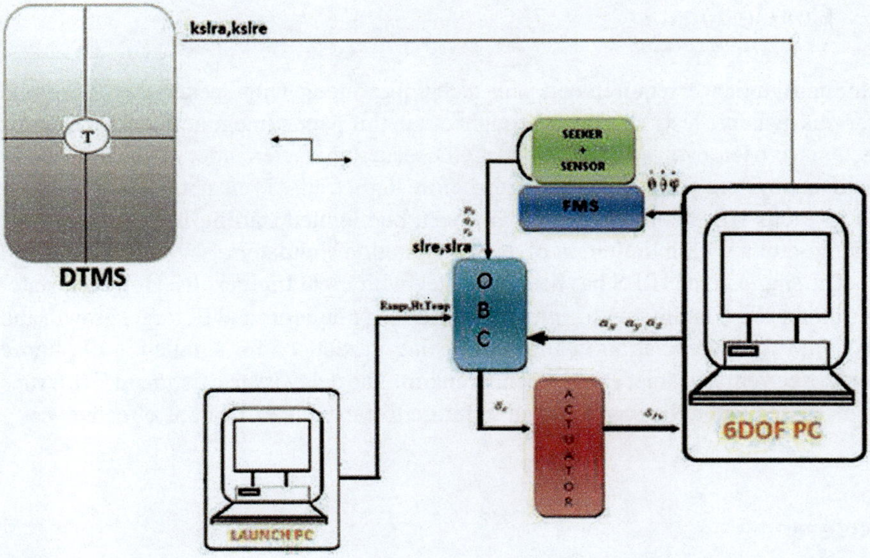

Fig. 14 Complete seeker HILS setup

However a full-scale 5- Axis closed-loop HILS facility has been evolved for future missions.

5 Delay Issues

The delay associated with different subsystems and motion simulators in HILS test-bed is very critical in conducting the HILS runs. As we said earlier the delay offered by TMS will cause the tracking delay of seeker. This tracking delay will enter into the track loop of the seeker and gives a delayed line of sight (LOS) errors and in turn gives delayed sight line rate (SLR) values. This delayed SLR enters into the guidance loop of Aerospace vehicle and gives delayed latex. This delayed latex will excite control algorithm and results into diverging oscillations.

This delay has to be compensated to conduct the HILS runs efficiently in order to validate the Embedded systems of Aerospace vehicle in real-time closed-loop environment much prior to the actual field tests.

6 Conclusions

Homing guidance requirements and techniques along with seeker modelling and various dynamic tests have been brought out in this paper. Integrated seeker dynamic tests have been evolved for IIR and CCD seeker characterization. It has helped in guidance system design and validation before flight trials. Typical modelling features and various HILS configurations have been highlighted starting from semi-natural configuration within limitation of available motion simulators.

The single plane HILS has been validated with flight trial results. However, state-of-the-art HILS facility with high fidelity motion simulators and IR target growth and dynamic scene generation facilities are being introduced for simulating the future aerospace vehicle-target engagement scenario. The delay issue existing in HILS runs is discussed and to be compensated in future to carry out HILS runs effectively.

References

1. Chaudhuri SK, Venkatachalam G, Prabhakar M et al (2001) Validation of guidance and control systems for tactical aerospace vehicles. In: AIAA modelingand simulation technologies conference, Montreal, Canada
2. Zarchan "Tactical and Strategic Missile Guidance"
3. Chaudhuri SK, Venkatachalam G, Prabhakar M (1997) Hardware-in-loop simulation for missile guidance & control systems. Defence Sci J 47(3)

4. Cosic K, Kopriva I, Kostic T, Slamic M, VolareviC M (1999) Design and implementation of a hardware-in-the loop simulator for a semi-automatic guided missile system. Simul Pract Theory 7(2):107–123
5. Underwood RC, Mc Millin BM, CrowML (2008) An open framework for highly concurrent hardware-in-the-loop simulation. In: 2008 32nd annual IEEE international computer software and applications conference, CD-ROM. https://doi.org/10.1109/COMPSAC.2008.165, ISBN: 978–0–7695–3262–2
6. Matar M, Karimi H, Etemadi A, Iravani R (2012) A high performance real-time simulator for controllers hardware-in-the-loop testing. Energies 5:1713–1733. https://doi.org/10.3390/en5061713
7. Lizarraga MI, Dobrokhodovy V, Elkaimz GH, Curryx R, Kaminer I (2009) Simulink based hardware-in-the-loop simulator for rapid prototyping of UAV control algorithms. In: Unmanned...Unlimited conference, Seattle, Washington
8. Duman E (2014) FPGA Based Hardware-in-the-Loop (HIL) simulation of induction machine model. In: 16th international power electronics and motion control conference and exposition electronic, Antalya, Turkey. ISBN: 978–1–4799–2060–0, https://doi.org/10.1109/EPEPEMC.2014.6980564
9. Zhang H, Zhang W, Wu Y, Wang J (2014) Background modeling in infrared guidance hardware-in-loop simulation system. In: Guidance, navigation and control conference (CGNCC), 2014 IEEE Chinese, pp 553–557
10. Shen N, Su Z, Wang X, Li Y (2009) Robust controller design and hardware-in-loop simulation for a helicopter. In: 4th IEEE conference on industrial electronics and applications, 2009. ICIEA 2009, pp 3187–3191

Acoustic Feedback Cancellation Using Optimal Step-Size Control of the Partition Block Frequency-Domain Adaptive Filter

S. Siva Prasad and C. B. Rama Rao

Abstract Acoustic feedback control is very critical in systems like hearing aids and public address systems. In hearing aid, lack of control of acoustic feedback can lead to instability of the system and causes annoying sound to the user. To control the acoustic feedback adaptive filtering approach is widely used. Limited tracking capability and slow convergence of the adaptive algorithms are also important limitations to be considered. However in hearing aid applications, for improving battery life, avoiding high computational complexity and long delays with appreciable feedback control is important. In this paper we proposed an AFC algorithm combining with decorrelation properties, by means of the prediction-error method (PEM), low delay and low computational complexity by means of a partitioned-block and frequency-domain implementation (PBFDAF), and included the optimal step-size (OSS) technique for achieving the fast convergence and low steady-state misalignment. Also we incorporated feedback path change detector (FPCD) in the proposed algorithm to improve convergence in the scenario of non-stationary feedback paths.

Keywords Acoustic feedback cancellation · Partition block frequency-domain adaptive filter · Optimum step-size · Hearing aids

1 Introduction

Hearing aids encounter the problem of acoustic feedback phenomenon by the virtue of the outflow of acoustic signal from loudspeaker to microphone. Input signal is the original voice signal which is feed to the microphone, and then processed by the processor and given to the loudspeaker. In acoustic feedback process the loudspeaker signal fed back to microphone thus forming a closed-loop from loudspeaker and microphone, i.e., (forward path from microphone to loudspeaker and feedback path

S. S. Prasad (✉) · C. B. R. Rao
Department of Electronics and Communication Engineering, National Institute of Technology, Warangal 506004, India
e-mail: sivaphd.nitw@gmail.com

T. Laxminidhi et al. (eds.), *Advances in Communications, Signal Processing, and VLSI*, Lecture Notes in Electrical Engineering 722,
https://doi.org/10.1007/978-981-33-4058-9_13

from loudspeaker to microphone). Thus according to barkhassen criteria this causes the system more oscillating and unstable when the magnitude of loop gain is greater than unity and phase angle is an integral multiples of 360 degrees. This also reduces total gain attained by the hearing aid [1, 2].

There are many feedback cancellation techniques among these adaptive filtering is vital, which includes feedback signal estimation and its subtraction from microphone signal. The major objective here is to enhance the instruments maximum usable gain and to safeguard intelligibility of speech [3]. On the basis of adaptation procedure the feedback cancellation systems are classified into continuous and non-continuous adaptation systems. In general non-continuous adaptation systems under certain instances use white noise as a probe signal for the estimation of acoustic feedback path. Continuous adaptation system majorly consists of adaptive filters, whose function is to continuously adapt to the transfer function of feedback path and here we need not depend on any training signal. The major defect in this continuous adaption technique is the desired signal is the summation of feedback as well as the input signal [4]. In general the correlation between the input signal and output signal results in the correlation of feedback signal with the input signal. Due to this reason the adaptive filter is not efficient in the estimation of feedback path properly. In this situation normal LMS algorithm produces a biased output because of its inability to converge [5].

There are many potential approaches in the literature to prevent this bias problem. Some of these techniques are employing a delay on feedback path or in forward path on the basis of statistical property of speech signal, adding a non-linear processing to input or feedback signal so as to reduce the correlation between them, addition of artificial noise, introducing constrained adaption, reducing the speed of adaptation, using a filtered X feedback cancellation with decorrelated adaptive filters, feedback path clipping, PEM-based approach [6, 7]. PEM-based approaches achieve an unbiased model of cancellation path by introducing a decorrelation by pre-filtering the loudspeaker primarily and then the signal of the microphone with its inverse model of near-end signal, before the signals feed to adaptive algorithm. Thus feedback path and near-end signal models can be collectively estimated using PEM and by this PEM produces an unbiased estimate of coefficient vectors of feedback path [8]. Similarly, other techniques like all-pass filters with time-varying poles [9], a band limited linear prediction coding based approach [10], two signal model AFC based on PEM [11], addition of low-frequency probe noise signal [12], improved prediction filter [13], formant and pitch-based estimation method [2], two microphone approach [14, 15], have been developed to reduce the bias. The development of an effective feedback canceller therefore requires a agreement between good steady-state performance with unbiased estimation of feedback path and the convergence of adaptive algorithm.

The poor convergence rate of LMS algorithm for coloured signals and also computational complexity is high when the modelling of feedback path with large number of filter coefficients [16]. The improvement in convergence is obtained by block frequency-domain adaptive filters (BFAF) for estimating the feedback path [17], but BFAF has long block processing delay when modelling the feedback path with many taps. To lessen the processing delay, Spriet et al. [7] proposed partioned-block frequency-domain implementation of AFC and can unite with time-domain pre-whitening filter so as to configure the PEM-based partition block frequency-domain adaptive filter (PBFDAF-PEM-AFC). The PBFDAF substitutes large point convolution task by adding up the outputs of smaller convolutions obtained by the use of smaller size block length, hence the processing delay is reduced. To achieve a trade-off between convergence rate and low steady-state misalignment, the selection of step-size mainly become a vital troublesome issue in the PBFDAF-PEM-AFC algorithm. In order to overcome the problems with fixed step-size, generally in the implementation of AFC or AEC, we adopt variable step-size, that is time varied in according to the algorithmic state. The objective of variable step-size is, when the system distance is large, i.e., at the beginning of the adaptation process $\mu(n)$ is large to allow fast adaptation, and $\mu(n)$ becomes low when the system distance is small to reduce the steady-state misalignment. The adaptively estimated step-size using variable step-size algorithms were proposed in [18–20]. An optimal step-size (OSS) technique for PBFDAF is proposed in [21], for AEC application. In this paper, to achieve fast convergence rate and low steady-state misalignment, PBFDAF with OSS technique is proposed for PEM-based AFC system (PBFDAF-OSS-PEM-AFC).

Further, the contents of the paper are organized as follows: Sect. 2.1 outlines implementation of PBFDAF-PEM-AFC with fixed step-size and Sect. 2.2 describes the proposed PBFDAF-OSS-PEM-AFC. Section 3 describes about the implementation of feedback path change detector(FPCD) for PBFDAF-OSS-PEM-AFC system. Simulation results are given in Sect. 4, and in the last section conclusion of the work is provided.

2 PBFDAF AFC with OSS

In this section, first we discuss the fixed step-size PBFDAF-PEM-AFC and then PBFDAF-OSS-PEM-AFC follows.

2.1 PEM-Based AFC Using PBFDAF with Fixed Step-Size

The prediction-error method is commonly used in acoustic feedback cancellation due to its decorrelation property and the PEM-AFC system is illustrated in Fig. 1. In PEM method, to reduce the correlation between microphone and loudspeaker signals, these signals are prefiltered and then used in the estimation of feedback path rather than original signals. The prefilter can be inverse of source signal model estimation. In this paper, the following notation is used, feedback path model is $F(q,n)$, source signal model $H(q,k)$, $A(q,k)$ is the prefilter response, such that the condition $H(q,k).A(q,k) = 1$ is satisfied, i.e., $A(q,k)$ is the inverse of $H(q,k)$. The source signal generator system can be modelled using AR model, and $H(q,k)$ is

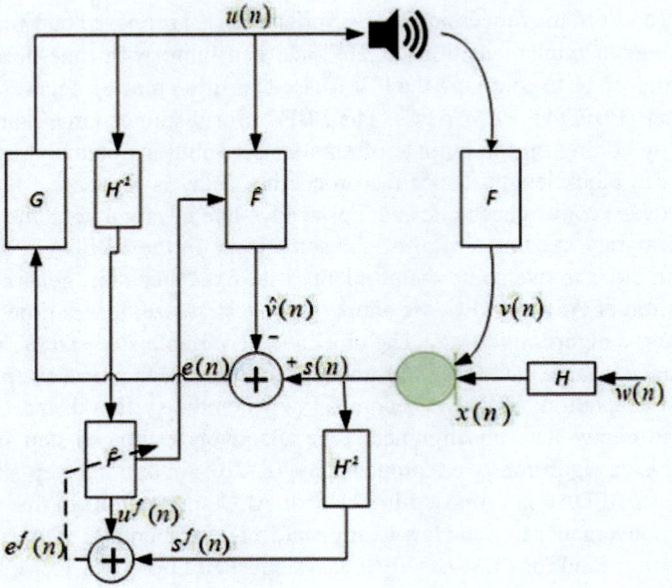

Fig. 1 PEM-based AFC system

estimated using levinson-durbin algorithm. The microphone signal and pre-whiten error signal are, respectively, given as

$$s(n) = \mathbf{u}^T(n)\mathbf{f}_o + x(n), \tag{1}$$

$$e^f(n) = s^f(n) - \mathbf{u^f}^T(n)\hat{\mathbf{F}}(n), \tag{2}$$

where $(\cdot)^T$ stands for transpose, $\mathbf{f}_o = [f_0, \ldots, f_{N-1}]^T$ is the feedback path coefficient vector of length N, $\mathbf{u}(n) = [u(n), \ldots, u(n-N+1)]^T$ is the sequence vector of loudspeaker signal, and $x(n)$ has both the near-end speech and background noise, $\hat{\mathbf{f}}(n)$ denotes an adaptive filter for the estimation of feedback path, $u^f(n)$, $s^f(n)$, and $e^f(n)$ denotes the pre-whiten signals of loudspeaker, microphone, and error signals, respectively.

In the PBFDAF algorithm, the weights of adaptive filter model $\hat{\mathbf{f}}(n)$ is partitioned into P smaller partitions as $\hat{\mathbf{f}}(n)$, where $\hat{\mathbf{f}}_p(n) = [\hat{f}_{pM}(n), \ldots, \hat{f}_{(p+1)M-1}(n)]^T$ is the weight vector of p-th subfilter with $M = N/P$ taps, and $p = 0, 1, ..P-1$. Thus, the pre-whiten error signal for PBFDAF-PEM-AFC is given as

$$e^f(n) = s^f(n) - \sum_{p=0}^{P-1} \mathbf{u}_p^{f^T}(n)\hat{\mathbf{f}}_p(n), \tag{3}$$

where $\mathbf{u}_p{}^f(n) = [u^f(n - pM), \ldots, u^f(n - (p+1)M + 1)]^T$. The frequency-domain pre-whiten loudspeaker signal vector of the p-th partition is $\mathbf{U}_p^f(k) = \text{diag}\{[U_{p,0}^f(k), \ldots, U_{p,2M-1}^f(k)]^T\} = \text{diag}\{\mathcal{F}\mathbf{u}_p^f(n)\}$, where the sequence $\mathbf{u^f}_p(n) = [u^f((n-p-1)M), \ldots, u^f((n-p+1)M-1)]^T$, $\mathcal{F}$ represents the FFT matrix of size $2MX2M$, and $diag$ represents diagonal matrix. The weights of the p-th partition is $\hat{\mathbf{F}}_p(k) = \mathcal{F}[\hat{\mathbf{f}}_p^T(kM), \mathbf{0}_{1\times M}]^T = [\hat{F}_{l,0}(k), \ldots, \hat{F}_{p,2M-1}(k)]^T$, where 0_{1XM} is a zero matrix. The frequency-domain representation of the pre-whiten error signal vector is $E^f(k) = [E_0^f(k), \ldots, E_{2M-1}^f(k)]^T$ and is given by

$$\mathbf{E}^f(k) = \mathbf{S}^f(k) - \mathcal{G}^{01}\sum_{p=0}^{P-1}\mathbf{U}_p^f(k)\hat{\mathbf{F}}_p(k), \tag{4}$$

where $\mathbf{S}^f(k) = \mathcal{F}[\mathbf{0}_{1\times M}, s^f(kM), \ldots, s^f((k+1)M-1)]^T$, is the frequency-domain vector of the pre-whiten microphone signal, and $\mathcal{G}^{01} = \mathcal{F}\begin{bmatrix}\mathbf{0}_M & \mathbf{0}_M\\ \mathbf{0}_M & \mathbf{I}_M\end{bmatrix}\mathcal{F}^{-1}$ is a windowing matrix. The weight update equation of PBFDAF-PEM-AFC is,

$$\hat{\mathbf{F}}_p(k+1) = \hat{\mathbf{F}}_p(k) + \mu\mathcal{G}^{10-1}(k)\mathbf{U_p^f}^H(k)\mathbf{E}^f(k), \tag{5}$$

where $(\cdot)^H$ indicates hermitian of a matrix, μ represents the fixed step-size, $\mathcal{G}^{10} = \mathcal{F}\begin{bmatrix}\mathbf{I}_M & \mathbf{0}_M\\ \mathbf{0}_M & \mathbf{0}_M\end{bmatrix}\mathcal{F}^{-1}$ is the constraint matrix, and the PSD matrix of the pre-whiten load speaker signal $\boldsymbol{\Lambda}(k) = \text{diag}\{[\Phi_{u^fu^f,0}(k), \ldots, \Phi_{u^fu^f,2M-1}(k)]^T\} = \mathcal{E}[\mathbf{U}_0^{f^H}(k)\mathbf{U}_0^f(k)]$, ($\mathcal{E}[.]$ represents expectation operator). The PSD of prefiltered loudspeaker can be recursively estimated as

$$\hat{\Phi}_{u^fu^f,i}(k) = \lambda\hat{\Phi}_{u^fu^f,i}(k-1) + (1-\lambda)\left|U_{0,i}^f(k)\right|^2, \tag{6}$$

where λ is a smoothing factor, $0 < \lambda < 1$.

2.2 PEM-Based AFC Using PBFDAF with OSS

In the weight update Eq. (5), of PEM-PBFDAF the step-size μ is fixed, this makes the algorithm either slow convergence with small step-size or high steady-state misalignment when the step-size is high. To overcome this we proposed an optimal step-size technique for PEM-PBFDAF acoustic feedback control algorithm. In [21], the derivation for estimation of optimal step-size in the scenario of acoustic echo cancellation(AEC) is given, and we followed the same approach of estimating OSS for the PEM-PBFDAF AFC algorithm. To represent the variable step-size weight update equation of PEM-PBFDAF, replace fixed step-size μ in Eq. (5) with $\mu_p i$, for the p-th partition's, i-th frequency bin.

$$\hat{\mathbf{F}}_p(k+1) = \hat{\mathbf{F}}_p(k) + \mathcal{G}^{10}\boldsymbol{\mu}_{p,i}(k)(k)^{-1}\mathbf{U}_p{}^{f^H}(k)\mathbf{E}^f(k), \tag{7}$$

where $\boldsymbol{\mu}_p = \text{diag}\{[\mu_{p,0}, \ldots, \mu_{p,2M-1}]^T\}$, is the variable step-size diagonal matrix. The following assumptions are made for convergence analysis (i) the sequences $U_p^f(k)$ and $x(k)$ are statistically independent and zero-mean stationary random processes ; (ii) the weight vector $\hat{\mathbf{F}}_p(k)$ is statistically independent of $U_p^f(k)$ and $x(k)$; (iii) $\mathcal{E}[U_{p,i}(k)U_{q,j}^*(k)] = 0$, for $p \neq q$. By performing the convergence analysis, we get the expression for recursive frequency-domain system distance $\delta_{l,i}(k)$, then for minimum system distance, the OSS $\mu_{l,i}(k)$ satisfies the expression $\frac{\partial \delta_{l,i}(k+1)}{\partial \mu_{l,i}(k)} = 0$. By solving, we get the optimal step-size and system distance equations as follows

$$\mu_{p,i}(k) = \frac{\Phi_{u^f u^f,i}(k)\delta_{p,i}(k)}{\Phi_{u^f u^f,i}(k)\left[\delta_{p,i}(k) + \frac{1}{2}\sum\limits_{l=0,l\neq p}^{P-1}\delta_{l,i}(k)\right] + 2\Phi_{xx,i}(k)}, \tag{8}$$

$$\begin{aligned}\delta_{p.i}(k+1) = &\left[1 - \frac{1}{2}\mu_{p,i}(k) + \frac{1}{4}\mu_{p,i}^2(k)\right]\delta_{p.i}(k) + \frac{1}{8}\mu_{p,i}^2(k)\\ &\times \sum_{l=0,l\neq p}^{P-1}\delta_{l,i}(k) + \frac{1}{2}\mu_{p,i}^2(k)\frac{\Phi_{xx,i}(k)}{\Phi_{u^f u^f,i}(k)},\end{aligned} \tag{9}$$

The initial value of the system distance could set to a constant value, as the performance of the algorithm convergence is not much sensitive to $\delta_{l,i}(0)$ [21]. Also, Eqs. (8) and (9) need the estimation of noise PSD, which consists of both stationary and non-stationary signals and this noise can be approximated to PEM error signal, when the adaptive filter converges to some degree. Hence, the PSD estimation of noise is $\Phi_{xx,i}(k) \approx \Phi_{e^f e^f,i}(k)$ and PSD estimation of PEM error signal as

$$\Phi_{e^k e^k,i}(k) = \alpha\Phi_{e^k e^k,i}(k-1) + (1-\alpha)\left|E_i^f(k)\right|^2, \tag{10}$$

where α is the smoothing factor and it lies between 0 and 1.

3 Feedback Path Change Detector (FPCD)

The PEM-FDAF-OSS has the fast convergence rate and low steady-state misalignment, but the tracking capability of change in feedback path is poor due to the decreasing property of system distance Eq. (9) with time and the overestimation of noise PSD when the feedback path changes. Thus due to the above reasons the step-size becomes very small and the behaviour of algorithm becomes unpredictable when a

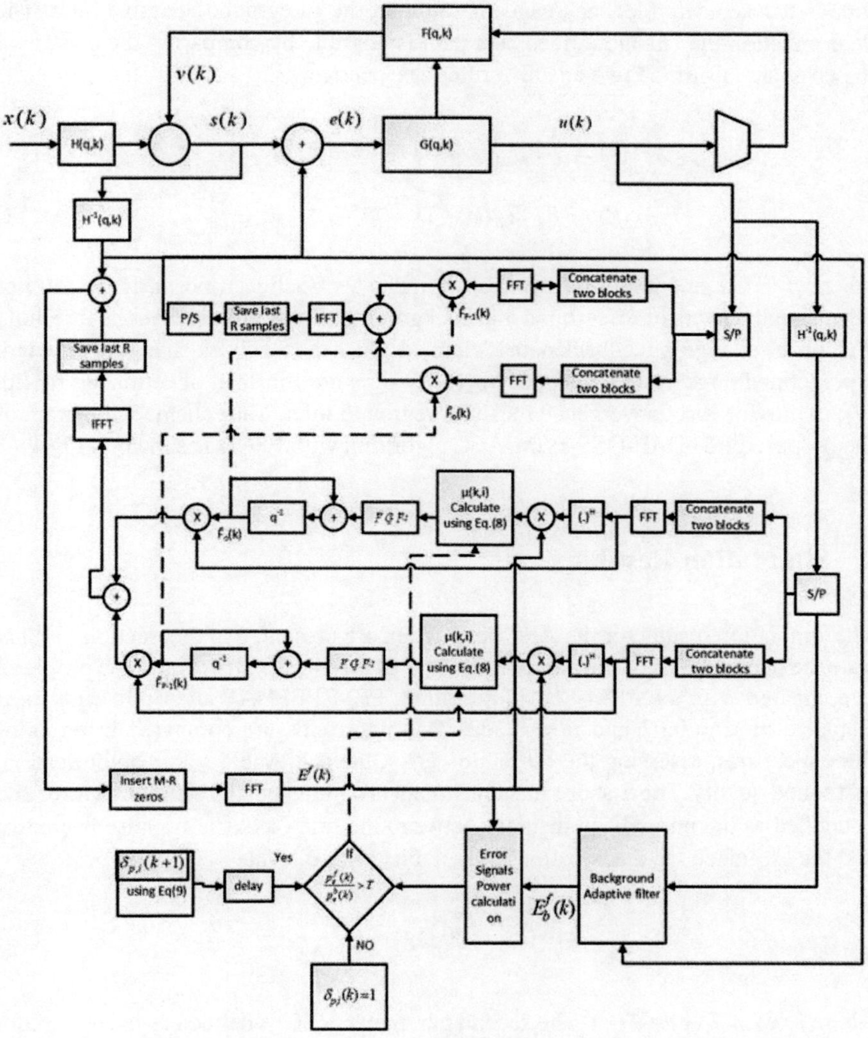

Fig. 2 Block diagram of proposed algorithm for AFC system

change in feedback path happens. In reference [21], to improve re-convergence capability of AEC, two independent parallel adaptive filters are adapted independently. The same approach is implemented in this paper, the foreground filter is the proposed PEM-PBFDAF-OSS which produces feedback cancellation output. The backward filter is the PEM-PBFDAF with large and fixed step-size $\mu = \mu_b$. The background filter length is shorter than the primary filter in order to reduce the complexity. Both the parallel adaptive filters are continuously adapted. Due to the large step-size of background adaptive filer, it exhibits fast convergence, but it has high steady-state misalignment. Thus, in the case of change in feedback path, backward filter can provide better acoustic feedback cancellation than the foreground adaptive filter. The determination the change in feedback path is possible by comparing the powers of the error signals of the two adaptive filters, respectively,

$$P_{e^f,f}(k) = \beta P_{e^f,f}(k-1) + (1-\beta)\left\|\mathbf{E}_f^f(k)\right\|^2, \tag{11}$$

$$P_{e^f,b}(k) = \beta P_{e^f,b}(k-1) + (1-\beta)\left\|\mathbf{E}_b^f(k)\right\|^2, \tag{12}$$

where β is the smoothing factor, and $\|\cdot\|$ indicates Euclidean norm. If the ratio of error signal powers of foreground and background filter is greater than the threshold T, then the change in feedback path is claimed. Once the feedback change is detected, a predefined large system distance $\delta_{p,i}(k) = \delta_c$ is used instead of estimated by Eq. (9), to provide fast convergence for the foreground filter. The schematic diagram of the proposed PBFDAF-OSS-PEM-AFC algorithm with FPCD is shown in Fig. 2.

4 Simulation Results

The simulation results for the AFC application are presented in this section to illustrate the performance of the proposed PEM-PBFDAF-OSS algorithm. The proposed is compared with the PEM-FDAF algorithm, PEM-PBFDAF algorithm with fixed step-size of $\mu = 0.02$ and $\mu = 0.002$. The algorithms are compared in terms of three measures, assessing the estimation error, the achievable stable amplification, and sound quality. The first one misadjustment is nothing but the estimation error and is defined as the normalized distance between the true (acoustic impulse response) and the estimated feedback path usually defined in dB level.

$$\text{Mis}(\kappa) = 20\log_{10}\frac{\|\mathbf{f}_{\text{r}}(\kappa)\|}{\|\mathbf{f}_{\text{t}}(\kappa)\|} \tag{13}$$

where $\mathbf{f}_{\text{r}}(\kappa) = \mathbf{f}_{\text{t}}(\kappa) - \hat{\mathbf{f}}(\kappa)$. The second parameter MSG is defined as the maximum achievable stable gain at a given time provided the the forward path is spectrally constant.

$$\text{MSG}(\kappa) = -20\log_{10}\left[\max_{l\in\mathcal{P}(\kappa)} |F_{\text{r}}(\kappa, l)|\right] \tag{14}$$

The third one is an objective evaluation for assessing the sound quality by means of frequency-weighted log-spectral signal distortion (LSD) and this distance measure proven to correlate well with subjective evaluation of AFC algorithms. The feedback paths considered for simulation are given in the figure. The results are obtained by averaging different trails with 10 different speech signals taken from TIMIT database, each one is of 1 minute duration. The sampling rate is 8Khz. The other

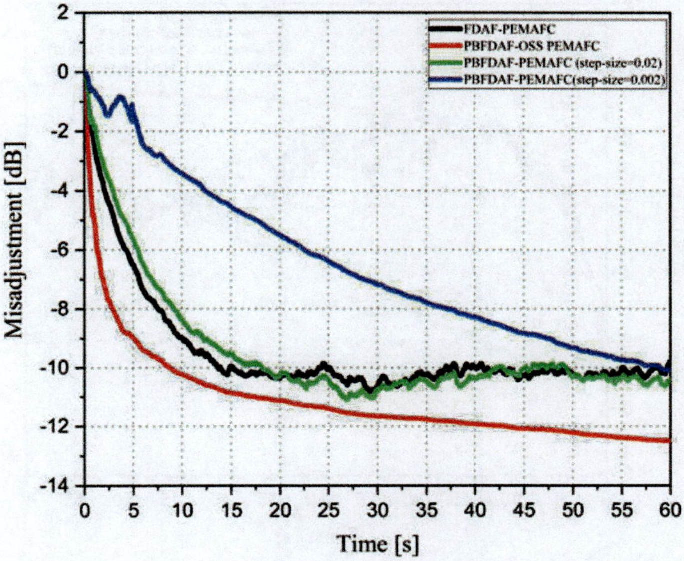

Fig. 3 Misadjustment performance of FDAF-PEMAFC, PBFDAF-PEMAFC with $\mu = 0.02$ and $\mu = 0.002$ and PBFDAF-OSS-PEMAFC using the AIR1Misadjustment performance of FDAF-PEMAFC, PBFDAF-PEMAFC with $\mu = 0.02$ and $\mu = 0.002$ and PBFDAF-OSS-PEMAFC using the AIR1

parameters of the proposed algorithm are $\delta_c = 1.0$, $\lambda = 0.85$, $\alpha = 0.9$, $\beta = 0.75$. In Fig. 3, we compared the convergence rate of four algorithms FDAF, PBFDAF with large step-size $\mu = 0.02$ and small step-size $\mu = 0.002$, and PBFDAF-OSS algorithms in PEMAFC configuration. The proposed algorithm has the convergence rate as fast as the PBFDAF algorithm with larger step-size and has better steady-state misalignment than the PBFDAF with small step-size. Figure 4 shows the ASG (Added stable gain) performance, PBFDAF-OSS manage to provide stable ASG compared to other algorithms. Under non-stationary environment, with feedback path change at 30 s, the algorithms performance is given in Fig. 5. The proposed algorithm without FPCD fails to track the change in feedback path. Also it can be observed that with including the FPCD in proposed algorithm, we have overcome the divergence problem for non-stationary feedback path. In Fig. 6, comparison of ASG performance is given when the feedback path is non-stationary and it can be concluded that PBFDAF-OSS-PEM-AFC with FPCD performs better compared to other algorithms. The mean and max LSD measure values of the four algorithms are given in Table 1.

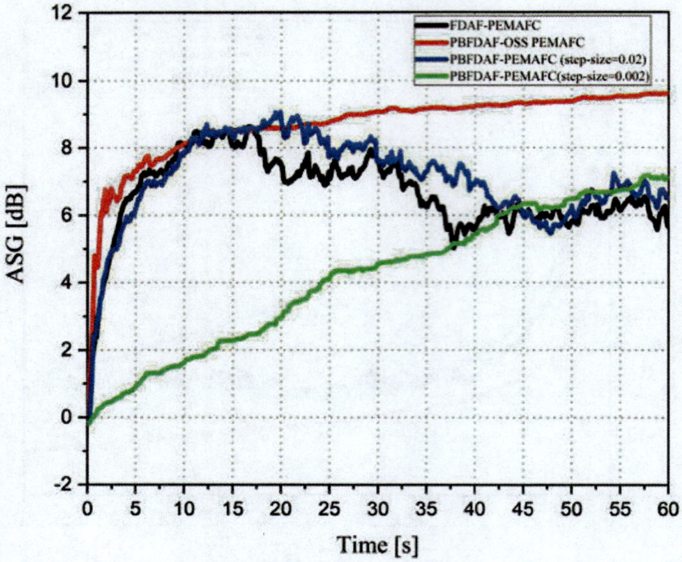

Fig. 4 ASG performance of FDAF-PEMAFC, PBFDAF-PEMAFC with $\mu = 0.02$ and $\mu = 0.002$ and PBFDAF-OSS-PEMAFC using the AIR1

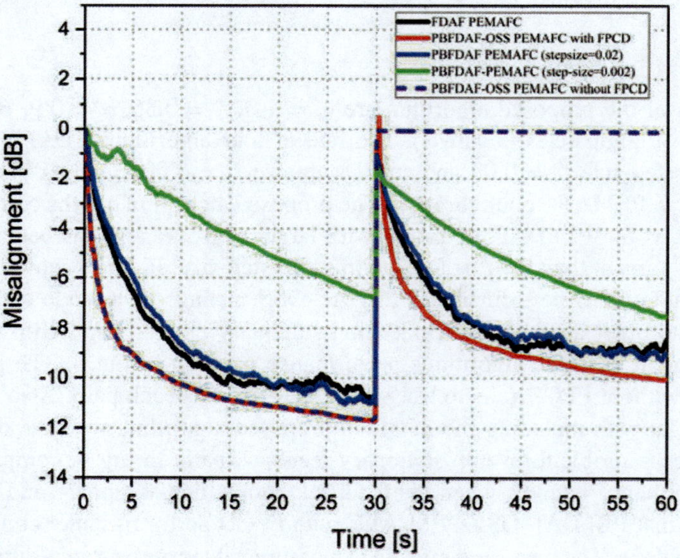

Fig. 5 Misadjustment performance of FDAF-PEMAFC, PBFDAF-PEMAFC with $\mu = 0.02$ and $\mu = 0.002$ and PBFDAF-OSS-PEMAFC, with change in Feedback Path from AIR1 to AIR 2 at 30 Sec

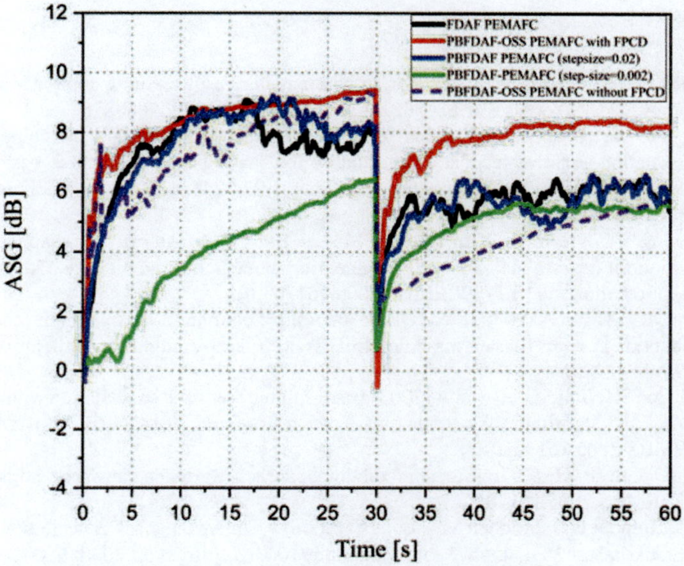

Fig. 6 ASG performance of FDAF-PEMAFC, PBFDAF-PEMAFC with $\mu = 0.02$ and $\mu = 0.002$ and PBFDAF-OSS-PEMAFC, with change in Feedback Path from AIR1 to AIR 2 at 30 Sec

Table 1 Mean and Maximum LSD values for the different algorithms in the two different simulated scenarios

	Stationary feedback path		Non-stationary feedback path	
	Mean	Max	Mean	Max
FDAFPEM AFC	0.63952	4.31936	0.72097	4.42553
PBFDAF-PEMAFC (μ = 0:02)	1.18837	4.58607	1.20183	4.09932
PBFDAF- PEM AFC (μ = 0:002)	2.81473	5.90118	2.71255	5.70194
PBFDAF-OSS-PEM AFC	0.26449	2.6882	0.82895	4.20533

5 Conclusion

We have proposed a robust optimal step-size PBFDAF-based PEM-AFC based on the framework presented in [21]. Simulation results indicated that the proposed PBFDF-OSS-PEM-AFC algorithm has faster convergence and lower steady-sate misalignment in stationary feedback path environment but fails to converge when the feedback path is non-stationary. We have incorporated FPCD technique in the proposed algorithm in order to achieve the better tracking capability in non-stationary scenario and simulation results show that the proposed algorithm has improved performance in both stationary and non-stationary feedback paths.

References

1. Kates JM (1991) Feedback cancellation in hearing aids: results from a computer simulation. IEEE Trans Signal Process 39:553–562. https://doi.org/10.1109/78.80875
2. Chiang YF, Wei CW, Meng YL, Lin YW, Jou SJ, Chang TS (2014) Low complexity formant estimation adaptive feedback cancellation for hearing aids using pitch based processing. IEEE/ACM Trans Audio, Speech Lang Process 22:1248–1259. https://doi.org/10.1109/TASLP.2014.2327300
3. Ji YS, Jung SY, Kwon SY, Kim IY, Kim SI, Lee SM (2006) An effcient adaptive feedback cancellation for hearing aids. In: IEEE engineering in medicine and biology 27th annual conference. https://doi.org/10.1109/IEMBS.2005.1617030
4. Siqueira MG, Alwan A (2000) Steady-state analysis of continuous adaptationin acoustic feedback reduction systems for hearing-aids. IEEE Trans Speech Audio Process 8. https://doi.org/10.1109/89.848225
5. Chi HF, Goa SX, Soli SD, Alwan A (2003) Band-limited feedback cancellation with a modified filtered-X LMS algorithm for hearing aids. Speech Commun 39:147–161. https://doi.org/10.1109/IEMBS.2005.1617030
6. Markel JD, GrayJr AH (1976) Linear prediction of speech. Springer, NewYork. https://doi.org/10.1109/IEMBS.2005.1617030
7. Spriet A, Rombouts G, Moonen M, Wouters J (2006) Adaptive feedback cancellation in hearing aids. J Franklin Inst 343:545–573. https://doi.org/10.1016/j.jfranklin.2006.08.002
8. Panda G, Puhan NB (2016) An improved block adaptive system for effective feedback cancellation in hearing aids. Digital Signal Process 48:216–225. https://doi.org/10.1016/j.dsp.2015.08.016
9. Boukis C, Mandic DP, Constantinides AG (2006) Bias reduction in acoustic feedback cancellation systems with varying all-pass filters. IET Electron Lett 42:556–558. http://orcid.org/10.1049/el:20060470
10. Guilin M, Gran F, Jacobsen F, Agerkvist FT (2011) Adaptive feedback cancellation with band-limited LPC vocoder in digital hearing aids. IEEE Trans Audio, Speech, Lang Process 19:677–687. https://doi.org/10.1109/TASL.2010.2057245
11. Van Waterschoot T, Moonen M (2009) Adaptive feedback cancellation for audio applications. Signal Process 89:2185–2201. https://doi.org/10.1016/j.sigpro.2009.04.036
12. Guo M, Jensen SH, Jensen J (2012) Novel acoustic feedback cancellation approaches in hearing aid applications using probe noise and probe noise enhancement, IEEE Trans Audio, Speech Lang Process 20:2549–2563. https://doi.org/10.1109/TASL.2012.2206025
13. Ngo K, van Waterschoot T, Christensen MG, Moonen M, Jensen SH (2013) Improved prediction error filters for adaptive feedback cancellation in hearing aids. Signal Process 93:3062–3075. https://doi.org/10.1016/j.sigpro.2013.03.042
14. Nakagawa CRC, Nordholm S, Yan WY (2015) Analysis of two microphone method for feedback cancellation. IEEE Signal Process Lett 22:35–39. https://doi.org/10.1109/LSP.2014.2345571
15. Pradham S, George NV, Albu F, Nordholm S (2018) Two microphone acoustic feedback cancellation in digital hearing aids: a step size controlled frequency domain approach. Appl Acoust 132:142–151. https://doi.org/10.1016/j.apacoust.2017.11.015
16. Derkx RMM, Egelmeers GRM, Sommen P (2002) New constraining method for partitioned block frequency-domain adaptive filters. IEEE Trans Signal Process 50:2177–2186. https://doi.org/10.1109/TSP.2002.801932
17. Estermann P, Kaelin A (1994) Feedback cancellation in hearing aids: results from using frequency-domain adaptive filters. In: IEEE international symposium on circuits and systems, vol 2, pp 257–260. https://doi.org/10.1109/ISCAS.1994.408953
18. Rotaru M, Albu F, Coanda H (2012) A variable step size modified decorrelated NLMS algorithm for adaptive feedback cancellation in hearing aids. In: Proceedings of the 10th international symposium on electronics and telecommunications. IEEE, pp 263–266. https://doi.org/10.1109/ISETC.2012.6408070

19. Gil-Cacho JM, Van Waterschoot T, Moonen M, Jensen SH (2014) A frequency-domain adaptive lter (FDAF) prediction error method (PEM) framework for double-talk-robust acoustic echo cancellation. IEEE/ACM Trans Audio, Speech, Lang Process 22:2074–2086. https://doi.org/10.1109/TASLP.2014.2351614
20. Strasser F, Puder H (2015) Adaptive feedback cancellation for realistic hearing aid applications. IEEE/ACM Trans Audio, Speech, Lang Process 23:2322–2333. https://doi.org/10.1109/TASLP.2015.2479038
21. Yang F, Yang J (2017) Optimal step-size control of the partitioned block frequency-domain adaptive filter. IEEE Trans Circuits Syst II: Express Briefs 814–818. https://doi.org/10.1109/TCSII.2017.2780880

A Compact IPD Based on-Chip Bandpass Filter for 5G Radio Applications

M. V. Raghunadh and N. Bheema Rao

Abstract In this paper, the model and simulation of compact on-chip bandpass filter (BPF) based on the Silicon-based integrated passive device (IPD) technology for 5G radio front end (RFFE) application is presented. A high performance 180 nm CMOS series stacked multilayer (ML) inductor with a novel double split structure is developed with minimal chip area. A high quality planar spiral capacitor is also designed and simulated. These two passive components are connected in a series resonator configuration to realize the BPF. The filter is simulated using high frequency structural simulator (HFSS). The simulated BPF demonstrated a quality factor (Q) of 13.68, fractional bandwidth of only 7.31%, an inband insertion loss of −1.28 dB along with −13.68 dB return loss at the center frequency of 8.2 GHz. The on-chip area occupied by the filter is only $400 \times 360\ \mu m^2$. Thus, it exhibits a narrow spectral response to yield high quality signals, yet occupying smallest footprint. Hence, this proposed compact resonator BPF is more suitable for various 5G RFFE applications such as cellular networks, navigation, and other IOT services. We simulated the filter by focusing around 8 GHz, as this spectral band is yet to be explored for licensed or unlicensed 5G broadband applications.

Keywords 5G · RFFE · IPD · HFSS · Multilayer · BPF · Quality factor

1 Introduction

Massive on demand deployment of networked smart devices had tremendous impact on global RFFE architecture evolutions. FCC approved 3.1 to 10.6 GHz unlicensed spectrum for UWB and 5G radio networking. Around 30 RFIC on-chip passive

M. V. Raghunadh (✉) · N. B. Rao
Department of ECE, National Institute of Technology, Warangal, India
e-mail: raghu@nitw.ac.in

N. B. Rao
e-mail: nbr@nitw.ac.in

T. Laxminidhi et al. (eds.), *Advances in Communications, Signal Processing, and VLSI*, Lecture Notes in Electrical Engineering 722,
https://doi.org/10.1007/978-981-33-4058-9_14

inductors and capacitors occupy large chip area (~70%) for smartphone architectures [5]. LTCC filters yielded low losses but suffer from large size and heat problems [18]. Microwave strip lines and stub resonators enable multiband responses with low losses but had large size and complexity [11]. MEMS technology permits low loss millimeter size component with good component integration. But MEMS components suffer low frequency (<2 GHz), power consumption, and instability problems [17]. IPD technology permits compact integrated passive circuits to meet the high performance micromachining needs of the 5G systems on a chip. IPD circuits are most preferred to implement multiple blocks like BPFs, LNAs, etc. in a single package [8, 9].

BPFs realized using CMOS technology possess low power consumption, lower footprints, and enhanced integration capabilities. Many low loss silicon substrate CMOS technology-based miniaturized RFFE circuits were successfully implemented using quality passive devices, to satisfy the system on-chip (SOC) requirements [7]. A 65 nm wideband CMOS receiver is developed for 2 and 10 GHz RF-band using a three-stage LC resonator BPF [18]. Flip chip IPD based 7–9 GHz UWB BPF was fabricated with 10 dB return loss and large are 1.68 mm^2 [2]. Glass substrate IPD dual-band BPF designed at 4.8 GHz with 2.5 dB insertion loss occupied chip size of 0.9 mm^2 [3]. Si IPD UWB filter designed at 7.656 GHz had Q value of 7.65, but it occupied large area of 1.67 mm^2 [4]. A 2.5 GHz Si IPD BPF showed −2.8 dB insertion loss but occupied 4 mm^2 chip space [5]. A triple-band microstrip UWB filter with 2.6 to 9.6 GHz passband had insertion loss of 1.5 dB but its size is 14 mm^2 [6].

Hence, it is proposed to design and simulate a compact high Q BPF to meet narrow spectral demands of the next Gen 5G RFICs, employing the low loss spiral passive devices. In this paper, we have designed a spiral ML inductor and a metal insulator metal (MIM) capacitor to develop resonator-based BPF. We simulated the filter S-parameters in HFSS by focusing the filter design around 8 GHz, as this spectral band is yet to be explored for licensed or unlicensed 5G broadband needs.

The paper presents the design, simulation, and analysis of the proposed BPF as given in Sects. 2, 3, and 4.

2 BPF Design and Analysis

We designed and simulated a miniaturized BPF with a spiral multilayer split inductor and a planar spiral capacitor structures on the Si substrate. Rectangular-shaped spiral is chosen as it possesses uniform current distribution and permits easier fabrication. Filter Integration is done in the Si-based IPD Technology. The structure mainly is composed of the stack of—Si substrate, dielectric, and three metal layers. The metal surfaces, leads, test, and bond layers make the model very concise. A SiO_2 dielectric layer is sandwiched between Cu metal layer and a 0.2 m thick Cu/gold metal (for via connections) on the Si Substrate as depicted.

2.1 On-Chip Inductor

Si-based on-chip inductor designs in the literature, showed maximum Q value of 10 with large chip area. Multi-turn stacked inductor structures increase the inductance, with improved self-resonant frequency. Copper metal as conductor material reduces the inductor parasitic resistance and also improves its quality factor [1]. The inductor is designed on thick silicon substrate with copper conductor. The external diameter is 100 μm with the occupied chip area as 100 × 100 μm^2. This will decrease losses while enhancing the quality factor. We used a novel double split along the entire conductor length to yield larger inductance (*L*) because of mutual coupling between every two turns. The proposed inductor is designed with multiple turns in six layers and simulated in the high frequency structured simulation software (HFSS) using the lumped model to obtain *Q*-factor and impedance *L* values. Extracting the inductance is an important step in component analysis. May expressions were developed based on the Grover method, but yielded incorrect *L* values [14]. Fundamental expression for inductance extraction is due to Greenhouse method [13]. The net inductance of inductor is the total sum of the self, mutual (positive and negative) inductances of the device structure.

$$L_{\text{Total}} = L_{\text{Self}} + \sum M_{+\text{ve}} + \sum M_{-\text{ve}} \tag{1}$$

The shunt capacitances C_{shunt} and the substrate capacitance C_{Si} in a lumped model of inductor is given by [17]

$$C_{\text{Si}} = \frac{\text{wlC}_0}{2} \tag{2}$$

$$C_{\text{Shunt}} = C_{\text{oxide}} \left\{ \frac{1 + w^2 (C_{\text{Si}} + C_{\text{oxide}}) C_{\text{Si}} R_{\text{Si}}^2}{1 + w^2 (C_{Si} + C_{\text{oxide}})^2 R_{\text{Si}}^2} \right\} \tag{3}$$

Here *w* is width of metal, l is conductor length, and C_0 is substrate capacitance for unit area. The expression for total magnetic flux from the primary and secondary current components is [17].

$$////\varphi_{21} = \frac{\mu_0 I_1}{4\pi} \int_{-l_2/2}^{l_2/2} \int_{-l_1/2}^{l_1/2} \frac{e^{-j\beta_1 z\prime}}{\sqrt{d^2 + (z - z\prime)^2}} dz\prime dz \tag{4}$$

Mutual inductance between the conductor paths is given as

$$M_{21} = \frac{\mu_0 I_1}{4\pi I_2} \left\{ \frac{\varphi_{21}}{e^{-j\beta Z}} \right\} \tag{5}$$

The evaluated mutual and self-inductance values are substituted in greenhouse method to get the inductance value. The inductor Q-factor is expressed as [17]

$$Q = \frac{\omega L_s}{R_s} \frac{R_p}{R_p + \left\{ \left(\frac{\omega L_s}{R_s} \right)^2 + 1 \right\} R_s} \left[1 - \frac{R_s^2 (C_s + C_p)}{L_s} - \omega^2 L_s (C_s + C_p) \right] \tag{6}$$

We chose the series stacked multilayer double split structure, which possess large mutual inductance and hence maximum possible inductance. The 3D and normal views of the designed rectangular-shaped spiral inductor are depicted below in Figs. 1 and 2. Si is the substrate material and the conducting material is copper.

Inductor structure employs three turns in six layers. The first half turn is placed into layer 1, with next half turn into layer 2 and the next half turn into layer 3 and so on, as shown in Fig. 2. The real part and imaginary parts of the Y-parameters are mapped from the simulated S-parameters. The values of Q and L for on-chip inductor are computed from the s-parameters [15]. The effective inductance versus frequency

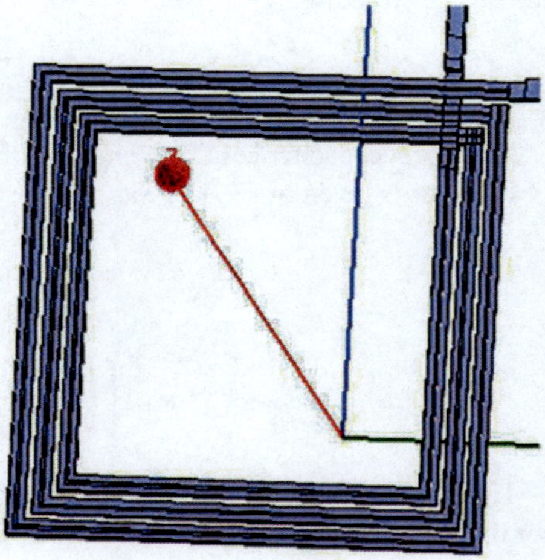

Fig. 1 The Planar view of the split structure spiral ML inductor used in BPF

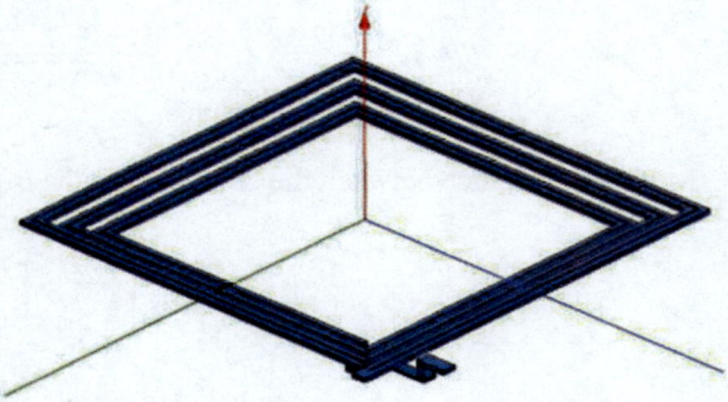

Fig. 2 The 3D view of the split structure inductor spiral ML inductor used in BPF

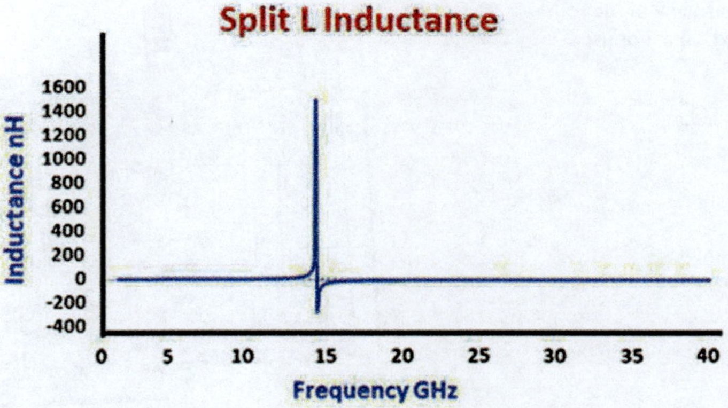

Fig. 3 Inductance (nH) versus Frequency (GHz) response of on chip Multilayer split spiral inductor

is extracted from the imaginary part of Y-parameters. So also the effective resistance (real part).

$$Q = \frac{\mathrm{Im}[Y_{11}]}{\mathrm{Re}[Y_{11}]} \tag{7}$$

$$L = \frac{-1}{2\pi f\{Im[Y_{11}]\}} \tag{8}$$

The following Fig. 3 indicates the inductance variation w.r.t. frequency for the proposed inductor.

The on chip inductor dimensions: Conductor width, thickness, and spacing are 4 μm, 2 μm, and 2 μm, respectively. The layer spacing is 2 μm with overall area occupied by the inductor is 180 × 180 μm^2. The simulated performance of this filter based on 0.18 μm RF CMOS technology yielded a high Q value of 26.3 and resonant frequency of 14 GHz [12]. Hence this inductor is suitable for C to X-band RF applications mainly due to the minimal area occupied on a chip.

2.2 *On-Chip Capacitor*

A planar rectangular-shaped spiral capacitor is also designed and simulated in HFSS. The structure uses Silicon substrate and the copper metal as conductor. All the turns are placed in a single layer. Conductor width is taken constant for all turns. The inductance and capacitance (C) values are mainly determined by the material properties, the conductor width, length, spacing, and the number of turns between the adjacent annular rings. Figure 4 shows the normal planar view of the on-chip spiral capacitor.

Fig. 4 Planar view of the spiral-shaped capacitor used in BPF

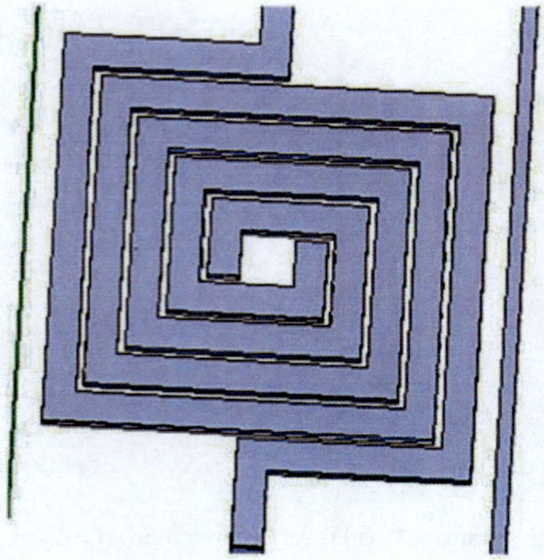

The capacitance variation with frequency is shown in Fig. 5. The Q and C values of the capacitor are determined from the S-parameters expressions shown below [15].

$$C = \frac{-\text{Im}[Y_{21}]}{2\pi f} \quad (9)$$

$$Q = \frac{\text{Im}[Y_{11}]}{\text{Re}[Y_{11}]} \quad (10)$$

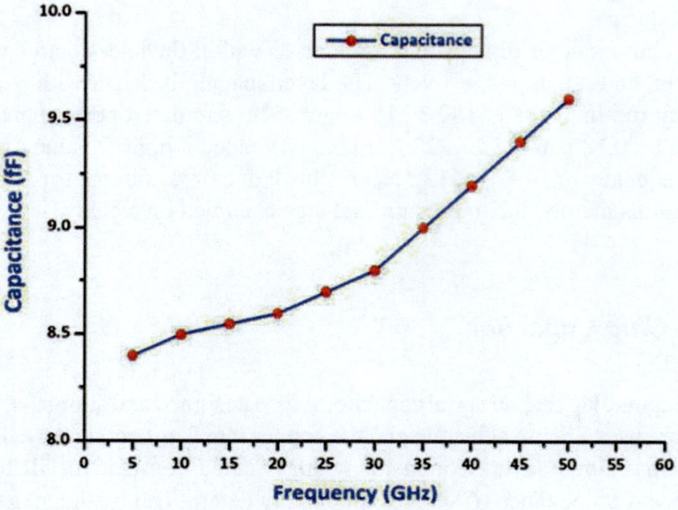

Fig. 5 The capacitance (fF) versus Frequency (GHz) response of on-chip spiral capacitor

The on-chip spiral capacitor dimensions are Conductor width, thickness, and spacing are 4 μm, 2 μm, and 1 μm, respectively. The overall area occupied by the capacitor is $50 \times 50\ \mu m^2$. The number of conductor turns is 2. The simulated capacitor based on 0.18 μm RF CMOS technology yielded improvements in terms of the *Q*-factor and capacitance. Hence this is suitable for *C*- to *X*-band RF applications because of minimal area occupied on a chip.

2.3 The BPF Circuit

Bandpass filters are the most essential components placed in between the antenna and RF amplifier. They decide the receiver selectivity to produce the desired high quality RF signals. Out of the several available approaches to design the BPF, we selected the simple LC resonant circuit as it is easy to implement and compare the performance to prove its effectiveness.

The schematic for the LC filter proposed is given in Fig. 6. It consists of the *L*, *C*, and parasitic lumped elements as a resonator. Filter simulation replaces the LC components with the designed rectangular-shaped ML spiral inductor and a planar spiral capacitor [17].

The simulation and optimization were performed in HFSS software. The *L* and *C* values are stable across the entire passband. Figures 7 and 8 show the normal and 3D views of the proposed series LC filter, respectively. High *Q* passive devices would

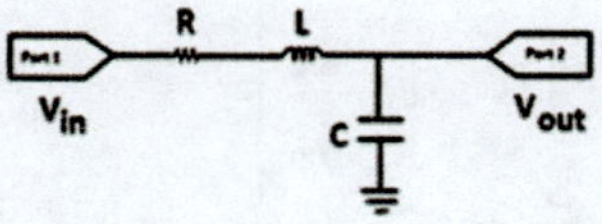

Fig. 6 Equivalent circuit schematic for the LC resonator BPF in radio receivers

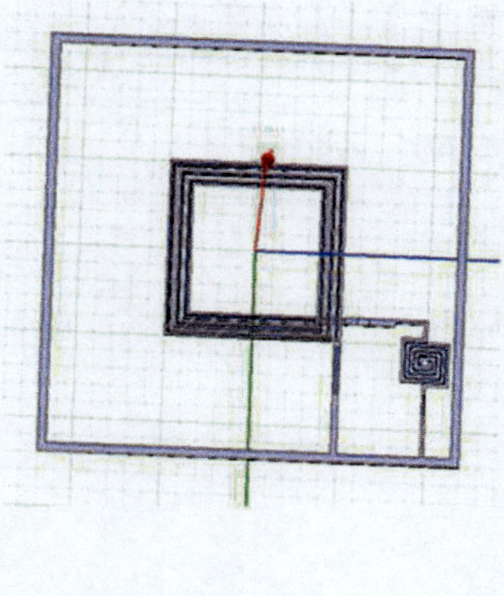

Fig. 7 Normal view of Series LC resonant circuit connection for BPF

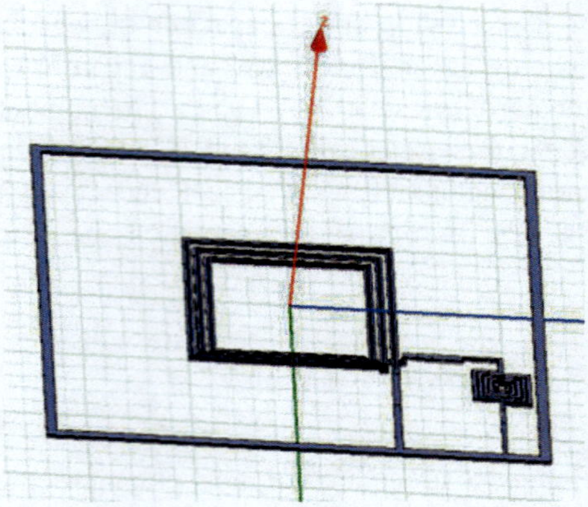

Fig. 8 3D view of the Series LC resonant circuit connection for BPF

only possess the narrowband spectrum, as the bandwidth and Q-factor vary inversely proportional. Filter designs in general involve in a trade-off between the loss and bandwidths.

Q values for the capacitor and the inductor employed in above BPF are shown in Figs. 9 and 10 given below.

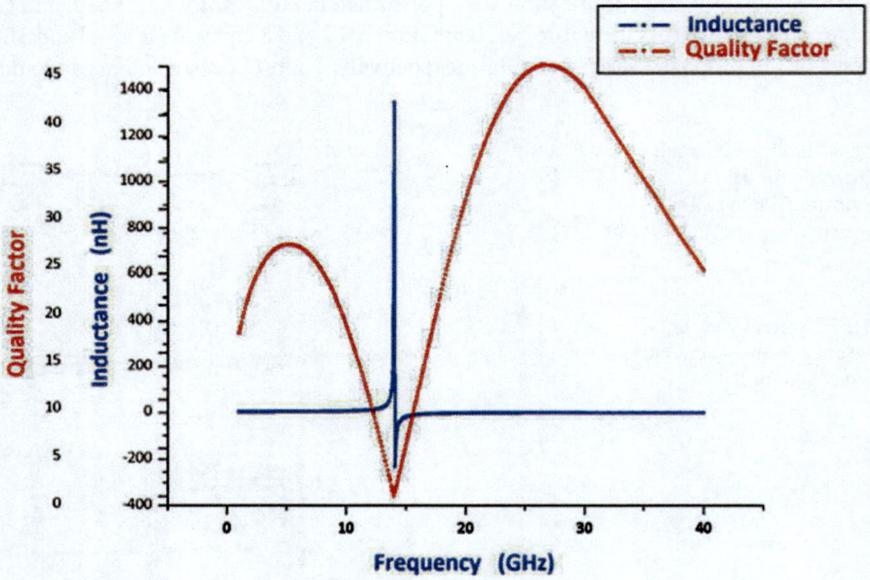

Fig. 9 Quality factor versus frequency for the inductor in the frequency range (1–40) GHz

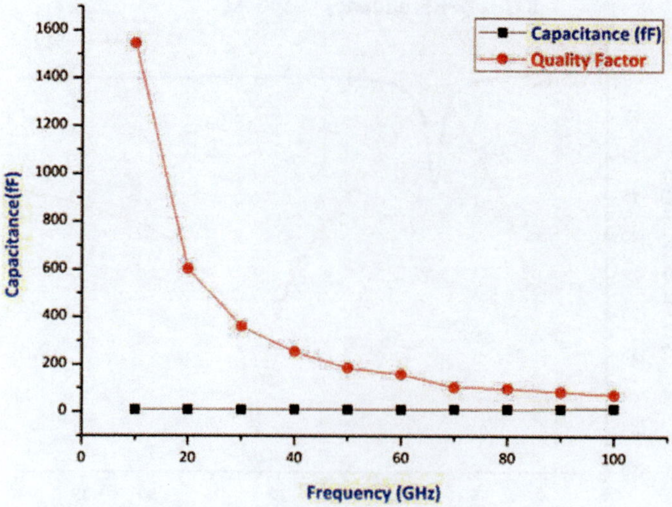

Fig. 10 Quality factor versus frequency for the capacitor in the frequency range (1–40) GHz

2.4 *Simulation*

A first-order LC BPF is chosen for the design to suit the requirements for 5G RFFE micro-components. The −10 dB line of insertion loss is selected to compute the filter bandwidth. We analyzed the filter using LC lumped model using HFSS to obtain *S*-parameters. *Y*-parameters are obtained from the simulated *S*-parameters. After many repeated simulations for different materials and dimensions' combinations, the filter losses are reduced to suit the 5G RFFE spectral needs. Theoretical predicted values are well matching with the simulation results.

3 Simulation Results

The maximum inband Q-factor attained by the inductor is 26.3 and by the capacitor is 480, as revealed from Figs. 9 and 10. The filter response shows −14.71 dB return loss (maximum S_{11}) and −1.28 dB insertion loss (minimum S_{21}) at the center frequency of 8.2 GHz as shown in Fig. 11. It is clearly identified that both the losses satisfy the minimum performance needed for a typical 5G bandpass filter. Finally, the simulated 8.2 GHz BPF resulted in a bandwidth of 600 MHz from 7.85 to 8.35 GHz.

Filter bandwidth = f_H-f_L = 8.35–7.85 = 600 MHz;

Loaded Q = fc/bandwidth = 8.2/0.6 = 13.68.

Fractional bandwidth = bandwidth/fc = 0.6/8.2 = 7.31%

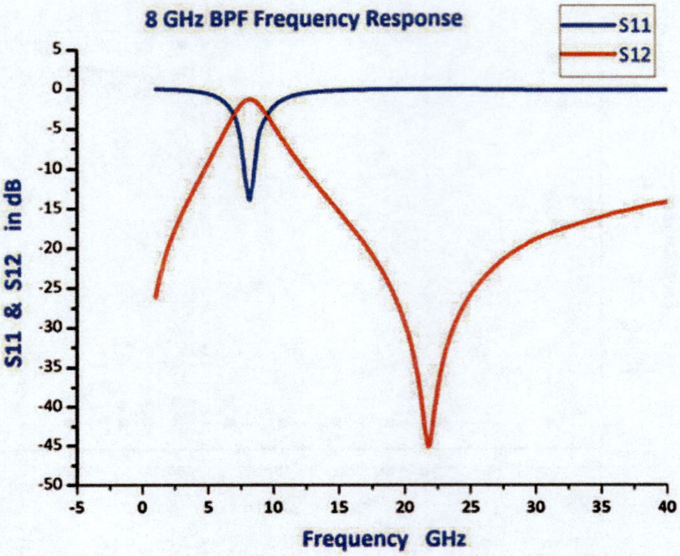

Fig. 11 Simulated BPF frequency response in the frequency range (1–40) GHz

The filter performance parameters like the insertion and return losses, loaded Q, Bandwidth, and fractional bandwidth from the simulation results are summarized in Table 1.

Any filter is referred to as a narrowband one if its fractional bandwidth is smaller than 20%. Our simulated single-stage series resonant BPF had exhibited smallest value of 7.31%. Hence this narrowband response definitely matches the stringent 5G BPF spectral needs. The maximum insertion loss makes this filter to operate efficiently with larger signal propagation. It also yielded narrow spectral bandwidth

Table 1 Summary of the designed BPF parameters

	Design specifications	Simulation results
Center frequency f_0—GHz	8	8.2
Bandwidth—MHz	500	600
Fractional bandwidth—%	5	7.31
Quality factor—Q	15	13.68
Return loss S_{11}—dB	<−15	−14.72
Insertion loss S_{12}—dB	<−1	−1.28
On-chip area—mm^2	<0.025	0.144

and smaller on-chip area. Hence, the proposed 8.2 GHz BPF filter is definitely suitable for realizing the RFICs for 5G applications near the *C* and *X* radio bands.

4 Conclusion

We proposed a very simple IPD-based 5G compatible BPF for the extended C- and X-band applications. We have designed a compact BPF successfully by employing IPD-based multilayer design and performed filter analysis in the simulation tool HFSS. Apart from the above-enhanced performance, the fabrication cost and time savings are possible due to IPD on multiple layers.

The return loss of 15 dB for our filter shows almost a minimum reflected signal in the filter passband. Also the best insertion loss of −0.24 dB is achieved among the reported BPFs. Thus our filter exhibited very good loss performance in the entire passband. Also the simulated 8.2 GHz filter size is less than $500 \times 500\ \mu\text{m}^2$. This is smallest among the other reported BPFs. By considering all above stated performance merits, our proposed BPF is highly suitable for the future 5G application requirements near the *C* and *X* radio bands. Our work demonstrates the feasibility to develop low loss highly selective compact narrowband 5G BPF fully integrated on Si substrate in IPD.

References

1. Wu CS, Chiu H-C, Lin Y-F (2008) Microwave band-pass filter and passive devices using copper metal process on Al_2O_3 substrate. https://www.microwavejournal.com/articles/5849
2. Lee Y-T, Liu K, Frye R, Kim H-T, Kim G, Ahn B (2009) Ultra-Wide-Band (UWB) band-pass-filter using integrated passive device (IPD) technology for wireless applications. In: 59th electronic components and technology conference 2009. IEEE. https://doi.org/10.1109/ECTC.2009.5074295
3. Chen C-H, Shih C-S, Horng T-S, Wu S-M (2011) Very miniature dual-band and dual-mode bandpass filter designs on an integrated Passive device chip. Prog Electromagn Res 119:461–476
4. Zhang X, Zhang W, Liguo Sun Lu, Huang, (2012) A compact UWB BPF design using silicon based IPD technology. IEEE Int Conf ISAPE. https://doi.org/10.1109/ISAPE.2012.6409005
5. Wang H, Pan J, Ren X, Liao A, Lu1 Y, Yu D, Shangguan D (2014) Optimization design and simulation for a bandpass-filter with IPD technology for RF front-end application. In: IEEE 15th international conference on electronic packaging technology, 2014. https://doi.org/10.1109/ICEPT.2014.6922620
6. Ruifang Su, Luo T, Zhang W, Zhao J, Liu Z (2015) A new compact microstrip UWB bandpass filter with triple-notched bands. Prog Electromagn Res 60:187–197
7. Zhou C, Guo P, Wu W (2016) Compact UWB BPF with a tunable notched band based on triple-mode HMSIW resonator. Int J Wirel Microwave Technol 5:1–9. https://doi.org/10.5815/ijwmt.2016.05.01
8. Li N, Li X-Z, Xing M-J, Chen Q, Yang X-D (2017) Design of super compact bandpass filter using silicon-based integrated passive device technology. In: IEEE 18th international

conference on electronic packaging technology, 2017. https://doi.org/10.1109/ICEPT.2017.8046627
9. Mao C, Zhu Y, Li Z (2018) Design of LC bandpass filters based on silicon based IPD Technology. In: IEEE 19th international conference on electronic packaging technology, 2018. https://doi.org/10.1109/ICEPT.2018.8480419
10. Shin KR, Eilert K, Compact low cost 5G NR n78 bandpass filter with silicon IPD technology. IEEE.https://doi.org/10.1109/WAMICON.2018.8363892, 978–1–5386–1267–5/18
11. Zhi-JiWang E-S, Liang J-G, Qiang T, Kim N-Y (2018) A high-frequency-compatible miniaturized bandpass filter with air-bridge structures using GaAs-based integrated passive device technology. Micromachines 9:463. https://doi.org/10.3390/mi9090463
12. Raghunadh MV, Bheema Rao N (2019) High performance series stacked multilayer on-chip inductor for wireless applications. In: Springer 3rd international conference OWT 2019, pp. 1–5 (proceedings in print)
13. Greenhouse HM (1974) Design of planar rectangular microelectronic inductors. IEEE Trans Parts Hybrids Packag 10(2):101–109. https://doi.org/10.1109/TPHP.1974.1134841
14. Mohan SS, Del Mar HM, Boyd SP, Lee TH (1999) Simple accurate expressions for planar spiral inductances. IEEE J Solid State Circuits 34(10):1419–1424. https://doi.org/10.1109/4.792620
15. Rao Vanukuru VN, Godavarthi N, Chakravorty A (2014) Miniaturized millimeter-wave narrow bandpass filter in 0.18 μm CMOS technology using spiral inductors and inter digital capacitors. IEEE Int Conf Signal Process Commun (SPCOM) 2014:1–4. https://doi.org/10.1109/SPCOM.2014.6983960
16. Chen Ji, Liou JJ (2004) On-chip spiral inductors for RF applications: an overview. IEEE J Semicond Technol Sci 4(3):149–167
17. Nagesh Deevi BVNSM, Bheema Rao N (2016) Miniature on-chip bandpass filter for RF Applications. Springer Microsyst Technol 23(3). https://doi.org/10.1007/s00542-016-3052-7
18. Jeng Y-H, Chang S-FR (2006) A high stopband-rejection LTCC filter with multiple transmission zeros. IEEE Trans Microw Theory Tech 54(2)

An 18-Bit Incremental Zoom ADC with Reduced Delay Overhead Data Weighted Averaging for Battery Measurement Systems

Manoj Katta Venkata and Veeresh Babu Vulligaddala

Abstract The paper presents a systematic implementation of 18-bit incremental zoom ADC for stacked lithium-ion battery voltage monitoring system with a reduced dynamic range analog front-end that comprises a resistive level shifter. A comparative study of the incremental zoom ADC and its incremental ADC counterpart, for a given modulator order, is presented in detail. The choice of incremental zoom ADC is demonstrated to be a best fit for an 18-bit resolution of the target application. Multi-bit DAC non-linearity is proven to be major bottleneck for the ADC performance. The performance of data weighted averaging (DWA) dynamic element matching (DEM) technique for improvising modulator linearity with 0.1 percent DAC component mismatch is analyzed. DWA DEM is confirmed to meet an in-band SNDR of 108 dB, operating on a 250 kHz oversampled clock frequency, required for 18-bit modulator linearity with a 5-bit DAC. An SNR of 114 dB is achieved with the first-stage integrator thermal noise limited to 3.3 μVrms over process voltage and temperature variations (PVT) using 0.35 μm CMOS process.

Keywords Feed-Forward · Sigma-Delta modulator · Incremental zoom ADC · Incremental ADC · SAR ADC · Data weighted averaging · Battery monitoring system

1 Introduction

To reduce the CO_2 emissions in automotive vehicles future is to use electric vehicles (EVs) and hybrid electric vehicles (HEVs). Batteries play important role in the EVs and HEVs. Li-ion battery chemistry makes an attractive choice for the EVs and HEVs [1]. Depending upon the system requirements, li-ion cells will be stacked to

M. K. Venkata (✉) · V. B. Vulligaddala
AMS Semiconductors India Pvt Ltd., Madhapur 500081, India
e-mail: manoj.kattavi@gmail.com; manoj.kattavi@ams.com

V. B. Vulligaddala
e-mail: VeereshBabu.Vulligaddala@ams.com

T. Laxminidhi et al. (eds.), *Advances in Communications, Signal Processing, and VLSI*, Lecture Notes in Electrical Engineering 722,
https://doi.org/10.1007/978-981-33-4058-9_15

get the required voltage ranges up to 400 V. In stacked system, each cell needs to be monitored individually with the help of proper battery management system (BMS).

Based on the IC technology, 5 to 12 stacked cells will be monitored using a single device. However, it is not so efficient to incorporate a dedicated ADC for every channel. In the current work, an area efficient solution for DC battery voltage monitoring system (BMS) of stacked lithium-ion (Li-ion) cells is proposed. It comprises a single ADC multiplexed across several channels as shown in Fig. 1. However, a resistive analog level shifter is used to attenuate the cell voltages into a single 5 V domain. This attenuated input dynamic range from analog front-end presents very high accuracy and ADC resolution requirements for a given system resolution.

Incremental data converters (IADCs) are the traditional means of achieving high accuracy and high-resolution systems for instrumentation and measurement systems [2–4] as these applications require absolute accuracy and cannot tolerate offset and gain errors. The static input nature, high resolution requirements and need to multiplex, altogether, best suits an incremental ADC (IADC) for the target application. A recently proposed incremental zoom ADC [5] is chosen for the target application due to its high energy efficiency. This paper presents the design of an ADC for 18-bit resolution at 8 ms throughput rate to suit the requirements of targeted BMS.

The rest of the paper is organized as follows. Section 2 presents a comparative study that presents the pros and cons of incremental zoom ADC over its incremental ADC counterpart. The measures taken to address ADC design challenges are summarized in Sect. 3. Section 4 describes the circuit implementation of the ADC. Simulation results are discussed in Sect. 5. Finally, Sect. 6 concludes the paper.

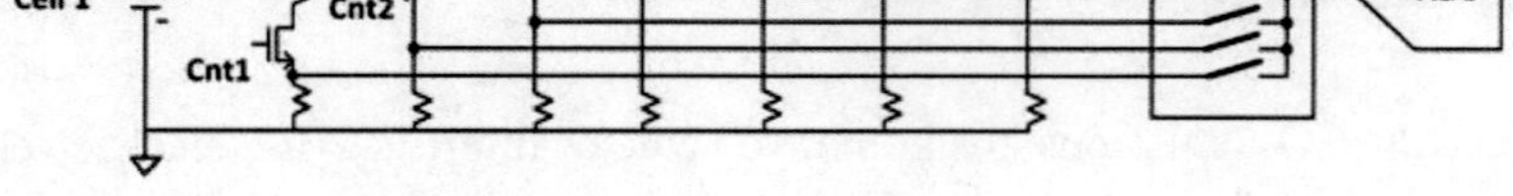

Fig. 1 Block diagram of the proposed battery measurement system

2 Comparative Study

2.1 Conventional Incremental ADC

The conventional incremental ADC comprises a front-end sigma-delta modulator (SDM) and a decimator at back-end, like sigma-delta ADCs. It is the incremental operation of IADCs that distinguishes them from their free running sigma-delta ADC counterparts. The architecture of IADCs establishes input and output mapping to ensure high absolute accuracy. The mapping process involves resetting of all the memory elements that includes integrators of the modulator and back-end decimator. As a result, the IADCs can be readily multiplexed which makes it a suitable choice for portable sensor applications due to its compact realization.

The low distortion feed-forward sigma-delta modulator (FFSDM) topology is chosen for the modulator implementation. FFSDM topology offers several advantages over conventional SDM. It presents relaxed distortion requirements on the integrators, for a given resolution, as it eliminates the need for integrators to operate on high-pass filtered input signal [6]. Moreover, the FFSDM relaxes the integrator output swings, compared to conventional modulator where integrators are to operate on superposition of high-pass filtered input and quantization error of the loop comparator. It offers reduced area overhead due to elimination of dedicated feedback DAC for each integrator in the modulator.

The second-order IADC (IADC2) can be considered for the target 18-bit resolution as first-order implementation severely limits the ADC throughput rate, whereas the higher order implementations present severe stability concerns. The second order FFSDM modulator incorporated into the design of IADC2 is shown in Fig. 2. The oversampling ratio (OSR) required for IADC2 for 18-bit operation can be best derived from time-domain analysis. Equation (1) expresses the output of

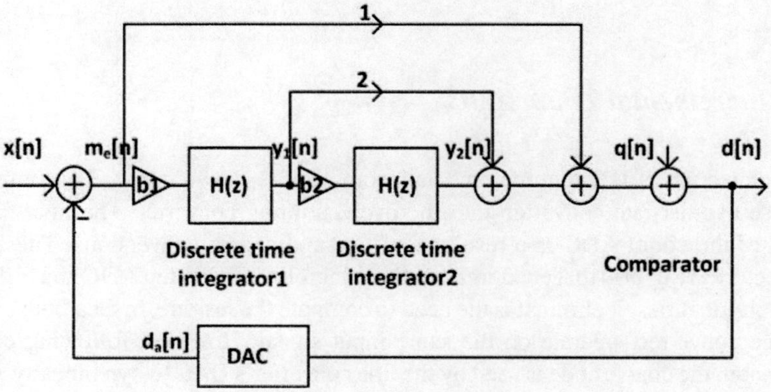

Fig. 2 Block diagram of second-order FFSDM of IADC2

$$y_2(N) = b1 \cdot b2 \sum_{m=0}^{N-1} \sum_{n=0}^{m-1} (x[n] - d[n]) \tag{1}$$

$$\left| ({}^{N}C_2)\bar{x} - \sum_{m=0}^{N-1} \sum_{n=0}^{m-1} d[n] \right| = \frac{V\text{max}}{b1 \cdot b2} \tag{2}$$

second-stage integrator of modulator on its Nth cycle, which is double summation of the weighted modulator error (m_e). Through the choice of proper integrator gain coefficients (i.e. b1, b2), the incremental operation can exercise a maximum bound on the second-stage integrator output (i.e. ± Vmax). Equation (2) outlines the conversion principle of an IADC2 as it defines a bound on error between unknown input signal and an expression made out of known digital bit stream values (i.e. d[n]) of the modulator. Thus by choosing an higher OSR(N) the quantization error associated with a sample can be brought down to realize higher resolutions. For an 18-bit resolution, with the ADC operating on a maximum input of 80 percent of 2.5 V reference, the OSR required is calculated to be about 1280, with following numericals (i.e. V_{max}= 1 V, b1 = .33 and b2 = .5). This shows IADC2 results only in a moderate throughput rate.

The critical building blocks that affect the overall performance of IADC2 include operational amplifier (op-amp) used in the first-stage integrator (Int1). The impact of op-amp non-idealities such as offset, finite gain and bandwidth on the modulator linearity is analyzed. It is found that in addition to auto-zeroing for Int1, minimum op-amp gain of 100 dB and a bandwidth that corresponds to a settling error of one-fourth LSB is necessary for target 18-bit resolution. The thermal noise of the op-amp of the first-stage integrator is another anomaly that can potentially limit [7] the resolution attainable from IADC2. In fact, the fractional gain coefficients (i.e. attenuation factors) make the constraints on output referred noise of the integrator more severe due to amplified input-referred noise. This in turn requires more power to meet the noise requirements for a given resolution.

2.2 *Incremental Zoom ADC*

The architecture of 1st order incremental zoom ADC is shown in Fig. 3. It comprises a coarse nyquist-rate converter and fine over-sampling converter. The final digital output of the zoom ADC is a resultant of fine and coarse conversions. The zoom architecture is a hybrid that encompasses the principles of two-step ADC and is robust to its non-idealities. It eliminates the need to compute the residue. In fact, both coarse and fine converters operate on the same input signal. However, following coarse conversion, the coarse code is used by the fine converter's DAC to dynamically zoom into a coarse LSB reference range around the input signal to eliminate the residue computation. Moreover, this considerably relaxes the resolution requirements of the

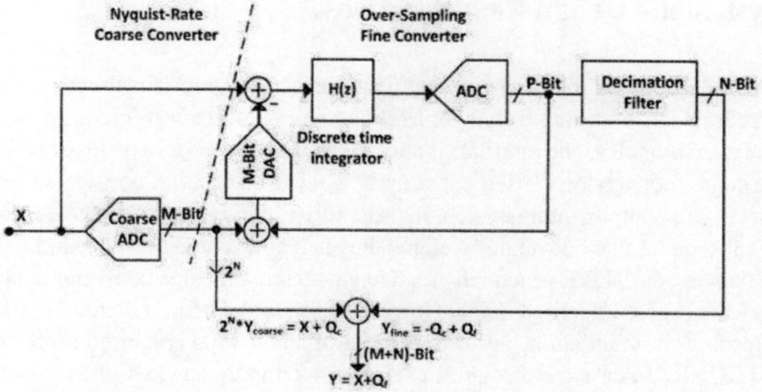

Fig. 3 Block diagram of 1st order incremental zoom ADC

back-end over-sampling converter and further improves its energy efficiency due to reduced voltage swings at internal nodes.

The second-order incremental zoom ADC (IZADC2) is considered for the target requirements of 18-bit resolution. The coarse ADC is realized with energy-efficient successive approximation register (SAR) architecture and the fine over-sampling converter is chosen with architecture of IADC2 discussed in the earlier section. However, the resolution of the IADC2 of zoom architecture is quite relaxed that alters the design constraints on its critical buildings blocks and also provides higher throughput rate. More importantly, the reduced internal node swings enable the op-amp to be better optimized for linearity and noise performance.

A 5-bit resolution (i.e. $M = 5$) is chosen for coarse ADC, which is sufficient enough to relax the swing and resolution requirements of back-end IADC2. The circuit level behavioural implementation is built to analyze the impact of non-idealities on the performance of IZADC2. The simulations substantiate that only a gain of 60 dB for the op-amp of the first-stage integrator is sufficient to ensure 18-bit resolution of IZADC2. The thermal noise constraints of the first-stage integrator are also considerably reduced as the zoom architecture no longer requires the IADC2 to employ fractional gain coefficients for its first-stage integrator. Simulations show that the IZADC2 requires an over-sampling ratio (OSR) of about 280 for the target SQNR of 118 dB, thereby, to get at least thermal noise limited SNR of about 108 dB.

Hence, the IZADC2 is chosen for the target application as it is more robust and energy efficient solution compared to its conventional counterpart, IADC2. However, it presents different set of challenges due to the presence of multi-bit feedback DAC of over-sampling converter, which can limit the overall resolution of the zoom ADC. This is usually addressed with dynamic-element-matching (DEM) techniques that significantly reduce the non-linearity associated with DAC component mismatch.

3 Systematic Design Considerations

The IADC2 discussed in the earlier section requires adaptation for the requirements of the zoom ADC with subtle changes as shown in Fig. 4. It doesn't require the input feed-forward branch as the internal signal swings are reduced as a result of zooming.

The coarse conversion carried out with SAR ADC does suffer non-idealities such as offset due to component mismatch. In order to avoid the effect of the non-idealities on the subsequent fine conversion, redundancy is incorporated between and fine and coarse conversion. This is practically realized by making the fine converter to operate on two LSB coarse reference range. However, this results in an increased OSR for a given resolution. Simulations indicate an OSR of about 380 is required for the target 118 dB SQNR. To ease the design of back-end decimator, an OSR of 512 is chosen [4]. The final ADC output incorporating a one-bit overlap between coarse and fine conversion results is expressed as follows:

$$Y = 2^{N-1} * Y_{Coarse} + Y_{fine} \tag{3}$$

The back-end decimation filter is chosen to be a sinc^2 filter, rather than cascaded-integrator-comb (CIC) filter, due to its symmetrical transfer function which makes it more resilient to the ripple introduced in the output bit stream as a result of DEM techniques. However, it requires the modulator to work for double the theoretical OSR (i.e. 1024).

Auto-zeroing (AZ) technique is incorporated into first-stage integrator as an offset correction technique. However, simulations indicate that AZ in itself is not sufficient to achieve the targeted accuracy of 16-bits. Thus, chopping at system level is incorporated into the design which requires the ADC to resolve the same input twice with swapped input polarities and take the final result as a difference average of the two results. This makes the required OSR to about 2048.

The multi-bit feedback DAC incorporated into IADC2 for fine reference generation can potential limit the overall linearity of the modulator due to component mismatch. Thus, data weighted averaging (DWA) is incorporated into the design as

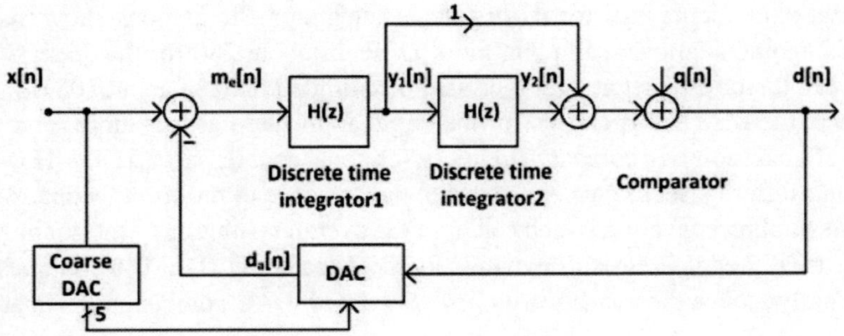

Fig. 4 Block diagram of modified 2nd order FFSDM for IZADC2

a dynamic element matching technique which averages the component mismatch by a rotation methodology, as will be described later, thereby improving the linearity of the modulator. DWA introduces a propagation delay in the modulator loop, however, this is minimal as compared to multi-bit sigma-delta implementations, as the fine converter operates on only two consecutive LSB reference range throughput its operation.

4 Circuit Implementation

The resourced shared circuit implementation proposed in [5] is employed due to its compact and efficient realization as shown in Fig. 5. It enables the reconfiguration of the same building blocks as coarse SAR ADC and fine IADC2 to optimize on-chip area by eliminating the redundant circuitry.

During coarse conversion, the first-stage integrator of the modulator is made to operate as a sample and hold (S/H) amplifier, which is facilitated by the same reset signal required for the incremental operation of the modulator. The output of the S/H amplifier is bypassed to the input of the comparator, which drives the SAR logic that completes the loop as required by SAR ADC. It resolves one bit at a time using binary search algorithm. Hence, it requires M clock cycles to perform coarse conversion. This conversion time is far negligible as compared to over-sampling required for subsequent fine conversion.

At the end of SAR conversion, one more guard band conversion (GBC) cycle is carried out in zoom architecture. During which one more bit (i.e. GBC Bit) is resolved which indicate whether the input lies in the lower-half or upper-half of the

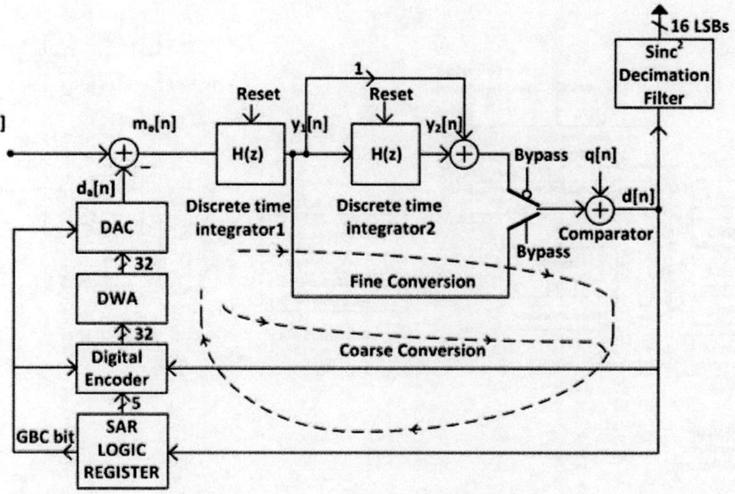

Fig. 5 Block diagram of resource shared 2nd order incremental zoom ADC

coarse LSB. This helps us make a better choice of two LSB reference range for subsequent fine conversion that always safely keep the input at mid-reference range. For example, if the coarse SAR conversion gives an output code that corresponds to a k*th* LSB of the reference. Then, if GBC bit is 0 (i.e. current input lies in lower-half LSB), the two LSB reference range for fine conversion is chosen as (k-1)*th* LSB and (k + 1)*th* LSB. Otherwise, if GBC bit is 1, then k*th* LSB and (k + 2)*th* LSB is chosen. Thus, the guard band correction avoids input overloading and prevents the modulator output from saturation.

The component mismatch induced non-linearity of DAC is the major bottleneck that limits the overall linearity of the modulator. There are various DEM techniques that are used to improve the linearity of DAC. Tree-structured DEM is most area efficient solution available in the literature [8], with second-order DAC noise shaping. However, it does not suit applications that target resolutions as high as 18-bit. Data weighted averaging (DWA) DEM technique particularly suits resolutions greater than 18-bit [9] with just first-order DAC noise shaping.

DWA algorithm involves cyclic rotation of the unary elements in the DAC by choosing the next available unused element in it. This requires the knowledge of past history of input codes to the DAC, which is often kept track of using a DWA pointer. An efficient implementation [10] of the DWA DEM is shown in Fig. 6. It comprises a pointer generator (PGR) and a logarithmic shifter. The accumulator used for the pointer generation performs the binary addition. Hence, it requires a thermometer-to-binary (T2B) decoder to change the thermometer code, which feeds the unary-weighted DAC, into binary format. T2B decoder is realized using an efficient ROM-based implementation. The cyclic rotation is implemented using an area efficient 5-stage logarithmic shifter. The PGR of DWA is usually decoupled from the signal feedback path to avoid delay overhead on feedback signal path.

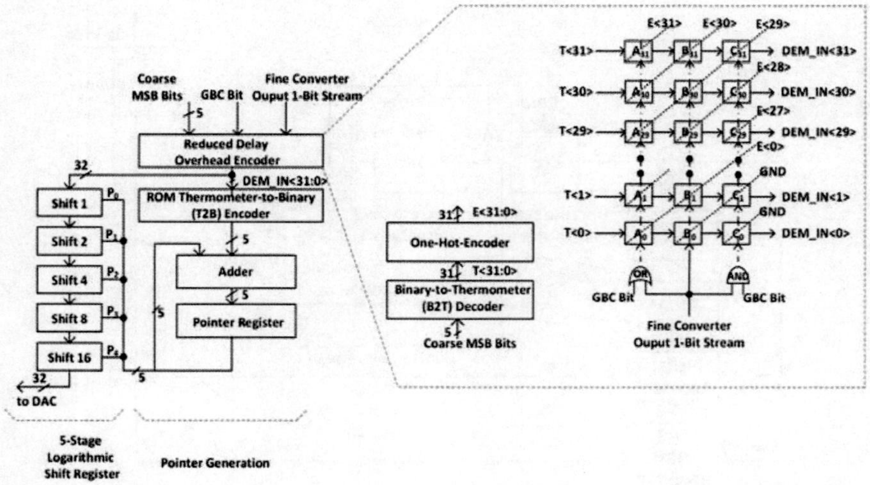

Fig. 6 Block diagram of proposed reduced delay overhead DWA realization

However, for the IZADC2, the thermometer code for reference selection is not readily available, unlike in multi-bit sigma-delta ADCs. The back-end fine converter references are selected based on single-bit modulator output and previous coarse conversion result along with GBC bit. This reference selection circuitry is now part of the feedback signal path. If not addressed, the associated delay overhead can stretch the required integration period, thereby, limit the modulator throughput. This reference selection can possibly be made using an n-bit binary Adder/Subtractor followed by binary-to-thermometer code generator. However, this is not an efficient way due to the huge propagation delay associated and area overhead.

A novel reduced delay overhead encoder for reference selection, shown in Fig. 6, is incorporated into DWA of IZADC2. It comprises a binary-to-thermometer (B2T) encoder, one-hot-encoder and three sequentially connected layers of multiplexers. Each of these layers comprises a stack of thirty-two two-input analog multiplexers to enable true thermometer encoding required for DWA. The output of the B2T decoder is held constant throughout the fine conversion due to the fact that back-end references just straddles across known coarse code. The one-hot-encoder generates the control signals, E < 31:0>, for the multiplexers such that the vertical inputs are feed forward, whenever it generates a control signal high, otherwise, horizontal inputs are feed forward. Table 1 gives a numerical example for a input binary code of 4. The LSB of the thermometer code T <0> is always tied to zero to facilitate binary search algorithm used for SAR coarse conversion. Thus, for any arbitrary input code k, on a range of 1–29, the encoder maps the output to a true thermometer code of either $k-1$, k, $k+1$ or $k+2$ depending on the GBC and BitStream inputs.

A fully differential circuit implementation of the second-order incremental zoom ADC is shown in Fig. 7. It requires two phase non-overlapping clock generation for its operation. Due to relaxed op-amp gain requirements of only about 60 dB, for first-stage integrator, it can easily be realized with simple folded-cascode architecture. Moreover, the relaxed integrator swings is another reason to limit the op-amp architecture just to a single-stage implementation.

The first-stage integrator sample capacitor C_{s1} in itself requires to be operated as a 5-bit DAC in the integration phase followed by sampling phase. This requires the sample capacitor C_{s1} to be split into 32 identical unit elements of 375 fF each which is sufficient enough to ensure atmost 0.1 percent mismatch required for DWA. The differential input to the modulator (i.e. $V_{IN}= V_{INP}-V_{INN})$ is sampled on to all the 32 elements in the sampling phase. In the following integration phase, reference (i.e. $V_{REF}= V_{REFP}- V_{REFN}$) is differentially sampled on to k-elements of

Table 1 Reduced delay overhead encoder functionality

GBC	BitStream	Coarse MSBs	T < 0:31>	E < 0:31>	D < 0:31>
0	0	00100	01111000.. (4)	00001000..	01110000.. (3)
0	1	00100	01111000.. (4)	00001000..	01111100.. (5)
1	0	00100	01111000.. (4)	00001000..	01111000.. (4)
1	1	00100	01111000.. (4)	00001000..	01111110.. (6)

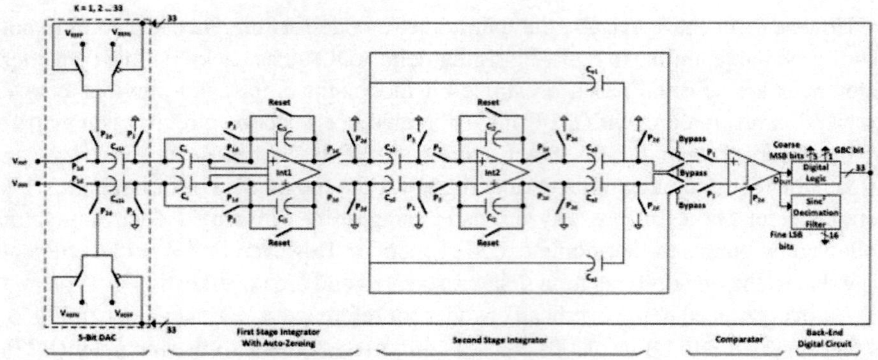

Fig. 7 Circuit implementation of 2*nd* order incremental zoom ADC

the DAC. The reference with opposite polarity (i.e$-V_{REF}$) is sampled on to rest of the (32−k) elements. As a result of these operations, the first-stage integrator accumulate the differential voltage represented by the following expression $V_{IN}+((2\ k-32)/32).V_{REF}$ on to its integration capacitor. To facilitate GBC cycle, an extra capacitance element is incorporated into the DAC that is made to sample only reference but not input.

As discussed in the previous section, the thermal noise performance of the first-stage integrator is critical for overall attainable resolution. Of the two components of thermal noise, op-amp thermal noise and KT/C of sample capacitors (C_{s1}) shown in Fig. 7, the op-amp noise constraints require the careful choice of op-amp architecture and sufficient quiescent power to be burned. Whereas, the KT/C_{s1} noise can only be limited by choosing a higher capacitance. Thus, the first-stage sample capacitance is chosen with a value of 12pF, in order to limit the KT/C_{s1} noise to -120 dB across temperature range of −40 °C to 125 °C with an OSR of 1024. This helps us achieve the op-amp thermal noise limited output of at least 18-bits. An auto-zeroing (AZ) technique is incorporated into the first-stage integrator (Int1) as a dynamic error correction technique. It eliminates the offset using a compensation capacitance (C_c) that is twice the sample capacitance (i.e. 24pF).

The gain settings of the first and second-stage integrator are chosen as 1.23 and 0.5, respectively, in order to limit the integrator swings as required for incremental operation of zoom ADC. In order to meet the throughput requirements of about 8 ms for overall ADC conversion, the modulator is chosen to operate with a sample frequency of 250 kHz. Both the integrators are designed with a unity-gain-bandwidth of about 4 MHz which is required to ensure the settling error of one-fourth LSB corresponding to overall ADC resolution of 18-bits. The Int1, with a scaling factor of four, is adopted for second-stage integrator (Int2). A latched comparator preceded by a preamplifier is chosen which can easily ensure atmost half-LSB comparator offset required for fine conversion.

5 Simulation Results

The second-order incremental zoom ADC is implemented in Cadence environment using AMS 0.35 μm process on a 5 V power supply. All the analog building blocks are implemented at transistor level. Whereas the design of digital building blocks is restricted to circuit-level behavioural implementation. The linearity of the op-amp is characterized by configuring the first-stage integrator as a sample and hold amplifier. Spectral analysis is carried out on the amplifier output, whose PSD density is shown in Fig. 8. The amplifier is found to have an SFDR of about 113 dB.

The noise analysis of first-stage integrator is carried out using PSS + PNOISE simulations by configuring it as S/H amplifier to facilitate PSS simulation. The integrated output RMS noise of the integrator for the two non-overlap phases is shown in Fig. 9 across corners with a maximum of 129 μV_{rms}. This output RMS noise, with an integrator gain of 1.23, if referred to the input gets about 104 μV_{rms}. This noise power corresponds to an SNR of 113.6 dB, with an OSR of 1024, with ADC operating on a differential input of $\pm$2.25 V. Thus, dominant first-stage integrator thermal noise limited ADC output realizes a resolution of about 18.3 bits.

In order to analyze the impact of component mismatch of the capacitive DAC, a random mismatch of maximum 0.1 percent is incorporated into the DAC. With DWA DEM turned off, the mismatch induced DAC noise results in a much elevated noise floor of 103 dB in the PSD of free running modulator shown in Fig. 10. With DWA averaging turned on, the in-band noise is spectrally shaped to ensure required 18-bit operation. The performance of DWA is observed to be coarse code or input dependent. DWA ensures superior performance at lower coarse codes as compared

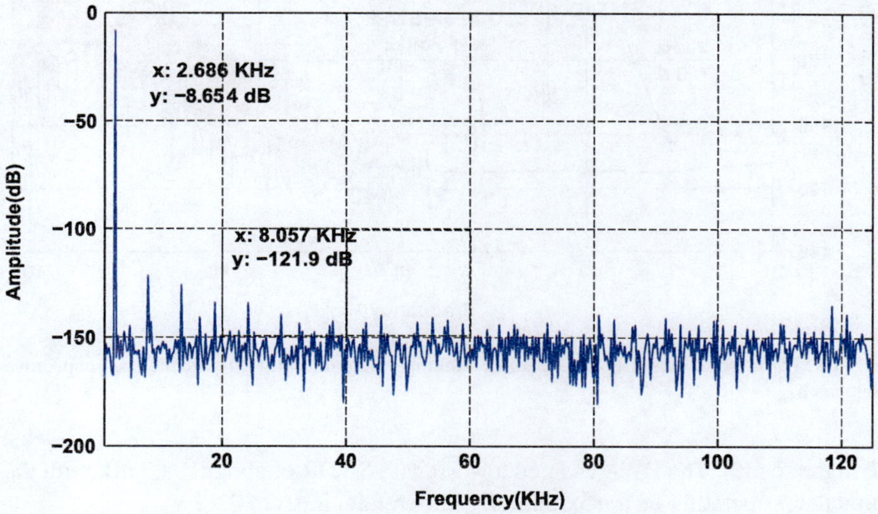

Fig. 8 1024-point output PSD of first-stage integrator configured as S/H amplifier

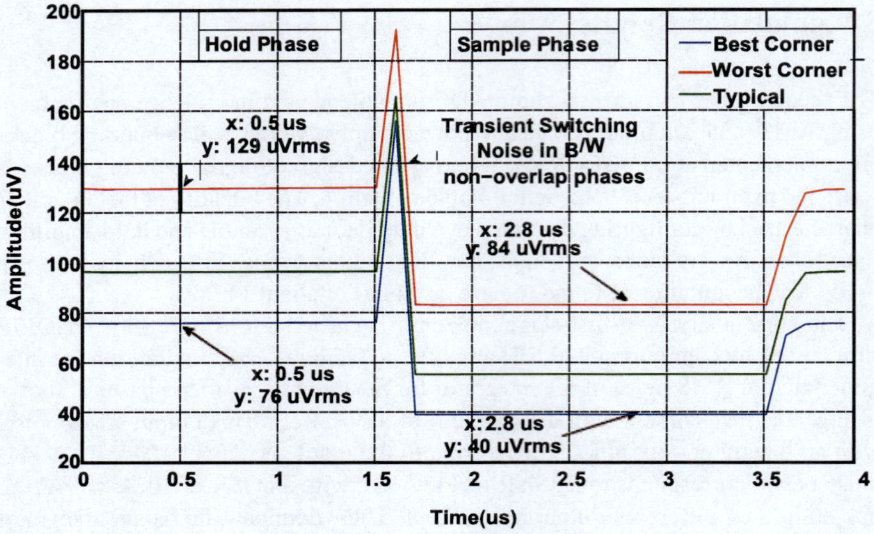

Fig. 9 Output integrated RMS noise of first-stage integrator across PVT corners

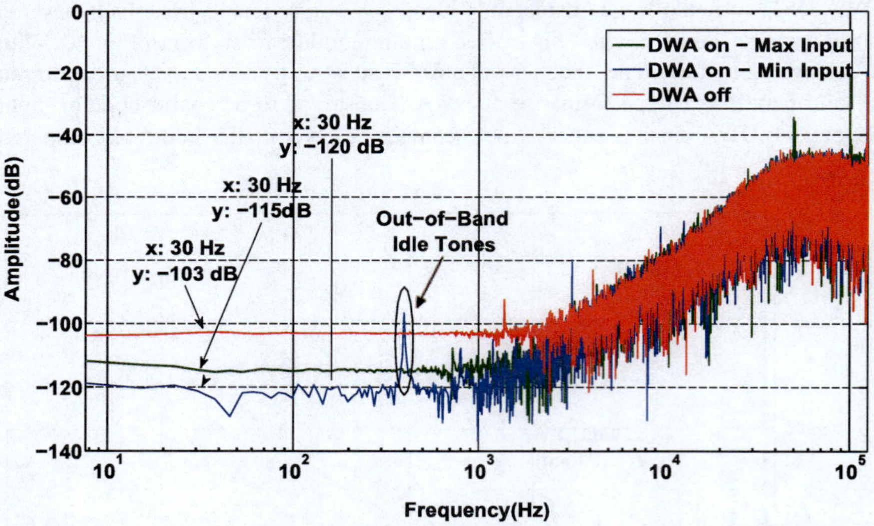

Fig. 10 32768-point output PSD of the free-running modulator with 0.1 percent DAC capacitive mismatch

to higher codes. The DWA ensured a worst case SNDR of about 108.3 dB, with the modulator operating on a maximum DC differential input of 2.25 V.

6 Conclusion

The design of an 18-bit incremental zoom ADC in AMS 0.35 μm CMOS process is presented. The choice of incremental zoom ADC is demonstrated to be a best fit for the target application. An SNR of 114 dB is achieved with the first-stage integrator thermal noise limited to 3.3 μV over PVT variations. The non-idealities that affect the ADC performance are analyzed in detail. Dynamic-error-correction techniques are incorporated to address the non-idealities. A novel DWA encoder realization, for incremental zoom ADC, with reduced delay overhead is proposed for high throughput required. DWA DEM ensured an in-band SNDR of 108 dB, with a maximum differential DC input of 2.25 V, operation on 250 kHz oversampled clock frequency.

References

1. Brandl M, H Gall et al (2012) Batteries and battery management systems for electric vehicles. In: Proceedings of the Design, automation and test in Europe, EDA Consortium, pp 971–976
2. Robert J et al (1987) A 16-bit low-voltage CMOS A/D converter. IEEE J Solid State Circuit 22(2):157–163
3. Robert J, Philippe D (1988) A second-order high-resolution incremental A/D converter with offset and charge injection compensation. IEEE J Solid State Circ 23(3):736–741
4. Quiquempoix V et al (2006) A low-power 22-bit incremental ADC. IEEE J Solid State Circ 41(7):1562–1571
5. Youngcheol C, Kamran S et al (2013) A 6.3uW 20-bit incremental zoom-ADC with 6 ppm INL and 1uV offset. IEEE J Solid State Circ 48(12):3019–3027
6. Silva J, Moon U, Steensgard J, Temes GC (2001) Wideband lowdistortion delta-sigma ADC topology. Electron Lett 37(12):737–738
7. Schreier R, Silva J et al (2005) Design-oriented estimation of thermal noise in switched-capacitor circuits. IEEE Trans Circ Syst I 52(11):2358–2368
8. Fogelman E, Jared W, Ian G (2001) An audio ADC delta-sigma modulator with 100-dB peak SINAD and 102-dB DR using a second-order mismatch-shaping DAC. IEEE J Solid State Circ 36(3):339–348
9. Baird RT, Terri SF (1995) Linearity enhancement of multibit A/D and D/A converters using data weighted averaging. IEEE Trans Circ Syst II Analog Digital Signal Process 42(12):753–762
10. Gabor CT (2015) ECE 627, class lecture, topic: modified DWA realization Cordley 2113, College of Engineering, Oregon State University, Mar. 30, 2015

A Non-Uniform Digital Filter Bank Hearing Aid for Aged People with Hearing Disability

K. Ayyappa Swamy, C. Sushma, Zachariah C. Alex, and S. Prathima

Abstract A hearing once lost cannot be regained naturally. One of the solutions to this problem is using of Hearing Aids. In this paper we are mainly concentrating on the solution for hearing loss in aged people (presbycusis). High frequencies are mainly affected in Presbycusis. In this paper we are going to implement Digital Filter Banks with non-uniform subbands so that we can adjust the gains precisely by referring to the particular audiogram. Here we are going to choose narrow/wide bands if there is a large/narrow variance of intensity of hearing loss in a narrow/wide band variance. This helps in precise gain adjustment leading to a good audiogram matching and hence less matching error. This is a major advantage in treatment of Presbycusis as high frequencies are majorly affected than low frequencies. This process helps in reducing the number of bands which leads to low complexity and cost. Thus, we can achieve a low cost and efficient hearing aid for elderly people. Audiograms with different degrees of hearing loss (Mild, Moderate, Mild to moderate and Severe) are chosen, and matching error is compared between the existing methods, proposed uniform, and non-uniform digital filter banks. The results have shown that the proposed method gives lower matching error values leading to an efficient hearing aid.

Keywords Digital filter banks · Hearing aids · Matching error · Non-uniform filter bank · Presbycusis

K. A. Swamy (✉) · C. Sushma · S. Prathima
Bio Signal Research Lab, Department of EIE, Sree Vidyanikethan Engineering College, Tirupati 517102 Andhra Pradesh, India
e-mail: ayyappa.kondru@gmail.com

K. A. Swamy · Z. C. Alex
SENSE, VIT University,, Vellore 632014 Tamil Nadu, India

T. Laxminidhi et al. (eds.), *Advances in Communications, Signal Processing, and VLSI*, Lecture Notes in Electrical Engineering 722,
https://doi.org/10.1007/978-981-33-4058-9_16

1 Introduction

Hearing Loss or Hearing impairment is the inability of a person to hear partially or completely [1]. It has become a major problem nowadays. It affects human life in many ways as it has an impact on the individual's mental health, working life, and social engagement. Major reasons for this problem are noise and aging. Various factors lead to the loss in hearing. Some of them are genetics, age effect, noise exposure, certain infections, medications, and trauma to the ear. A person with hearing loss faces several difficulties in his life such as finding difficulty to have conversation with family and friends, unable to understand the advice of a doctor, difficulty in hearing alarms, mobile phone rings, doorbell rings, etc. So a hearing problem becomes worse when it is ignored or not treated.

The hearing loss is not same in all the people. Table 1 shows the degrees of hearing loss [1, 2]. Depending on the cause of occurrence, there are different types of hearing losses [1–6] which includes Noise-induced hearing loss (NIHL), Sensorineural hearing loss (SNHL)[1], Conductive Hearing Loss, sensorineural hearing losses and Presbycusis hearing loss [3, 6].

For a hearing-impaired person, hearing ability is represented by an "Audiogram" [2, 7, 8]. The softest sounds one can hear also called as Hearing Thresholds [9] are graphically represented in a typical pure tone audiogram. Figure 1 shows the audiogram of a person with normal hearing (Solid line) and audiogram of a person with hearing impairment due to noise exposure (dashed line). Here the two marked curves "X" and "O" represent the fine hearing level of left and right ear, respectively. An audiogram [1, 10] shows the hearing capability in dB at different frequencies (250 Hz, 500 Hz, 1k Hz, 2k Hz, 4k Hz, and 8 k Hz). A hearing aid is an electro-acoustic device [1, 2, 4–8, 10, 11]. This serves the purpose of amplifying the sound in order to make the speech more intelligible. It is an electronic instrument you wear in or behind your ear [1].

The basic block diagram of a hearing aid algorithm is as shown in Fig. 2. Here using a microphone, a speech signal is given as input to the system. Then it is amplified using a preamplifier. As digital signal processing has much advantages over analog signal processing [2, 4, 9], the input analog audio signal is converted into digital audio signal. Then it is passed through a digital signal processor where the

Table 1 Degrees of Hearing Loss

Degree of Loss	Range of Hearing Loss (dB HL)
Slight	16–25
Mild	26–40
Moderate	41–55
Moderately Severe	56–70
Severe	71–90
Profound	91 and above

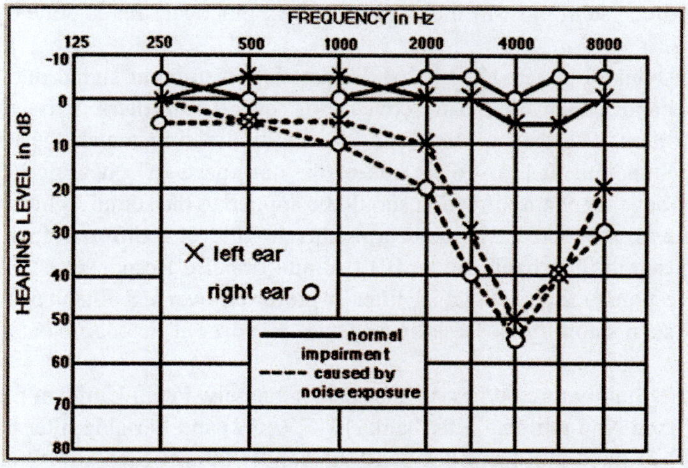

Fig. 1 Typical Audiogram of Normal Hearing and Hearing Impairment

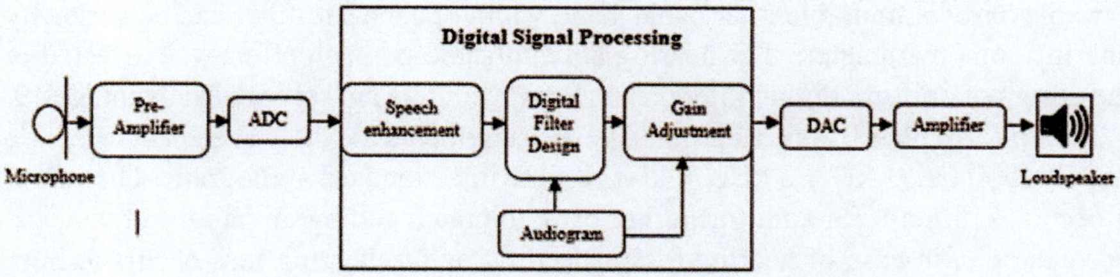

Fig. 2 Basic block diagram of hearing aid algorithm

audio signal undergoes speech enhancement to remove the unwanted noise. Then the enhanced signal is divided into subbands using a filter bank and then gain is adjusted according to the given audiogram. Finally, it is reconverted back into analog signal using an DAC converter and then post amplified. Then it is sent as output using a loudspeaker.

The main task [4, 9, 12] of the hearing aid is to selectively intensify the audio sounds and send the processed sound to the ear. Its main purpose is to provide accurate adjustment of gain [6] in the required frequency. Analog hearing aid is a low-cost device but it has a disadvantage of amplification of noise along with sound without discrimination. In order to program the gain and frequency settings, a programmable control circuitry is added to analog hearing aid. This increases the complexity. Because of using advanced DSP (Digital Signal Processing) algorithms, a Digital hearing aid is more advantageous than analog hearing aids. Digital hearing aids can be precisely tuned to the users hearing loss. A digital hearing aid must satisfy the following properties [13] to become ideal. This includes linear phase response, arbitrary and instantaneous adjustable magnitude response, less power consumption and the entire system must fit into a canal aid. In practical, this is done by using a filter bank.

A Filter bank [1, 8, 9, 12, 14–16] divides the given input signal into different frequency bands. Each filter bank consists of several sub-filters. It is an array of bandpass filters. The gain of each sub-filter is modified to match the audiogram of hearing impaired. It [15, 16] is one of the core parts of a hearing aid design. Frequency-dependent amplification should be applied to the sound signal in order to compensate for the increase in the hearing threshold level at different frequencies.

Filters used in filter bank can be IIR (Infinite Impulse Response) or FIR (Finite Impulse Response) filters. But FIR filter is preferred over IIR filter because of its properties such stablility, possess linear phase response if its coefficients are symmetric.

There are mainly three types of filter banks, namely, Fixed Uniform filter banks [13–18], Fixed Non-uniform filter banks [3–7, 9, 12] and Variable filter banks [10, 11, 19–24] In fixed uniform filter bank the band width of each sub-filter of filter bank is uniform or same and it is fixed,whereas in fixed non-uniform filter bank different sub-filters have different bandwidth of frequencies and it is unaltered. In variable or reconfigurable filter banks band width of each sub-filter can be varied by altering some parameters. For hearing aid applications much effort was invested in the design of uniform digital filter banks. However there are several disadvantages[9, 12] of uniform filter banks. Hearing level measurements are done at each octave, i.e., 250 Hz, 500 Hz, 1 kHz, 2 kHz, 4 kHz, 8 kHz in a standard audiogram. Therefore, it becomes difficult for a uniform filter bank to match audiogram at all frequencies. For example, in case of hearing loss caused by aging, hearing loss occurs at high frequencies. In such a case this uniform filter bank will not be much useful to reduce the matching error. So non-uniform filter bank will be more beneficial than uniform filter bank to reduce the matching error.

In case of non-uniform filter bank, different sub-filters have different bandwidths leading to efficiency in matching the audiogram of a person. However, fixed non-uniform filter banks are not much useful to match an audiogram precisely as each individual has their own audiogram values based on the frequencies at which they lost their hearing. One solution to this is to increase the number of subbands that reduces the matching error. But increasing the number of bands [1] will increase the cost and power consumption. So, it is important to choose frequency bands so that number of bands is less and matching the audiogram is precise. So, we can go for a variable filter bank where the band divisions done by adjusting some variable parameters according to the audiogram. This process leads to a smaller number of bands that reduce the power consumption as well as cost. In this paper, we are going to design both uniform and non-uniform filter banks using MATLAB Simulink Software. Then we are adjusting the gain values according to the audiogram using Gain adjustment block. This reduces the matching error. Finally, we are tabulating matching errors of four different audiograms having different severities of hearing loss (Mild, Moderate, Mild to Moderate and Severe Hearing loss) and compare the values of Uniform filter bank and non-uniform filter bank with the existing methods.

2 Proposed Method

Figure 3 shows the simulink model of the filter bank design along with gain adjustment. First, a speech signal from multimedia file is given as input to the filter bank. Using an audio device writer the audio samples are written to an audio output device. By this we can listen to the audio taken as input. Using spectrum analyzer the speech signal frequency domain spectral characteristics (Frequency vs Magnitude) are obtained. We use FIR filters in the filter bank, because of its advantages specified in [1, 21] compared to IIR, and the number of sub-filters chosen is twelve. Sampling frequency chosen is 16 KHz. Four different audiograms of elderly people having different severity of hearing loss are chosen. First, we design a uniform filter bank with 12 bands with 12 equal frequency divisions from 0–8 KHz. Then we adjust the gains of each subband using gain adjustment block according to the audiogram. Finally, we calculate the matching error for each audiogram and tabulate those values.

Now we design a non-uniform filter bank with 12 bands having more bands in high frequencies than in the lower frequencies as we are designing this hearing aid for the case of Presbycusis. Then audiogram matching is done by adjusting the gain values using gain adjustment block according to the audiogram. Then the matching errors are evaluated and tabulated for the four audiograms. By comparing the tabular values of uniform and non-uniform we can observe that with the same number of bands non-uniform filter bank has lower matching error compared to uniform filter bank. Also,

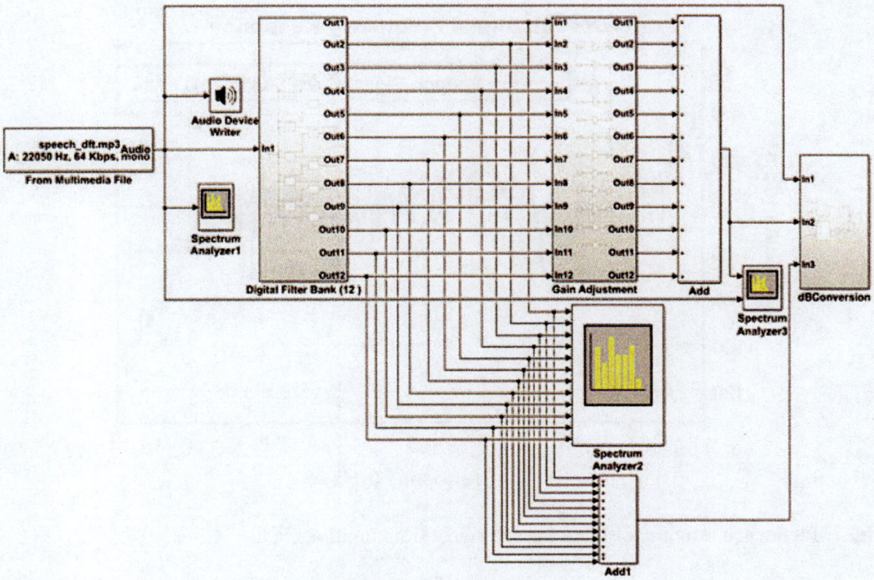

Fig. 3 Simulink model of Filter Bank design for hearing aid application

comparison of matching errors of proposed method with existing methods is done. This shows that there is a close matching to audiogram of hearing-impaired person in proposed non-uniform filter bank leads to qualitative hearing of the hearing disabled person.

3 Results

The output of the speech signal from multimedia file is as shown in Fig. 4.

3.1 Audiograms

As we are mainly focusing on the elderly or aged people hearing loss who's hearing is lost at mostly high frequencies, we have chosen audiograms with four different cases of hearing loss in high frequencies. The first audiogram depicts the mild hearing loss, the second audiogram depicts the moderate hearing loss, the third audiogram depicts the mild to moderate hearing loss, and the fourth audiogram depicts the severe hearing loss. These four audiograms are shown in Figs. 5, 6, 7, 8. Let us consider the right ear audiogram2 details as shown in Table 2.

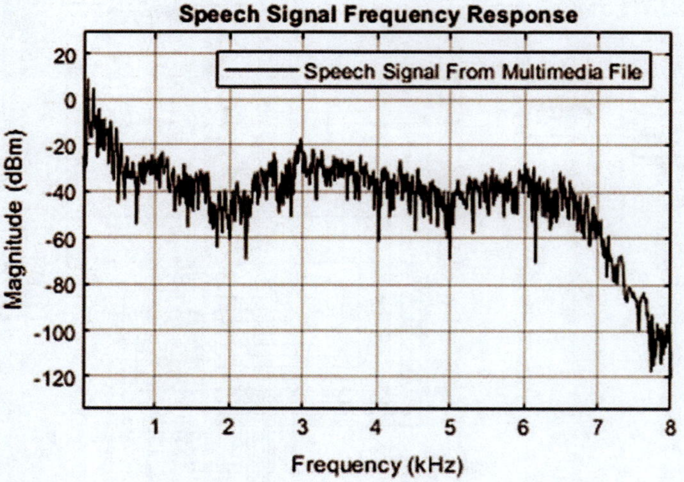

Fig. 4 Frequency response of Input Speech signal from multimedia file

Fig. 5 Audiogram for Mild hearing loss [8]

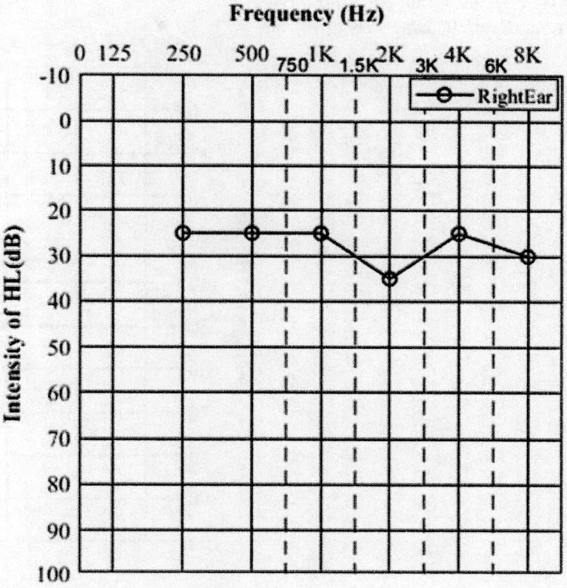

Fig. 6 Audiogram for Moderate hearing loss [25]

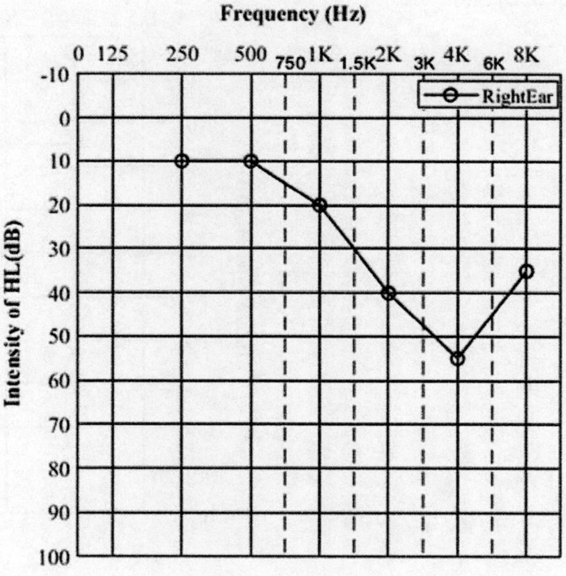

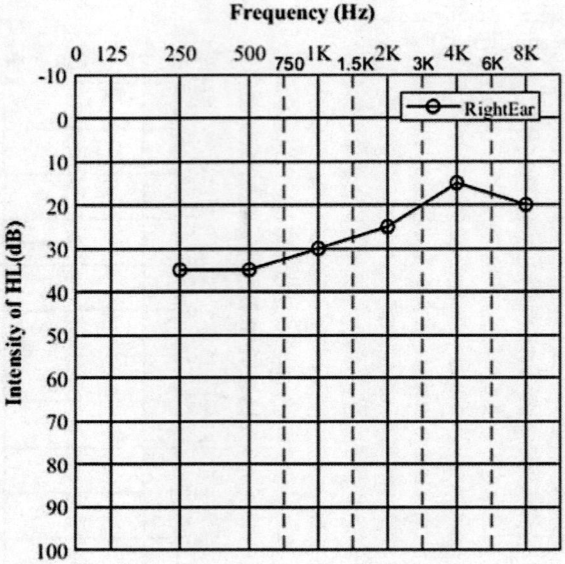

Fig. 7 Audiogram for Mild to moderate hearing loss [26]

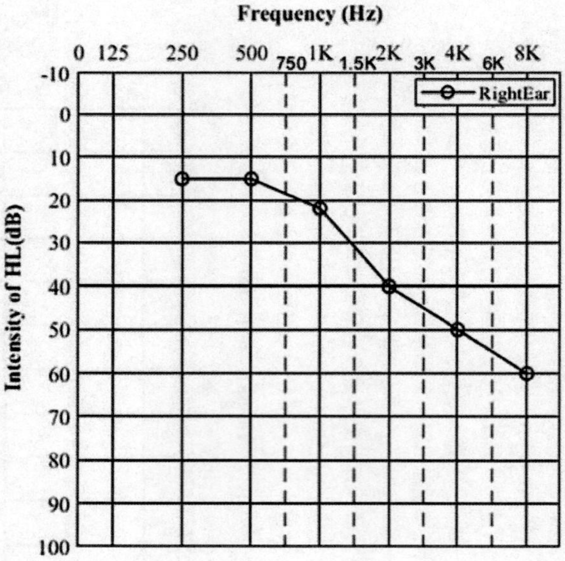

Fig. 8 Audiogram for Severe hearing loss [7]

Table 2 Frequency and hearing loss level details of Right ear of Audiogram2

Frequency (Hz)	Hearing level (dB)
250	10
500	10
1000	20
2000	40
4000	55
8000	35

Table 3 Band divison and gain adjustment details of uniform filter bank for Audiogram2 (Moderate Hearing Loss)

Band	Frequency Range (Hz)	Gain
1	0–666.6	11.67
2	666.6–1333.3	20
3	1333.3–2000	33.3
4	2000–2666.6	42.5
5	2666.6–3333.3	47.5
6	3333.3–4000	52.5
7	4000–4666.6	53.25
8	4666.6–5333.3	49.75
9	5333.3–6000	46.5
10	6000–6666.6	43
11	6666.6–7333.3	39.5
12	7333.3–8000	36.5

3.2 *Uniform Filter bank*

The frequency division for uniform filter bank for 12 bands and its gain adjustment for right ear of audiogram2 are as shown in Table 3. The output of uniform filter bank with 12 bands is as shown in Fig. 9.

Now by adjusting the gains with the values specified in Table 3. For the respective bands the spectrum is as shown in Fig. 10. Finally, after adjusting the gain values filter bank response nearly matches to the audiogram2 curve. This audiogram matching is as shown in Fig. 11. The error between filter bank response (total response of subbands) and audiogram is known as matching error[12] and it is a quality measure parameter of filter design. The matching error graph is as shown in Fig. 12. Least is the matching error best is the hearing aid device. Similarly, we repeat the same process for remaining three audiograms and tabulate the gain values and matching error values.

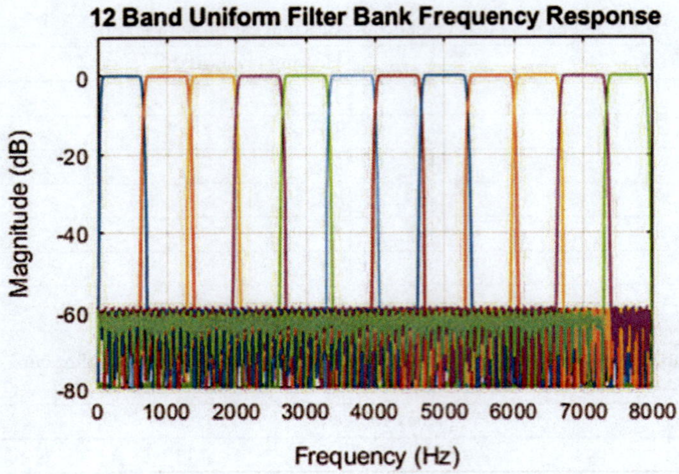

Fig. 9 Frequency response of Uniform Filter bank

Fig. 10 Gain adjustment of uniform filter bank(for audiogram2)

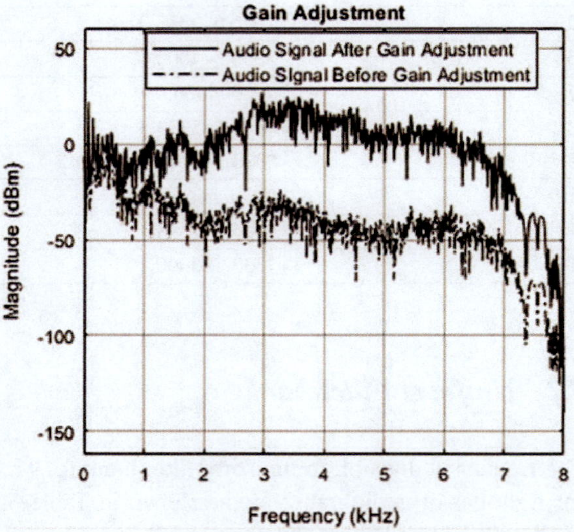

3.3 Non-Uniform Filter Bank

Here we adjust the frequency band values of the subbands in a non-uniform fashion to better match the audiogram for a given number of bands. Considering the same case of right ear of audiogram2 as taken in uniform filter bank, the frequency division for 12 bands and its gain adjustment are as shown in Table 4.

The output of non-uniform filter bank with above 12 bands is as shown in Fig. 13. Now by adjusting the gains with the values specified in Table 4. for the respective

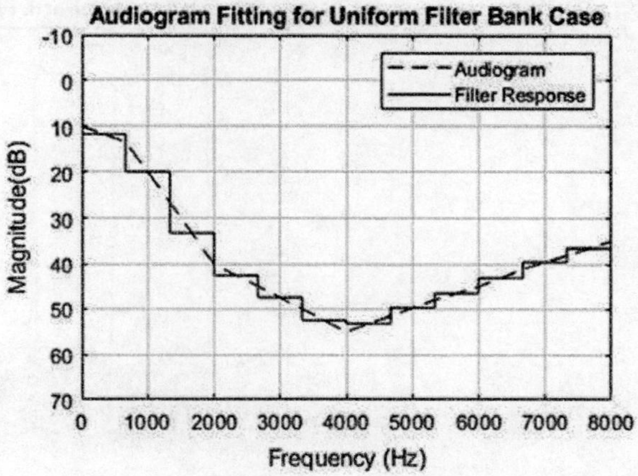

Fig. 11 Audiogram matching in uniform filter bank case for audiogram2

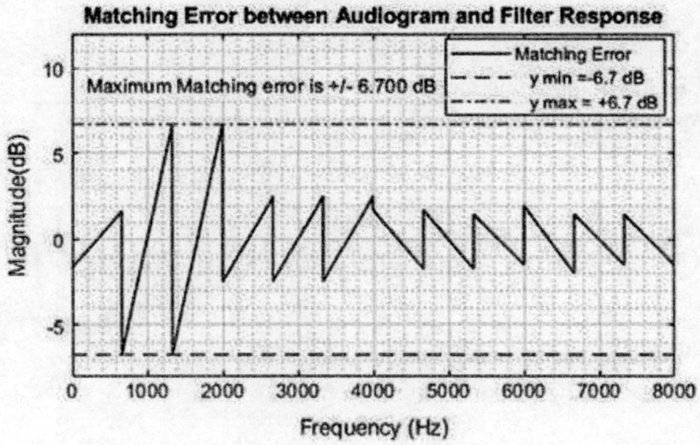

Fig. 12 Matching error between audiogram2 and filter response in uniform filter bank case

Table 4 Band divison and gain adjustment details of Non-Uniform Filter bank (for Audiogram2)

Band	Frequency Range (Hz)	Gain
1	0–750	12.5
2	750–1000	17.5
3	1000–1250	22.5
4	1250–1500	27.5
5	1500–1750	32.5
6	1750–2067	37.75
7	2067–2800	43.25
8	2800–3533	48.75
9	3533–5000	50.75
10	5000–6000	47.5
11	6000–7000	42.5
12	7000–8000	37.5

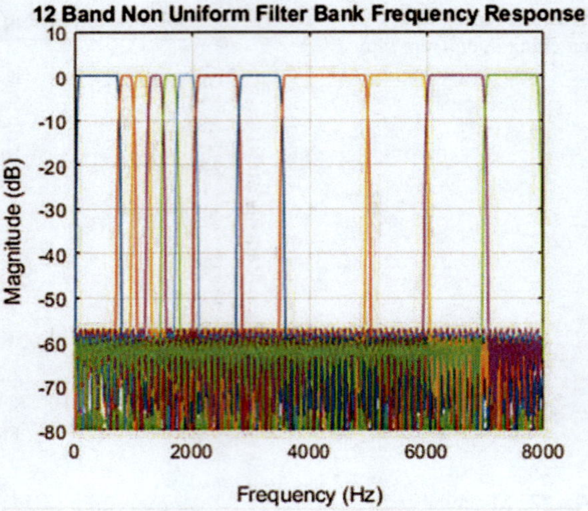

Fig. 13 Frequency response of non-uniform filter bank

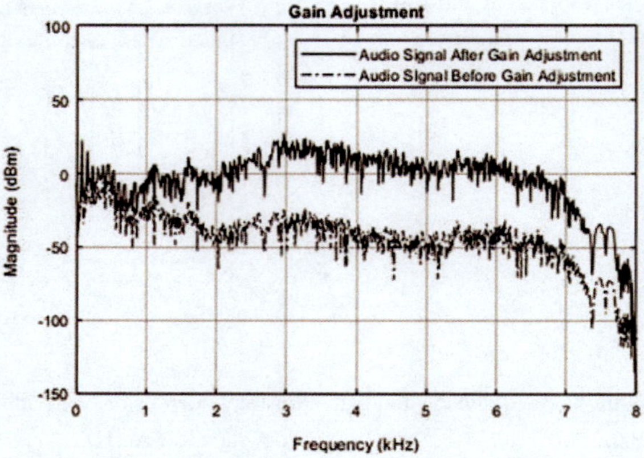

Fig. 14 Gain adjustment for non-uniform filter bank for audiogram2

bands the spectrum is as shown in Fig. 14. The audiogram matching and its matching error are as shown in Figs. 15 and 16.

Similarly we repeat the same process for remaining three audiograms and tabulate the gain values and matching error values. Figures 17, 18 show the audiogram matching and matching errors of audiogram1, Figs. 19, 20 show the audiogram matching and matching errors of audiogram3 and Figs. 21, 22 show the audiogram matching and matching errors of audiogram4 respectively.

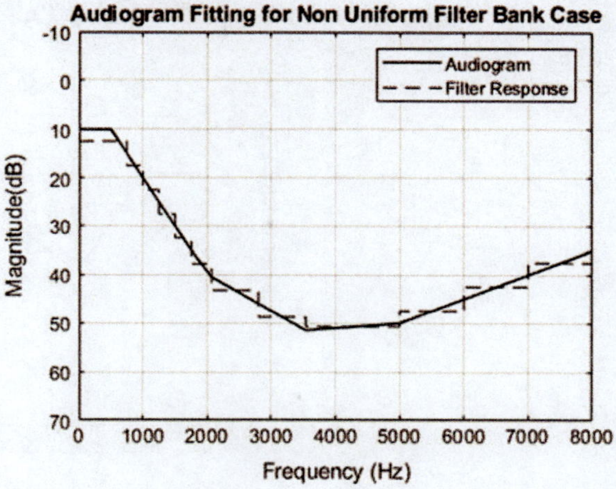

Fig. 15 Audiogram matching in non-uniform filter bank case for audiogram2

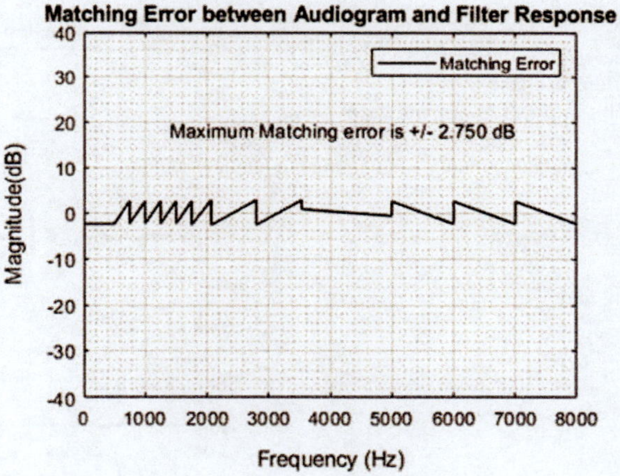

Fig. 16 Matching error between audiogram2 and filter response in non-uniform filter bank case

4 Comparison of Results

Once the results are obtained for uniform and non-uniform filter banks we compare them to justify the best filter bank for hearing aids. The matching error for each different audiogram in uniform case and non-uniform cases is compared. Figures 23, 24, 25, 26 show the matching error comparision graphs between uniform and non-uniform filter banks for the audiogram1,audiogram2 ,audiogram3, and audiogram4, respectively.

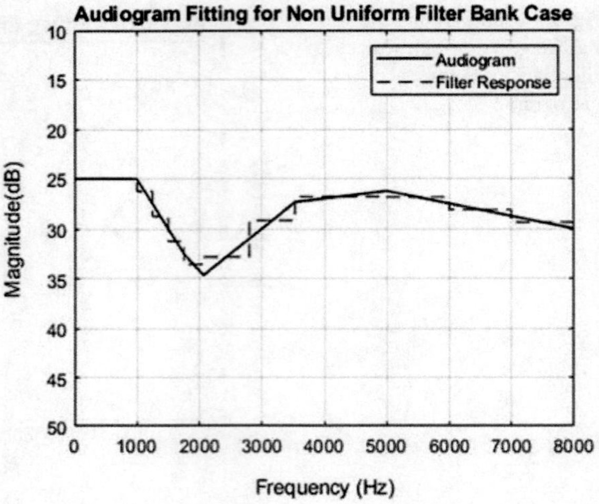

Fig. 17 Audiogram matching in nonuniform filter bank case for audiogram1

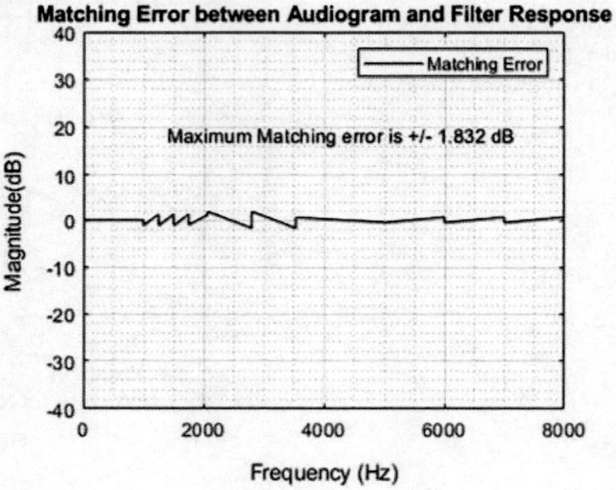

Fig. 18 Matching error between audiogram1 and filter response in non-uniform filter bank case

By tabulating the values of the matching errors of existing methods and the proposed uniform and non-uniform filter banks as shown in Table 5. It clearly shows that a non-uniform filter bank has closer values to audiogram than an uniform filter bank and the existing methods in all different cases of hearing loss.

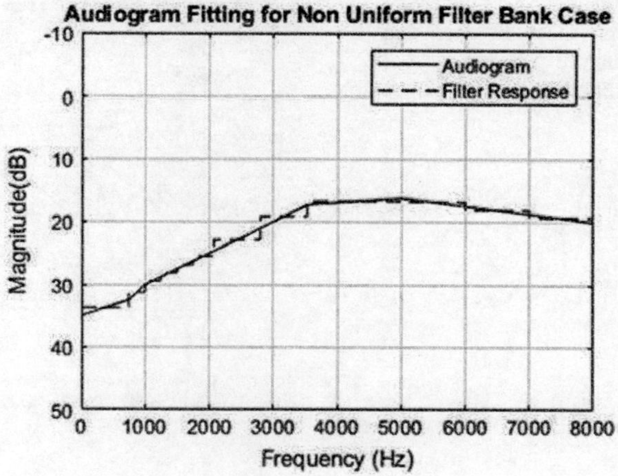

Fig. 19 Audiogram matching in non-uniform filter bank case for audiogram3

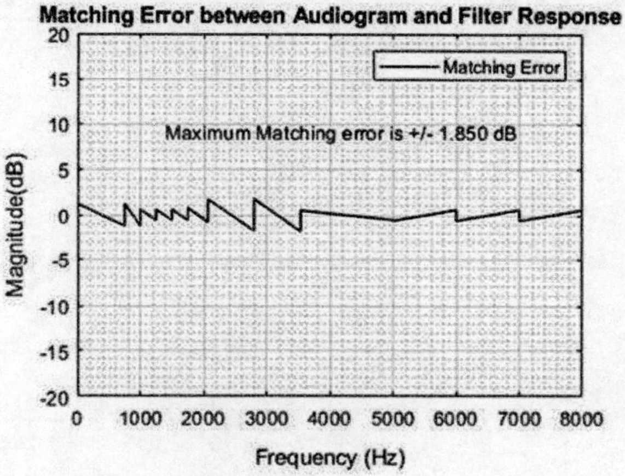

Fig. 20 Matching error between audiogram3 and filter response in non-uniform filter bank case

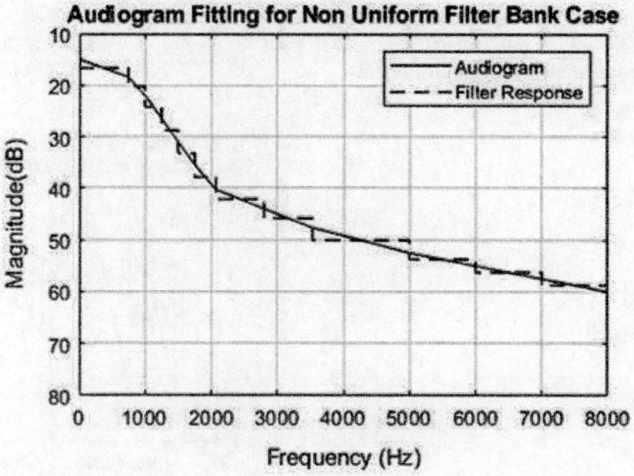

Fig. 21 Audiogram matching in non-uniform filter bank case for audiogram4 (Severe Hearing Loss)

Fig. 22 Matching error between audiogram4 and filter response in non-uniform filter bank case

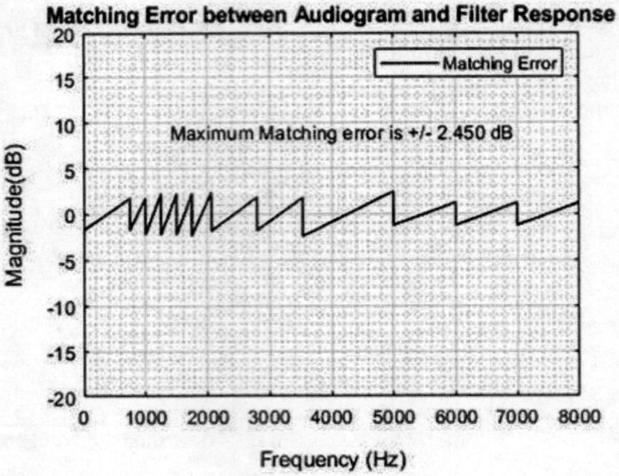

Fig. 23 Matching error comparision for audiogram1

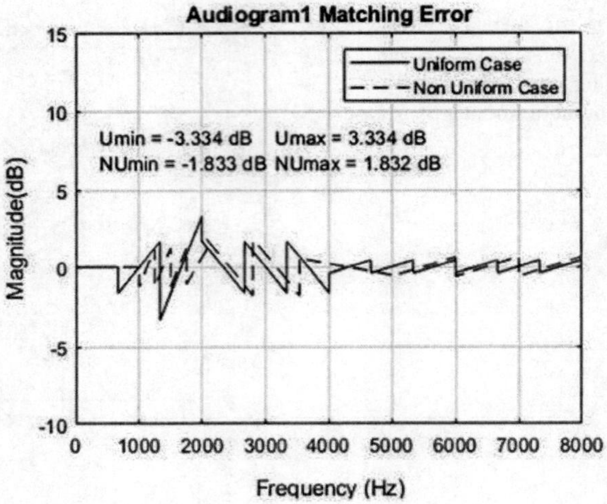

Fig. 24 Matching error comparision for audiogram2

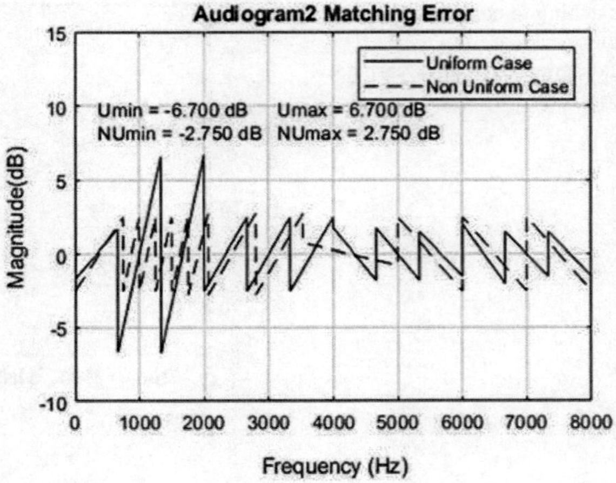

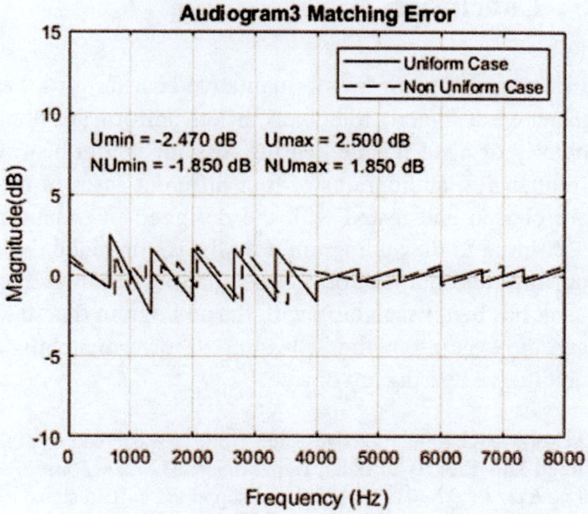

Fig. 25 Matching error comparision for audiogram3

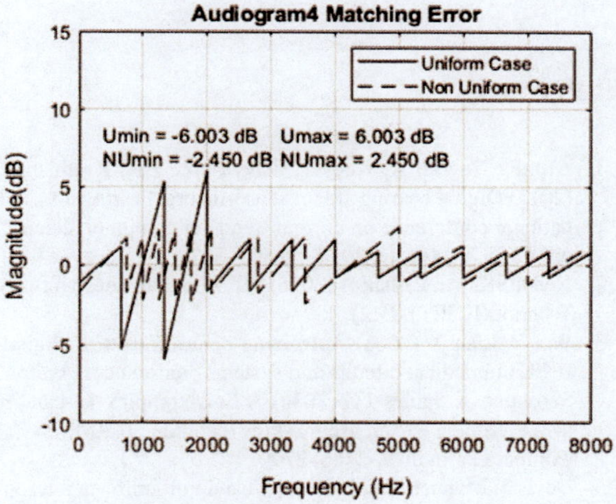

Fig. 26 Matching error comparision for audiogram4

Table 5 Comparison of matching error between Existing methods and Proposed Uniform and Non-Uniform Filter banks

Audiogram type	Existing method (dB)	Uniform Case (dB)	Non-Uniform Case (dB)
Audiogram1 [8]	+4	±3.334	±1.832
Audiogram2 [25]	–	±6.7	±2.75
Audiogram3 [26]	±1.9	±2.5	±1.85
Audiogram4 [7]	±5	±6.003	±2.45

5 Conclusion

In the aim of providing a qualitative hearing for a hearing-impaired person in this paper we designed a filter bank in both uniform and non-uniform models. Considering elderly or aged people hearing loss for whom hearing loss mainly occurs at high frequencies, audiograms of four different cases of hearing loss in high frequencies are chosen and tested with the designed filter banks along with gain adjustment according to the audiogram. Finally, we obtained the matching error values of both uniform and non-uniform cases that have shown that proposed non-uniform filter bank has better matching with the audiogram than the proposed uniform filter bank and the existing methods, leading to lower matching error values. This results in a qualitative hearing aid device.

Acknowledgements This research work is sponsored by **Technology Interventions For Disabled and Elderly (TIDE), Department of Science and Technology, Government of India**, #$SEED/TIDE/015/2017/G$ and supported by Bio-signal research center, Sree Vidyanikethan Engineering College, Tirupati, India.

References

1. Arpitha Nagesh K, KavyaP, Kavyashree BK, Kruthishree KS, Surekha TB, Girijamba TL (2017) Digital hearing aid for sensorineural hearing loss: (Ski-Slope Hearing Loss). 2017 international conference on current trends in computer, electrical, electronics and communication (CTCEEC), Mysore, 2017, pp 505–507
2. Nema SK, Mr. Pathak A (2016) FIR filter bank design for Audiogram Matching. Int Res J Eng Technol (IRJET) 03(2), 409–414
3. Wei Y, Lian Y (2006) A 16-band nonuniform FIR digital filterbank for hearing aid. (2006) IEEE biomedical circuits and systems conference. London 2006:186–189
4. Sebastian A, James TG (2014) A Low complex 12-band non-uniform FIR digital filter bank using frequency response masking technique (FRM) for hearing aid. Int Jo Rec Innovat Trend Comput Commun 2, 2786–2790
5. Devis T, Manuel M (2018) A 17-band non-uniform interpolated fir filter bank for digital hearing aid. 2018 International Conference on Communication and Signal Processing (ICCSP), Chennai, 2018, pp 0452–0456
6. Sebastian A, James TG (2015) Digital filter bank for hearing aid application using FRM technique. 2015 IEEE international conference on signal processing, informatics, communication and energy systems (SPICES), Kozhikode, 2015, pp 1–5
7. Sebastian A, Ragesh MN, James TG (2014) A low complex 10-band non-uniform FIR digital filter bank using frequency response masking technique for hearing aid. 2014 first international conference on computational systems and communications (ICCSC), Trivandrum, 2014, pp 167–172
8. Raj S, Shaji A (2016) Design of reconfigurable digital filter bank for hearing aid. Int J Sci Res (IJSR) 5(7):450–454
9. Ying W, Yong L (2004) A computationally efficient non-uniform digital FIR filter bank for hearing aid. IEEE international workshop on biomedical circuits and systems, Singapore, 2004, pp S1.3.INV.17-20

10. Haridas N, Elias E (2015) Efficient farrow structure based bank of variable bandwidth filters for digital hearing aids. 2015 IEEE international conference on signal processing, informatics, communication and energy systems (SPICES), Kozhikode, 2015, pp 1–5
11. Raj S, Shaji A (2016) Design and implementation of reconfigurable digital filter bank for hearing aid. 2016 International conference on emerging technological trends (ICETT), Kollam, 2016, pp 1–6
12. Lian Y, Wei Y (2005) A computationally efficient nonuniform FIR digital filter bank for hearing aids. IEEE Trans Circ Syst 52(12):2754–2762
13. Lunner T, Hellgren J (1991) A digital filterbank hearing aid-design, implementation and evaluation. 1991 International conference on acoustics, speech, and signal processing, Toronto, Ontario, Canada, vol 5, 1991, pp 3661–3664
14. Lim Y (1986) A digital filter bank for digital audio systems. IEEE Trans Circ Syst 33(08):848–849
15. Onat E, Ahmadi M, Jullien GA, Miller WC (2000) Optimized delay characteristics for a hearing instrument filter bank. Proceedings of the 43rd IEEE midwest symposium on circuits and systems (Cat.No.CH37144), Lansing, MI, vol 3, 2000, pp 1074–1077
16. Li H, Jullien GA, Dimitrov VS, Ahmadi M, Miller W (2002) A 2-digit multidimensional logarithmic number system filterbank for a digital hearing aid architecture. 2002 IEEE international symposium on circuits and systems, phoenix-scottsdale, AZ, USA, 2002, pp II760–II763
17. Parameshappa G, Jayadevapp D (2018) Efficient uniform digital filter bank with linear phase and FRM technique for hearing aids. Int J Eng Technol 7(1.9):69–74
18. Chun Zhu Yang and Yong Lian (2003) A new digital filter bank for digital audio applications. Seventh International Symposium on Signal Processing and Its Applications, Paris, France 2:267–270
19. Huang S, Tian L, Ma X, Wei Y (2016) A reconfigurable sound wave decomposition filterbank for hearing aids based on nonlinear transformation. IEEE Trans Biomed Circ Syst 10(2):487–496
20. Wei Y, Liu D (2013) A reconfigurable digital filterbank for hearing-aid systems with a variety of sound wave decomposition plans. IEEE Trans Biomed Eng 60(6):1628–1635
21. George GT, Elias E (2014) A 16-band reconfigurable hearing aid using variable bandwidth filters. Glob J Res Eng 14(1)
22. Reshma AS, Manuel M (2017) Reconfigurable digital FIR filter bank for hearing aids using minimax algorithm. 2017 international conference on trends in electronics and informatics (ICEI), Tirunelveli, 2017, pp 803–808
23. Deng T-B (2010) Three-channel variable filter-bank for digital hearing aids. IET Signal Process 04(02):181–196
24. McAllister HG, Black ND, Waterman N (1995) A body worn digital hearing aid. Proceedings of 17th international conference of the engineering in medicine and biology society, Montreal, Quebec, Canada, vol 2, 1995, pp 1613–1614
25. https://www.noisehelp.com/hearing-test-results.html
26. Wei a Y, Liu D (2011) A design of digital FIR filter banks with adjustable subband distribution for hearing aids. 2011 8th international conference on information, communications and signal processing, Singapore, 2011, pp 1–5

Optimal Resource Allocation Based on Particle Swarm Optimization

Naidu Kalpana, Hemanth Kumar Gai, Amgothu Ravi Kumar, and Vanlin Sathya

Abstract In Heterogeneous Networks (or HetNets), resources (both bandwidth and power) are apportioned by the Macro Base station to multiple User Equipments (UEs) with the intention of providing maximum possible date rate to all UEs. Concurrently, UE's received power has to be above its receiver sensitivity. In addition, cross-tier co-channel interference to the nearest Femto Base station must be limited. This resource allocation problem is solved using Particle Swarm Optimization (PSO) method. However, this paper procures optimal solution with lesser number of iterations by properly choosing minimum and maximum possible powers in PSO.

Keywords LTE · Heterogeneous networks · Resource allocation · Powers · Optimal capacity · PSO

1 Introduction

As per Huawei and Nokia-Siemens, 60% of voice and video traffic comes from the users operating in environments like a cafeteria, indoor stadium, airport, etc. But, these environments are subjected to low data rates due to poor coverage. Thus, it is essential for the cellular operator to enhance the data rate [1].

One solution could be deploying more small cell base stations in these environments by operating on the reuse factor one spectrum. Various challenges in denser

N. Kalpana (✉) · H. K. Gai · A. R. Kumar
Department of ECE NIT, Warangal, India
e-mail: kalpana@nitw.ac.in

H. K. Gai
e-mail: hemanth1997.gai@gmail.com

A. R. Kumar
e-mail: amgothuravikumar@gmail.com

V. Sathya
University of Chicago, Chicago, IL, USA
e-mail: vanlin@uchicago.edu

T. Laxminidhi et al. (eds.), *Advances in Communications, Signal Processing, and VLSI*, Lecture Notes in Electrical Engineering 722,
https://doi.org/10.1007/978-981-33-4058-9_17

small cells are non-seamless handover, resource allocation, and co-channel interference, etc. In this paper, resource allocation problem is addressed in Heterogeneous Networks (or HetNets), wherein Macrocell Base station (BS) and Femtocell BS operate on the same frequency channels.

Macrocell BS assigns channels to the Mobile Equipment (ME). Then, to those channels, appropriate powers are assigned. Simultaneously, while assigning resources to ME, Macrocell BS has to succeed in doing the following :

(a) Macrocell BS transmissions should reach the Mobile Equipment (ME) in such a way that ME's received signal strength is more than its receiver sensitivity.
(b) Additionally, Macrocell BS transmissions to ME should not create more interference to the Femtocell BS that is nearer to ME [2].
(c) Further, assigned powers need to increase the amount of data transmission between Macrocell BS and ME as much as possible [3].

This paper undertakes the aforesaid tasks. From here onwards, resource allocation problem handling all the above tasks is identified as "Resource Allocation by Macrocell Base station" (RAMB).

Reference [4] deals only with the above mentioned tasks (b) and (c). Besides, channel allocation alone is carried out in [5]. Likewise, [6, 7] implements both of the channel allocation and power allocation using iterative algorithms.

Conversely, [8, 9] reduces interference to the Femtocell BS by reinforcing that Macrocell BS does not assign power to the channel more than its utmost possible power. Moreover, [10, 11] untangles the number of positive powers at first for the resource allocation problem and following that uncovers those positive powers alone.

Particle Swarm Optimization (PSO) is applied to resolve some of the resource allocation problems as well. For instance, [12] assigns the radio resources efficiently in LTE downlink through PSO method. Similarly, transmission rate (or capacity) in LTE system is maximized by complying with minimum rate for each user in [13]. Likewise, [14] uses PSO to achieve Quality-of-service while assigning appropriate bandwidths for all the approaching traffic flows.

Unlike all the above algorithms, the proposed algorithm handles the received signal strength as well, which is very important in deciding the performance of the receiver. Additionally, while resolving RAMB using PSO, proposed algorithm fixes on the appropriate minimum and maximum values viable for the optimal powers. Thus, proposed algorithm gets optimal solution with very small number of iterations. Herein lies the novelty of the proposed solution.

Remaining paper is arranged as follows. Section 2 presents the HetNet system model we adopt for study. Thereafter, Sect. 3 describes the proposed PSO model. Following that Sect. 4 provides simulation results. Eventually, conclusion is inferred in Sect. 5.

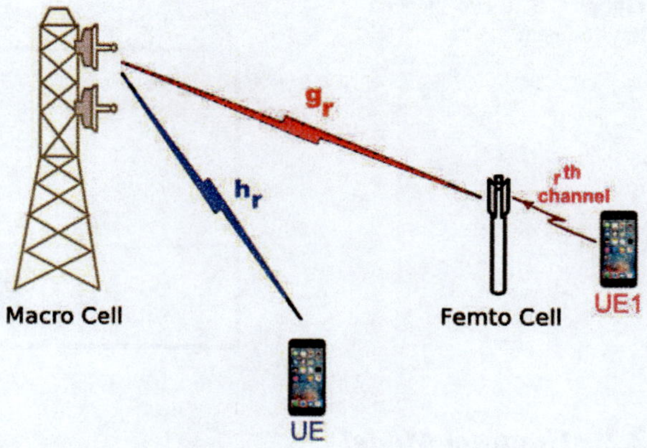

Fig. 1 HetNet Structure

2 System Model

In this paper, heterogeneous system is considered, wherein Macrocell and Femtocell are deployed in a different tier. Further, Macrocell Base Station (BS) and Femtocell BS operate on the same frequencies [2].

As conveyed in Fig. 1, Macrocell BS transmits to User Equipment (UE) on r = 1, 2, . . ., R channels. Besides, Femtocell BS also receives transmissions from other UEs (showed as UE1 in Fig. 1) on the same R channels. Hence, there is a cross-tier co-channel interference in between Macrocell BS and Femtocell BS.

h_r is the channel gain from Macrocell BS to UE on rth channel. Similarly, g_r is the channel gain from Macrocell BS to Femtocell BS on rth channel. This g_r creates interference to the Femtocell BS.

We further assume that the Macrocell BS and Femtocell BS are connected via Femto Gateway (F-GW). So, there will be a centralized co-ordination between Macrocell BS and Femtocell BS. The control plane (C-plane) and Data plane (D-plane) for the UE comes from the same Macro/Femto. In our work, Particle Swarm Optimization (PSO) model runs in the centralized controller (for example F-GW) which decides the optimal bandwidth chunks (i.e., channels) to be allocated to the connected user. Also, PSO resolves the optimal powers to be assigned to the channels.

2.1 *Environment*

Entire terrain region S is divided into smaller sub-regions s_i, where $s_i \in$ S. Each sub-region is of length z_x and width z_y as shown in Fig. 2, where brown circle represents the Macro BS, red circle represents Femto BS and blue color represents UE.

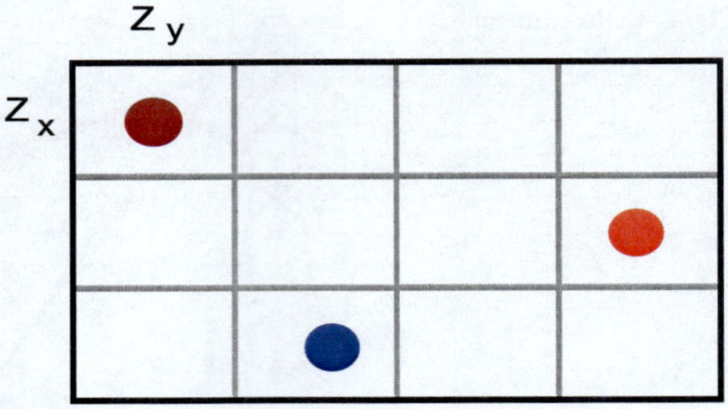

Fig. 2 A Bird-eye View on Environment

2.2 Channel Model

The path loss (P_L) from Macrocell BS to UE is given in Decibels (db) by [15]:

$$P_L = 40\log_{10}\frac{d}{1000} + 30\log_{10} f + 49 + t\ \rho \tag{1}$$

In (1), d is the distance from Macrocell BS to UE [15], t is number of walls/obstructions in between Macrocell BS and UE, f is the center frequency of the bandwidth chunk assigned by Macrocell BS to UE, and ρ is the penetration loss.

Additionally, thermal noise in UE receiver is Additive White Gaussian Noise (AWGN) with variance ξ decibels. Hence, UE encounters AWGN along with the path loss from Macrocell BS.

UE should receive the signal from Macrocell BS such that the average received signal strength is above its receiver sensitivity, Γ. Then only, UE can decipher the signal coming from Macrocell BS. If Γ (in db) = P_L (in db) + ξ (in db), then Γ (in Watts) = $\sigma^2 \times$ x, where $\xi = 10 \log_{10} \sigma^2$ and x = 10 $\log_{10} P_L$. This $\sigma^2 \times$ x decides the receiver sensitivity of UE (Table 1).

Table 1 Glossary

Notation	Definition
P_r	Power allotted by Macrocell to the User Equipment (UE)
B_r	Bandwidth allotted to the rth channel by Macrocell
S_i	Set of all sub-regions
h_r	Channel gain from Macrocell to UE on rth channel
g_r	Channel gain from Macrocell to Femto cell on rth channel
σ^2	Thermal noise in the receiver of UE
T_I	Interference Threshold for the Femto cell

2.3 Optimization Model

The main goal of the optimization model is to allocate the efficient spectrum by guaranteeing minimum SINR threshold.

For transmitting data to User Equipment(UE), Macrocell Base Station (BS) assigns R channels. Bandwidth of rth channel is B_r. Likewise, P_r is the power assigned to rth channel. Then, optimal amount of data gets transmitted from Macrocell BS to UE by assigning optimal powers and optimal bandwidths to R channels. Thus, optimal amount of data transmitted (or Optimal data Capacity) is [8, 10, 11]

$$C = \sum_{r=1}^{R} B_r \, log_2 \left(1 + \frac{P_r h_r}{\sigma^2}\right)$$

Besides, Macrocell BS has to transmit data to multiple UEs. Hence, Macrocell BS assigns limited bandwidth, B_T to UE so that remaining bandwidth can be used for other UEs. Therefore, below equation becomes valid.

$$\sum_{r=1}^{R} B_r \leq B_T \tag{2}$$

Moreover, following equation indicates that the average received signal strength by the UE should be above its receiver sensitivity as per Sect. 2.2.

$$\sum_{r=1}^{R} \frac{P_r h_r}{R} \geq \sigma^2 \times x \tag{3}$$

Additionally, Femtocell BS can sustain T_I amount of interference from Macrocell BS [2, 10, 11] as given subsequently.

$$\sum_{r=1}^{R} P_r \, g_r \leq T_I \tag{4}$$

Combining all the above, we get the following optimization problem:

$$\max_{P_r, B_r} C = \sum_{r=1}^{R} B_r \, log_2 \left(1 + \frac{P_r h_r}{\sigma^2}\right) \tag{5}$$

with constraints:

$$\sum_{r=1}^{R} B_r \leq B_T; \tag{6}$$

$$\sum_{r=1}^{R} \frac{P_r h_r}{R} \geq \sigma^2 \times x; \tag{7}$$

$$\sum_{r=1}^{R} P_r\, g_r \leq T_I; \tag{8}$$

$$\text{and } P_r \geq 0.r \leq R \tag{9}$$

Hereafter, resource allocation problem of (5)–(9) is recognized as "Resource Allocation by Macrocell Base station" (RAMB).

3 Solution Using PSO

Optimal capacity is achieved for the RAMB optimization problem, when equivalence is followed in (6) and (8) equations [8, 9]. Then, RAMB of (5)–(9) is transformed into

$$\max_{P_r, B_r} C = \sum_{r=1}^{R} B_r\, log_2 \left(1 + \frac{P_r h_r}{\sigma^2}\right) \tag{10}$$

with constraints:

$$\sum_{r=1}^{R} B_r = B_T; \tag{11}$$

$$\sum_{r=1}^{R} \frac{P_r h_r}{R} \geq \sigma^2 \times x; \tag{12}$$

$$\sum_{r=1}^{R} P_r\, g_r = T_I; \tag{13}$$

$$\text{and } P_r \geq 0.r \leq R \tag{14}$$

In addition, bandwidth for a channel (or sub-carrier) in LTE is taken as same [2, 16–18]. Consequently, B_r becomes same for all r. Thus, (11) gives B_r maximum value to be same as $\frac{B_T}{R}$. As B_r is constant now and optimal value for B_r is $\frac{B_T}{R}$; RAMB of (10)–(14) assumes the form of

$$\max_{P_r} C = \sum_{r=1}^{R} log_2\left(1 + \frac{P_r h_r}{\sigma^2}\right) \tag{15}$$

with constraints:

$$\sum_{r=1}^{R} \frac{P_r h_r}{R} \geq \sigma^2 \times x; \tag{16}$$

$$\sum_{r=1}^{R} P_r \, g_r = T_I; \tag{17}$$

$$\text{and } P_r \geq 0.r \leq R \tag{18}$$

Particle Swarm Optimization (PSO) is used to find the optimal capacity for the RAMB of (15)–(18).

3.1 Particle Swarm Optimization (PSO)

Particle Swarm Optimization (PSO) is explored briefly here. The parameter that needs to be altered to get the optimal solution is considered as a particle in PSO. Like that, many particles can be taken simultaneously and this collection of particles is called population.

All the particles are initialized with some random value satisfying the constraints mentioned. In each consecutive iteration, particles' present value is updated such that the constraints are met and the solution for this present value is computed. Till that iteration, if the solution obtained is the best of all the previous solutions of the particle, then that particle is considered to have achieved its personal best. Additionally, if the solution attained is the best of all the personal best solutions in the population, then it is considered as global best [19].

Velocity is the rate at which particles change their values. In each iteration, Particle's values are altered using velocity vector. Every time, particle's velocity is updated using [19]

$$\begin{aligned} v_r(i+1) = I_C \;\; v_r(i) + A_1 R_1 \;\; [b_r(i) - x_r(i)] + \\ + A_2 R_2 \; [g_b(i) - x_r(i)] \end{aligned} \tag{19}$$

where "r" is the particle index and "i" denotes the iteration number. Besides, "I_C" is the inertia coefficient that decides the velocity updation and its value lies in between 0.8 and 1.2 [19]. Moreover, A_1 and A_2 are acceleration coefficients, whose range lies in between $0 \leq A_1, A_2 \leq 2$. Further, R_1 and R_2 are random values taken in between 0 and 1 [19]. In addition, $v_r(i)$ is the particle's velocity in ith iteration and $x_r(i)$ is the particle's value in ith iteration. Also, $b_r(i)$ is the particle's personal best in ith iteration and $g_b(i)$ is the particle's global best.

Each particle's value is updated by using

$$x_r(i+1) = x_r(i) + v_r(i+1) \tag{20}$$

During each iteration, updation of particles, finding personal best for all particles and also finding global best for the entire population are performed. In every iteration, better optimal solution (i.e.,lobal best) is achieved when compared to previous iterations. The algorithm is terminated after a given number of iterations, or once the global best value is achieved [19].

In our problem, powers allotted to the channels are considered as particles.

3.2 Algorithm Finding the Optimal Solution

Generally, initial values for the particles are taken as random values in PSO. However, initializing the particles to random values compels PSO to have more number of iterations. But, in contrast to existing algorithms, proposed solution provides both the lower limit and the upper limit for all particles. Thus, proposed solution produces less number of iterations while implementing PSO.

Equation (18) specifies that minimum power to be allotted for rth channel is

$$P_{r,min} = 0; \forall r \tag{21}$$

Similarly, (17) implies that maximum possible power that can be assigned for rth channel is

$$P_{r,max} = \frac{T_I}{g_r}.\forall r \tag{22}$$

Until now, PSO starts with randomized values to get the optimal powers. This in turn may bring more iterations for the PSO algorithm to acquire optimal capacity. But, proposed solution unfolds the minimum and maximum values (= $P_{r,min}$ and $P_{r,max}$; ∀ r) that can be there for the optimal powers. By making use of these $P_{r,min}$ and $P_{r,max}$, optimal capacity is attained quickly with lesser number of iterations in Algorithm 1.

Algorithm 1 Algorithm that obtains the Optimal Capacity

Require: Inputs required are R, σ^2, $P_{r,min}$, $P_{r,max}$, h_r, x, g_r and T_I.

Ensure: Output is rth power P_r, $\forall$ r and optimal Capacity C.

Take the size of particles as R, i.e., $v_s = [1\ R]$.

Set Inertia Weight Damping Ratio as $W_{damp} = 1$.

Initialize $P_{r,min} = 0$ and $P_{r,max} = \frac{T_I}{g_r}$; r $\leq R$.
  Also, equate var_{min} to $P_{r,min}$, $\forall$ r.
  Besides, reset n_p (number of particles) to 50.

Further, initialize I_C (intertia weight) to 0.72984. Moreover, set both c_1 & c_2 (personal and global learning coefficients) to 1.4962.

Assign maximum and minimum velocities (that determine the rate at which particles change their values) as $v_{r,min} = 0$ and $v_{r,max} = 0.01 \times \left(P_{r,max} - P_{r,min}\right)$, $\forall$ r.

Find $L = \sum_{r=1}^{R} P_{r,min}\ g_r$; $K = \frac{1}{R} \sum_{r=1}^{R} P_{r,min}\ h_r$ and
  Receiver sensitivity $U = \sigma^2 \times x$.

while ($K \leq U$) **do**
  Update $P_{r,min}$ as $P_{r,min}$ = previous $P_{r,min}$ +
  { random value in between [0,1] } $\times 10^{-6}$ for r $\leq R$.
  Obtain $L = \sum_{r=1}^{R} P_{r,min}\ g_r$ and $K = \frac{1}{R} \sum_{r=1}^{R} P_{r,min}\ h_r$.
  if $L \leq T_I$ **then**
    break from the loop.
  else
    Update $P_{r,min}$ as $P_{r,min} = var_{min}$ +
    { random value in between [0,1] } $\times 10^{-6}$; $\forall$ r.
  end if
  Get $K = \frac{1}{R} \sum_{r=1}^{R} P_{r,min}\ h_r$.
end while

Initialize number of particle "i" to 1 and
  "GlobalBest.Cost" to 0.

for ($i \leq n_p$) **do**
  Fix P_{ri} = random number created from uniform
  distribution with lower and upper bounds
  specified by [$P_{r,min}$, $\frac{P_{r,max}}{3.52}$], $\forall$ r.
  Specify velocity $v_{ri} = v_{r,min}$; $\forall$ r.
  Determine $L = \sum_{r=1}^{R} P_{ri}\ g_r$ and $K = \frac{1}{R} \sum_{r=1}^{R} P_{ri}\ h_r$.
  if { $L \leq T_I$ } and { $K \geq U$ } **then**
    Don't change the powers.
  else
    Assign $P_{ri} = P_{r,min}$, $\forall$ r.
  end if
  Calculate $C_i = \sum_{r=1}^{R} log_2 \left(1 + \frac{P_{ri}\ h_r}{\sigma^2}\right)$.
  Particle i's personal best is $C_{ib} = C_i$.

29: **if** { C_{ib} > "GlobalBest.Cost" } **then**
30: Assign "GlobalBest.Cost" = C_{ib};
"GlobalBest.Sensitivity" = $\frac{1}{R} \sum_{r=1}^{R} P_{ri}\ h_r$ and
"GlobalBest.interference" = $\sum_{r=1}^{R} P_{ri}\ g_r$.
Take Particle's best powers as $P_{ir,best} = P_{ri}, \forall$ r.
Also, obtain "GlobalBest.powers" as $P_{Gr} = P_{ri}, \forall$ r.
31: **end if**
32: **end for**
33: Set particle number "i" to 1 and number of iteration "W" to 1.
34: **for** ($W \leq 100$) **do**
35: **for** ($i \leq n_p$) **do**
36: Find y_{1r} and y_{2r} to be random values in between 0 & 1.
37: ith particle velocity is changed as : $v_{ri} = I_C\ v_{ri} + c_1\ y_{1r}\ \left[P_{ir,best} - P_{ri}\right] + c_2\ y_{2r}\ \left[P_{Gr} - P_{ri}\right]$.
38: Set v_{ri} = maximum of present v_{ri} & $v_{r,min}$ and v_{ri} = minimum of v_{ri} & $v_{r,max}$; $\forall$ r.
39: Update particle's power as P_{ri} = previous P_{ri} + v_{ri}; $\forall$ r.
40: Further, do the updation as
P_{ri} = maximum of P_{ri} & $P_{r,min}$ and
P_{ri} = minimum of P_{ri} & $P_{r,max}$.
41: Assign $L = \sum_{r=1}^{R} P_{ri}\ g_r$ and $K = \frac{1}{R} \sum_{r=1}^{R} P_{ri}\ h_r$.
42: **if** { $L \leq T_I$ } and { $K \geq U$ } **then**
43: Don't change the powers.
44: **else if** { $L > T_I$ } or { $K < U$ } **then**
45: Set $P_{ri} = P_{ri} - v_{ri}, \forall$ r.
46: **else**
47: Assign $P_{ri} = P_{ri} - v_{ri}, \forall$ r.
48: **end if**
49: Evaluate the step 27.
50: **if** { $C_i > C_{ib}$ } **then**
51: Update personal best as $C_{ib} = C_i$.
52: **if** { C_{ib} > "GlobalBest.Cost" } **then**
53: Assign "GlobalBest.Cost" = C_{ib}. Also,
set "GlobalBest.powers" as $P_{Gr} = P_{ri}$.
54: **end if**
55: **end if**
56: **end for**
57: To decrease the velocities of particles, update I_C as
I_C = previous $I_C \times W_{damp}$.
58: **end for**
59: Optimized powers are $P_r = P_{Gr}, \forall$ r and
optimized capacity is C = "GlobalBest.Cost".

4 Simulation Results

Simulations are done using MATLAB R2010b software. Channel gains h_r & g_r, ∀ r are generated using Rayleigh fading. These h_r and g_r are produced by employing two normal distributions exhibiting the variance of $\frac{1}{15}$. Table 2 lists out the variables adopted for simulations.

Optimal powers allotted to $R = 5$ channels are depicted in Fig. 3. Further, Fig. 4 indicates that capacity is improved with the increase in number of iterations. Moreover, it can be observed from Fig. 4 that it takes lesser number of iterations to obtain optimal capacity in this paper. This is because of the initialized minimum (= $P_{r,min}$) and maximum (= $P_{r,max}$) bound levels for optimum powers.

Table 3 imparts the number of iterations that the Algorithm 1 takes to give out the optimal capacity. As per Table 3, number of iterations taken to obtain the optimal capacity is small in number even for different number of channels.

Table 2 Variables adopted for simulations

Variables	Value of the variable
Sustainable interference level at the Femtocell BS (= T_I)	25 μ Watt
AWGN noise variance (= σ^2)	0.5
Bandwidth of rth channel (= B_r, ∀ r)	5 MHz
Receiver Sensitivity (= $\sigma^2 \times x$)	7 μ Watt

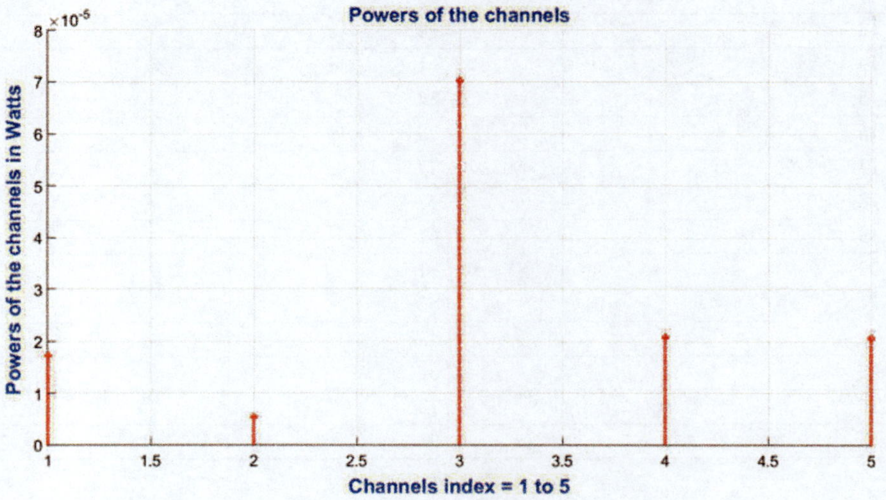

Fig. 3 Optimal Powers assigned to $R = 5$ channels

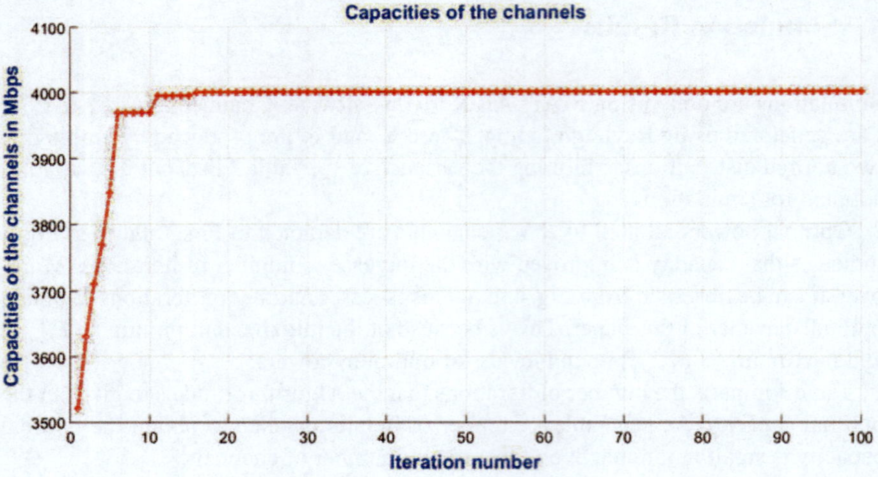

Fig. 4 Iterations vs Capacity

Table 3 Number of iterations taken by PSO algorithm to obtain optimal capacity

No. of channels (= *R*)	No. of iterations
6	30
7	34
8	80
9	56
10	80
11	89
12	88
13	84
14	61
15	34
16	70
17	77
18	47
19	67
20	77
21	52
22	42
23	61
24	73
25	93

5 Conclusion

In this paper, Particle Swarm Optimization is applied to obtain the optimal solution for RAMB of (5)–(9). Novelty of this paper lies in getting the optimal solution with lesser number of iterations. This is attained because of setting proper bounds for the optimal powers.

References

1. CISCO Visual Networking Index. https://www.cisco.com/c/en/us/solutions/service-provider/visual-networking-index-vni/index.html
2. Kalpana et al. (2018) Quick resource allocation in heterogeneous networks. Wir Netw, Spring J 24(8):3171–3188
3. Kalpana et al. (2018) Resource allocation at MAC to Provide QoS for cognitive radio networks. In: Janyani V, Tiwari M, Singh G, Minzioni P (eds) Optical and wireless technologies. Lecture Notes in Electrical Engineering, vol 472. Springer, Singapore
4. Kalpana and Mohammed Zafar Ali Khan (2015) Fast computation of generalized water-filling problems. IEEE Signal Process Lett 22(11):1884–1887
5. Konnov I et al (2017) dual iterative methods for nonlinear total resource allocation problems in telecommunication networks. Int J Mathe Comput Simul 11
6. Raja Balachandar S, Kannan K (2011) A heuristic algorithm for resource allocation/reallocation problem. J Appl Math
7. Chen L, Krongold B (2007) An iterative resource allocation algorithm for multiuser OFDM with fairness and QoS constraints. in Proc. IEEE VTC-Spring, Dublin
8. Kalpana, Zafar Ali Khan M, Hanzo L (2016) An efficient direct solution of cave-filling problems. IEEE Trans Commun 64(7), 3064–3077
9. Kalpana, Ali Khan Z (2016) A fast algorithm for solving cave-filling problems. In: Proceedings of IEEE VTC-Fall, 18–21 Sep 2016
10. Kalpana et al (2018) Swift resource allocation in wireless networks. IEEE Trans Veh Technol 67(7):965–5979
11. Kalpana et al (2018) quicker solution for interference reduction in wireless networks. IET Commun 12(14):1661–1670
12. Su L, Wang P, Liu F ()2012 Particle swarm optimization based resource block allocation algorithm for downlink LTE systems. In: Proceedings of IEEE APCC, pp 970–974
13. Dul J et al (2015) Using joint particle swarm optimization and genetic algorithm for resource allocation in TD-LTE systems. In: Proceedings of QSHINE, pp 171–176, Aug 2015
14. Vieira FH et al (2015) Dynamic resource allocation in LTE systems using an algorithm based on particle swarm optimization and βmwm network traffic modeling. In: Proceedings of IEEE LASCAS, pp 1–4, 2015
15. Sathya V et al (2014) On placement and dynamic power control of femtocells in LTE HetNets. In Proceedings IEEE GlobeCom, 2014
16. Kalpana et al (2011) Optimal Power allocation for Secondary users in CR networks. In: Proceedings of IEEE ANTS, Dec 2011
17. Kalpana et al (2017) The Fastest possible solution to the Weighted Water-Filling Problems. In: Proceedings of IEEE IACC, Jan 2017
18. Kalpana (2019) Simple Solution to Reduce Interference in Cognitive Radio Networks. In: Proceedings of IEEE IMICPW-2019, May 2019
19. Naik B et al (2012) "Cooperative swarm based clustering algorithm based on PSO and k-means to find optimal cluster centroids," in Proc. Durgapur, NCC, pp 1–5

Design and Simulation of a W-Band FMCW Radar for Cloud Profiling Application

Akash Tyagi, Bijit Biswas, G. Arun Kumar, and Biplob Mondal

Abstract Remote sensing for weather applications is gaining importance for accurate measurement of atmospheric parameters, especially in the higher microwave and millimeter-wave frequency ranges. In this work, a feasibility study on the design of a W-band FMCW radar has been carried out for cloud profiling application. Carrier frequency of 95 GHz has been chosen for the radar operation. Few practical examples of cloud radar have been investigated through literature survey and a design of FMCW-based cloud radar has been proposed. System-level simulation has been carried out in SystemVue and results have been produced.

Keywords Atmospheric radar · Cloud profiling radar · FMCW · W-band

1 Introduction

Cloud radars are used for observation and measurement of clouds and atmospheric parameters in the millimeter-wave frequency range [1]. The millimeter-wave frequency range includes the absorption and transmission band of the atmosphere and is, therefore, important for atmospheric remote sensing and research [2].

A. Tyagi · B. Mondal
ECE Department, Tezpur University, Tezpur, Assam, India
e-mail: akashtyagi087@gmail.com

B. Mondal
e-mail: biplob.tezu@gmail.com

B. Biswas
Circuits & Systems Department, SAMEER, Kolkata, WB, India
e-mail: bijit@mmw.sameer.gov.in

G. Arun Kumar (✉)
ECE Department, NIT Warangal, Telangana, India
e-mail: g.arun@nitw.ac.in

T. Laxminidhi et al. (eds.), *Advances in Communications, Signal Processing, and VLSI*, Lecture Notes in Electrical Engineering 722,
https://doi.org/10.1007/978-981-33-4058-9_18

Cloud radar is an active remote sensing instrument. Electromagnetic waves are emitted through an antenna into the atmosphere and the signals returned by hydrometeors, e.g., ice crystals and water droplets in the air mass provide the measurement of reflectivity, range profiles, and Doppler velocity [3]. These measurements give information on important atmospheric parameters such as precipitation, rainfall rate, and wind speed. [4]. Cloud radars are also used for weather prediction and early warning applications because they can measure differences in precipitation levels at a resolution of a few square kilometers or better and provide the ability to quickly monitor the evolving weather events, which are critical for declaring serious and dangerous weather early warnings [5].

Practically deployed cloud radars (e.g., RPG-FMCW-94 radar) are being used for atmospheric research and measurement of reflectivity, range profiles, and Doppler velocity. RPG-FMCW-94 radar operates at 94 GHz in the frequency range of W-band, which enables the instrument to reach high sensitivity with small sizes of the instrument [6]. Also, this radar utilizes frequency modulated continuous wave (FMCW) signals, where much lower transmit powers are required in comparison to radars using pulsed mode.

Practically, FMCW-based approach lowers the cost of development by eliminating the requirement of high power transmitter as in the case of pulsed radar [1]. FMCW-based radars also provide more flexibility in the operation.

Due to its elevated precision and time continuity, despite its attenuation problem during rain, nowadays cloud radar has become a popular and effective instrument for atmospheric remote sensing [5].

2 Design Configuration

There are five stages common to all radar models: a transmitter system, a target model system, a receiver system, a signal processing system, and a measurement system. In a radar system, a signal is generated from the transmitter and transmits via a target model to reach the receiver part. After mixer in the receiver part, signal processing takes effect and the processing result is ready for required measurement [7].

In a cloud radar system, the target will be hydrometeors, i.e., ice crystals and water droplets in the air mass and the measurement of interest will be reflectivity, range profiles, and Doppler velocity, which provide information on important atmospheric parameters such as precipitation, rainfall rate, wind speed, etc.

Figure 1 shows the basic design blocks of a cloud radar model, where FMCW waveform is generated at a carrier frequency of 95 GHz and transmitted via cloud target model defined by radar cross-section (RCS) value of hydrometeors to reach the receiver part [8]. Heterodyne methods are often used in the receiver end to transform the signal to lower frequencies allowing commercial amplifiers and signal processing to be used.

Fig. 1 Block diagram of FMCW cloud radar system

3 Simulation

In this section, the system-level simulation of W-band cloud radar at 95 GHz using FMCW approach in SystemVue environment is presented. The results of simulation give an impression on the output range and output velocity for the input range and input velocity of cloud hydrometeors, i.e., ice crystals and water droplets in the air mass. SystemVue setup of W-band cloud radar is shown in Fig. 2. Also, various parameterizations of simulation components are listed in Table 1 [9].

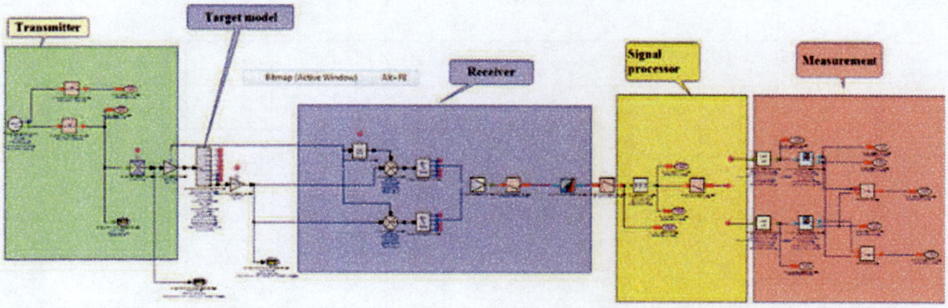

Fig. 2 SystemVue setup of FMCW cloud radar

Table 1 Parameters value of FMCW cloud radar simulation

Parameter	Value/Status
Operating frequency	95 GHz
Transmitted power	30 dBm
Input FMCW waveform	Triangular
Sampling frequency	100 MHz
Antenna gain	54 dBi
RCS	10^{-8} m^2
Scatter type	single
Measurements	Range, Velocity
Waveform bandwidth	50 MHz
Input range	1000 m
Input velocity	30 m/sec

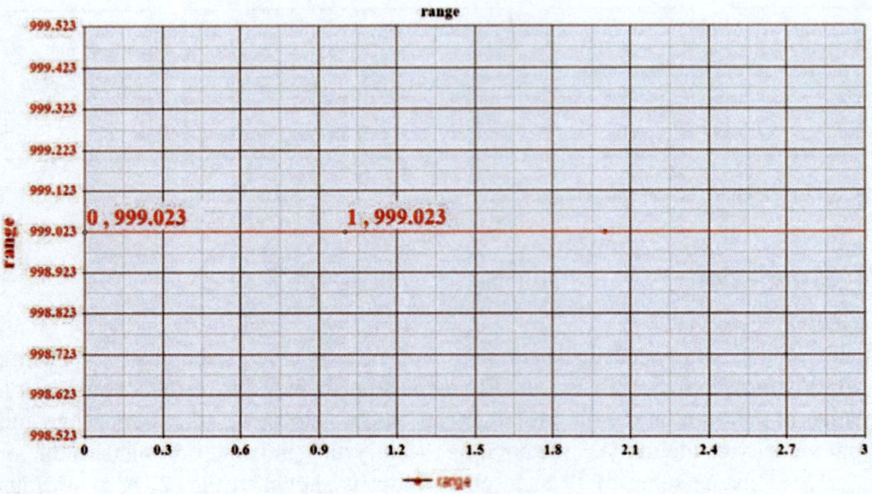

Fig. 3 Output range for input range (1000 m)

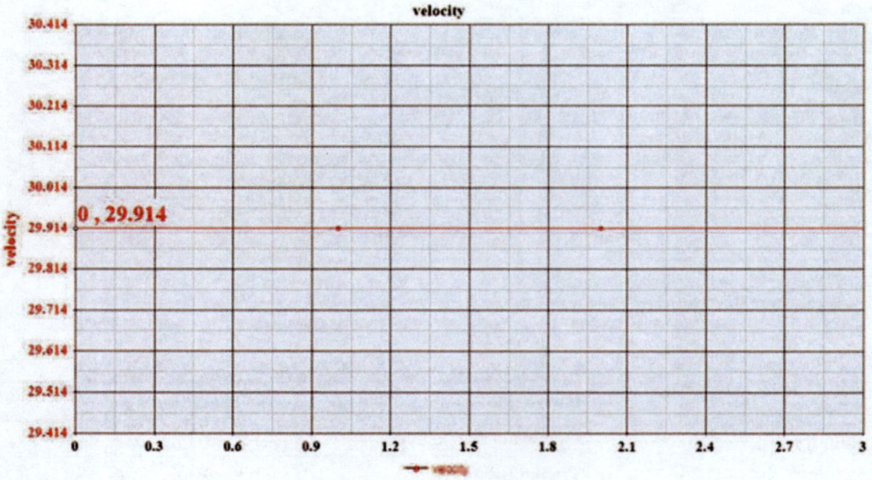

Fig. 4 Output velocity for input velocity (30 m/sec)

Simulation Results. In this section, simulated results of the FMCW cloud radar are presented in terms of output estimated range and estimated velocity for input range and input velocity (Table 2).

Table 2 Consolidated result for FMCW cloud radar at 95 GHz

Parameter	Input	Output	Figure
Range	1000 m	999.023 m	Figure 3
Velocity	30 m/sec	29.914 m/sec	Figure 4

4 Conclusion

In this work, cloud profiling radar has been investigated. The origin of different measurable parameters and their relation to the cloud parameters has been indicated. In simulation part, we have successfully modeled a FMCW-based low power could radar in W-band frequency range and completed the simulation in SystemVue environment. This work may be extended towards realization of the cloud radar RF front-end and the signal processing back-end.

Acknowledgements The authors express their gratitude to Mr. Arijit Majumder, Officer-In-Charge, SAMEER Kolkata Centre for allowing this material for publication.

References

1. Delanoë J et al (2016) BASTA: A 95-GHz FMCW doppler radar for cloud and fog studies. J Atmos Oceanic Technol 33(5):1023–1038
2. Huggard PG et al (2008) 94 GHz FMCW cloud radar. In: Millimetre Wave and Terahertz Sensors and Technology, vol 7117: 711704. International Society for Optics and Photonics
3. Küchler N et al (2017) A W-Band radar-radiometer system for accurate and continuous monitoring of clouds and precipitation. J Atmos Oceanic Technol 34(11):2375–2392
4. Wilson JW, Brandes EA (1979) Radar measurement of rainfall -a summary. Bull Am Meteor Soc 60(9):1048–1060
5. Ventura JFI (2009) Design of a high resolution X-band doppler polarimetric weather radar. Dissertation at Delft University of Technology
6. RPG-Radiometer Physics GmbH, https://www.radiometer-physics.de
7. ISO 19926-1:2019, Meteorology -Weather radar -Part 1: System performance and operation
8. Yamaguchi J et al (2006) Sensitivity of FMCW 95 GHz cloud radar for high clouds. In: Asia-Pacific microwave conference 1841–1846. IEEE, Yokohama
9. Laborie P et al (2002) Technical aspects of a new w-band cloud radar. In: 32nd European microwave conference, pp 1–4. IEEE

Single Layer Asymmetric Slit Circularly Polarized Patch Antenna for WLAN Application

Swetha Ravikanti and L. Anjaneyulu

Abstract This paper presents an asymmetric circularly polarized patch antenna for WLAN application is proposed. V-shaped slits are chopped on to the patch to create asymmetry and realize circular polarization. The performance parameters like impedance bandwidth, axial ratio bandwidths, and gain are 2.83%, 0.5%, and 4.11dBi, respectively.

Keywords Patch antenna · Circular polarization · WLAN

1 Introduction

Nowadays circular polarized antenna gain much attention in wireless communication device since it doesn't depend on the orientation of the transmitter or receiver for communication. It is very flexible, low power losses, less multipath fading mitigation, and useful for wireless communication. Patch antennas are low profile antenna with circular polarization resulted in better bandwidth, many applications need compact circularly polarized antennas like RFID reader, portable wireless devices. Compact circular polarization antennas can be used for wireless application such as WLAN (Wireless local area network) and the frequency fall under 2.4 GHz-2.484 GHz,5.15–5.25 GHz,5.25–5.35 GHz,5.47–5.725 GHz,5.725–5.850 GHz.

Here single layer feeding is taken into account in order to reduce the complexity than dual feeding [1]. The single feed structure slightly chopping the antenna at edges with respect to feed position resulted in orthogonal modes for circular polarization. Several types of unsymmetrical techniques like truncating corners, slits, notches create circular polarization radiation of the antenna [2–4]. The performance of the compact antenna geometry is discussed in Sects. 2, 3 gives the simulation results, and Sect. 4 discusses the conclusion of the circularly polarized antenna.

S. Ravikanti (✉) · L. Anjaneyulu
National Institute of Technology, Warangal, India
e-mail: swetha.rks@nitw.ac.in; swetha.rks@gmail.com

T. Laxminidhi et al. (eds.), *Advances in Communications, Signal Processing, and VLSI*, Lecture Notes in Electrical Engineering 722,
https://doi.org/10.1007/978-981-33-4058-9_19

2 Geometry of the Proposed Antenna

The asymmetric slits circularly polarized antenna for WLAN is shown in Fig. 1. The

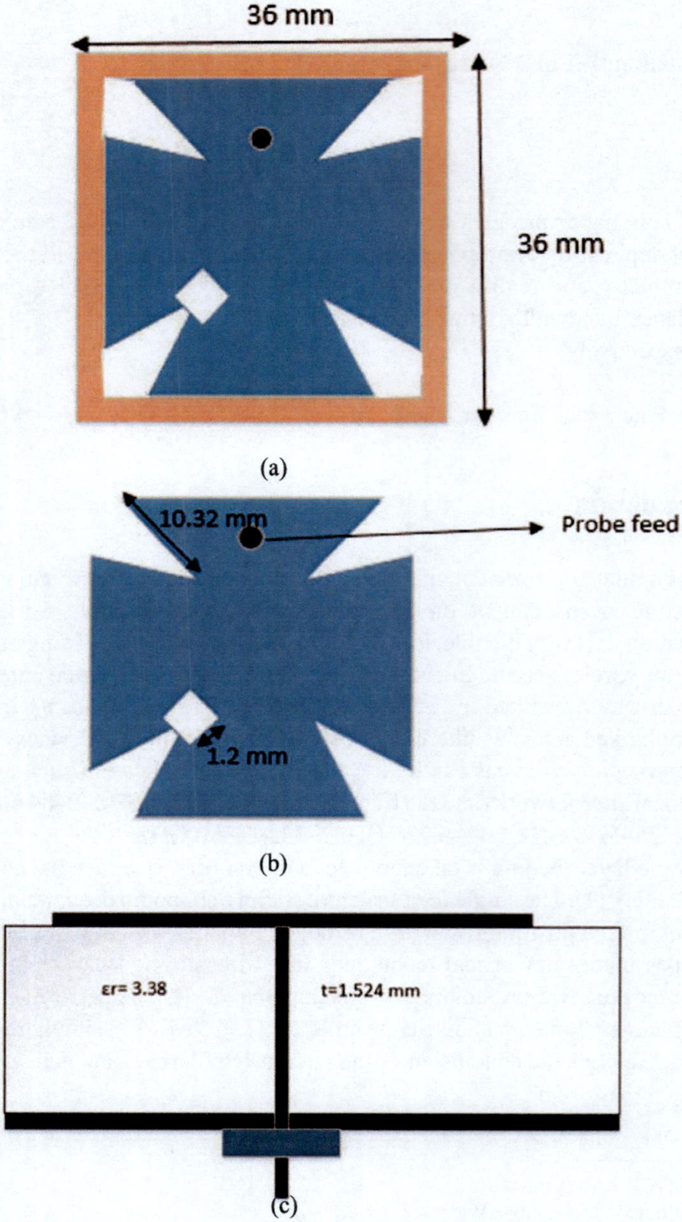

Fig. 1 (**a**) Antenna Geometry (**b**) top view of the patch (**c**) cross-sectional view

asymmetric slit patch size of 31.6 mm × 31.6 mm is etched onto a substrate with a thickness of 1.524 mm. The coaxial-feed location (5, 0) is along the orthogonal axis from the center of a patch for good impedance matching. The v-shaped asymmetric slit patch is designed on a RO4003C substrate ($\varepsilon_r = 3.38$, $\tan\delta = 0.0027$).

2.1 Parametric Analysis

The positioning (r1) of the square box chopped on one of the corners is 45° and the location is varied to obtain better performance are shown in Fig. 2 and tabulated in table 1

Corner square edge position	Resonant frequency(GHz)	Return loss bandwidth (%)	Axial ratio bandwidth (%)	Gain(dBi)
(−8.3,−8.3)	2.402	2.83	0.5	4.11
(−8.4,−8.4)	2.408	2.61	0.5	3.96
(−8.5,−8.5)	2.44	2.29	0.5	4.34
(−8.6,−8.6)	2.443	2.21	0.43	4.35

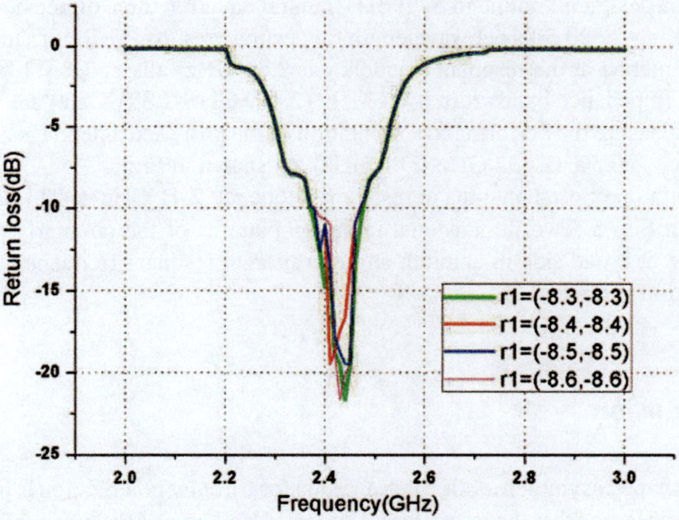

Fig. 2 Return loss for various positions of square box chopped at the fourth corner of the patch

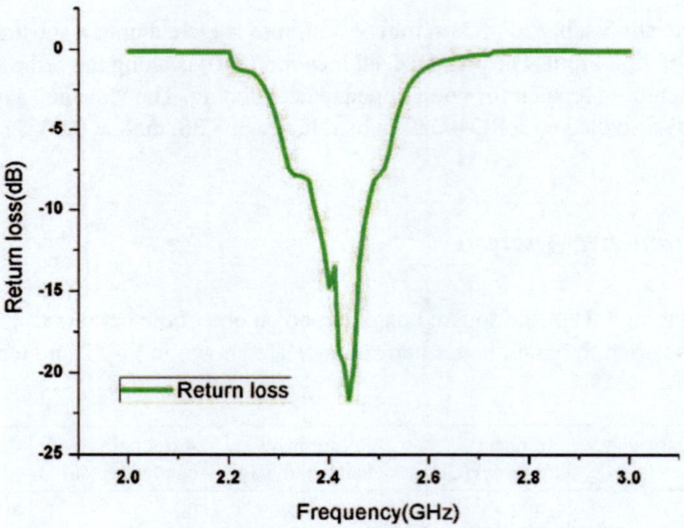

Fig. 3 Return loss of the proposed antenna geometry

3 Simulated Results

The antenna design is simulated by IE3D (Integral equation three dimensional) software by opting for 20 cells per wavelength for accurate results [5]. Return loss for the proposed antenna at the resonant frequency is 2.41 GHz falls under WLAN application and impedance bandwidth is 62 MHz (2.459 GHz–2.391 GHz) are shown in Fig. 3. To measure the performance orientation of the proposed antenna is axial ratio and its value is 0.5% (2.432 GHz–2.42 GHz) are shown in Fig. 4.

Gain of the proposed antenna at resonant frequency 2.41 GHz is 4.11dBi, which is shown in Fig. 5. Two-dimensional radiation patterns of the circularly polarized antenna are at broad side in azimuth and elevation at resonant frequency 2.41GHz are shown in Figs. 6 and 7.

4 Conclusion

A novel v-shaped asymmetric slit patch antenna for Circular polarization is proposed. It has been observed that the return loss, axial ratio, and gain of the antenna geometry does not depend on the slit shape, but rather depend on the area of the slits. The simulated gain, axial ratio, and return loss bandwidth are 4.11dBi, 0.5%, and 2.83%, respectively. A compact circularly polarized antenna of size 36 mm × 36 mm × 1.524 mm and an asymmetric slit technique is useful for circularly polarized compact patch antenna design.

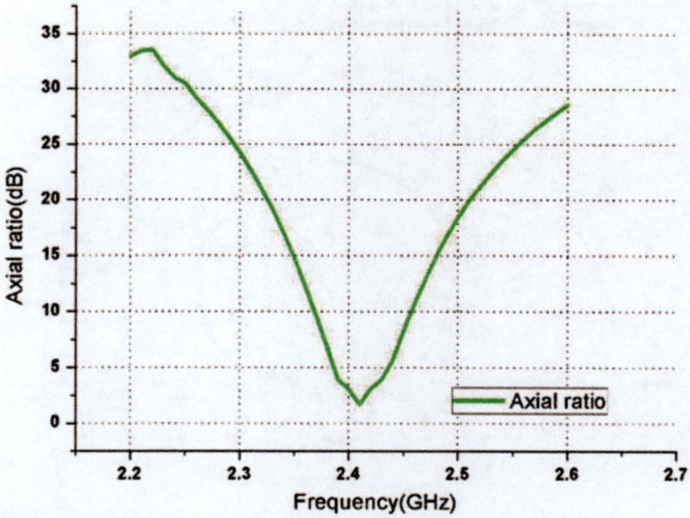

Fig. 4 Axial ratio of the proposed antenna geometry

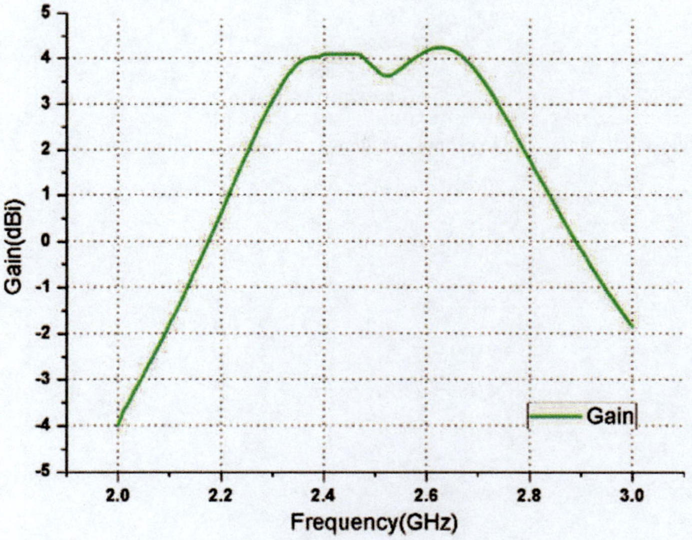

Fig. 5 Gain of the proposed antenna geometry

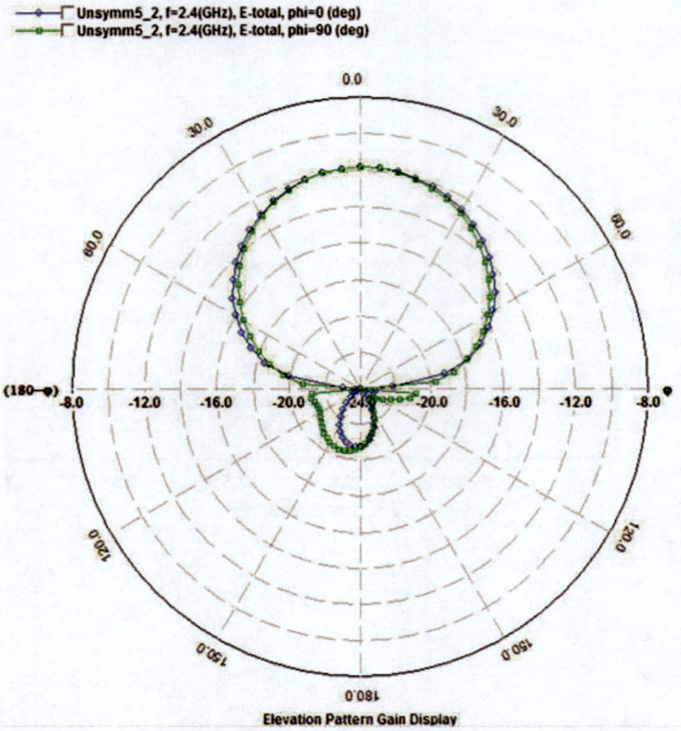

Fig. 6 2D elevation pattern of the proposed antenna geometry at resonant frequency 2.41 GHz

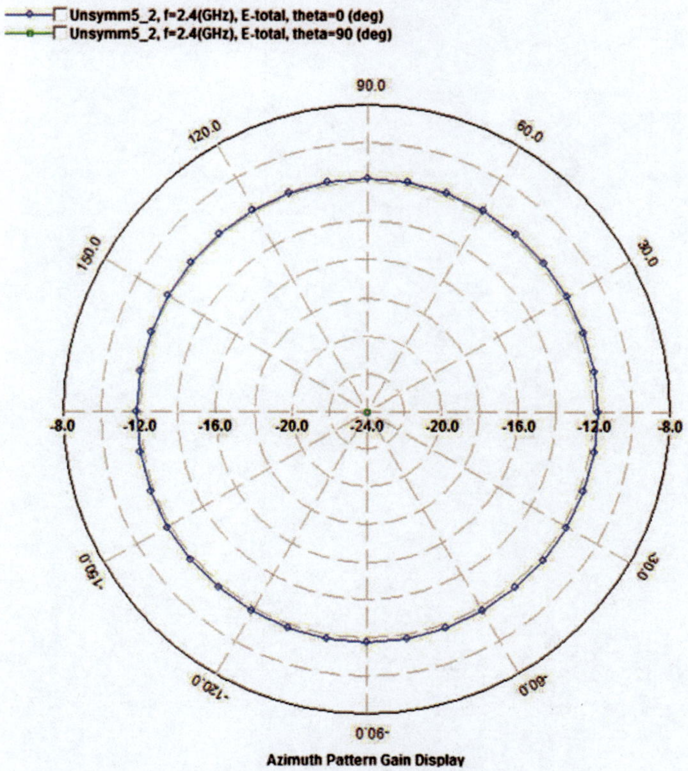

Fig. 7 2D Azimuth pattern of the proposed antenna geometry at resonant frequency 2.41 GHz

Acknowledgements The authors would like to thank the Department of Electronics and Communication Engineering, National institute of technology Warangal for providing the requisite software and Network analyzer.

References

1. Garg R, Bhatia P, Bahl I, Ittipboon A (2001) Microstrip antenna design handbook. Artech Huse, Norwood, MA
2. Sharma PC, Gupta KC Analysis and optimized design of single feed circularly polarized microstrip antennas, IEEE Trans Antennas Propag 29(6):949–955
3. Nakano H, Vichien K, Sugiura T, Yamauchi J (1990) Singly-fed patch antenna radiating a circularly polarized conical beam, Electron Lett 26(10):638–640
4. Chen WS, Wu CK, Wong KL (1998) Single-feed square-ring microstrip antenna with truncated corners for compact circular polarization operation. Electron Lett 34(11):1045–1047
5. Wong ML, Wong H, Luk KM (2005) Small circularly polarized patch antenna, Electron Lett 41(16):7–8 LNCS Homepage, https://www.springer.com/lncs, last accessed 2016/11/21
6. IE3D Manual Version (2006) Release 12, Zealand Software Inc.

Contour and Texture-Based Approaches for Dental Radiographic and Photographic Images in Forensic Identification

G. Jaffino, J. Prabin Jose, and M. Sundaram

Abstract In forensic odontology, the challenging task is to identity the decomposed and severely burnt corpse of individual person. In such a situation, dental records have been used as a prime tool for forensic identification. The main goal of this work, by comparing the analysis of contour shape extraction and texture feature extraction for both radiographic and photographic images, is used to identify a person. In this work, contourlet transform is used as a contour shape extraction; Local Binary Pattern (LBP), Center-Symmetric Local Binary Pattern (CS-LBP) are used as texture features. Both AM and PM images are used to identify the person more accurately by comparing different matching algorithms. In order to salvage better matching performance, Cumulative Matching Curve (CMC) is used for both radiographic and photographic images. Better matching is observed for radiographic images than photographic images by Hit rate performance metrics.

Keywords Contourlet transform · Local binary pattern · Hit rate · Person identification · Dental images

1 Introduction

The field of forensic dentistry or forensic odontology involves the application of dental science to those unidentified persons who are only to be identified by using dental evidence. In other words, it is the use, in legal proceedings, of dentistry deriving from any facts involving teeth. Dental identification can contribute to the major role for identification of person when there are some changes in postmortem like tissue

G. Jaffino (✉)
Aditya College of Engineering, Surampalem, India
e-mail: First.jaffino22@yahoo.com

J. P. Jose
Kamaraj College of Engineeing, Virudhunagar, India

M. Sundaram
VSB Engineering College, Karur, India

T. Laxminidhi et al. (eds.), *Advances in Communications, Signal Processing, and VLSI*,
Lecture Notes in Electrical Engineering 722,
https://doi.org/10.1007/978-981-33-4058-9_20

injury, DNA, and lack of fingerprint, etc. Dental identification is a main consequence in cases where the deceased person is decomposed, skeletonized, or burned. The principal part of forensic identification is identification of the individual that is affected by mass disaster, flood, Earthquake, bomb blast, and Tsunami, etc. The main significance of teeth in person identification is to identify deceased fatalities whose remains have been severely damaged for the above circumstances. David R.Senn et.al [1] clarified that a professional forensic dentist identification of an individual from dental records has long been established and accepted by the court as a way of providing the identity. Dental identification has proved that the most suited biometric for victim identification in Tsunami of December 26, 2004 and Thailand Tsunami attack in January 2005. Also for the most recent incident of Malaysia Airlines Flight was shot down in July 2014, among all 283 passengers 127 people were identified by victim identification. For these circumstances, all the other means of identification are less effective compared to dental identification. Dental identification is used for identifying an individual by comparing records from both Antemortem and Postmortem records.

Related Work:

There are limited literature papers available for person identification. For the initial paper, Anil.K.Jain et al. [2, 3] proposed the semi-automatic method for person identification. In this work, they explained the concept of semi-automatic-based contour extraction technique. This algorithm may not be suitable for blurred and occluded dental images. Automatic segmentation of the dents using active contours without edges has been explained by Samir et al. [4]. Handling of poor-quality images is the major issue for this work. Banumathi et al. [5] proposed the morphological corner detection algorithm and Gaussian mask to determine the shape of tooth contour. But, for all processes, the Gaussian test does not produce satisfying results. The identification of persons using tooth shape and appearance was explained by Nomir et al. [6]. Hofer et al. [7] demonstrated dental work extraction. Dental work is one of the major issues in person identification. Yet to determine the dental work alone may not be efficient for exact matching, in which this may be taken into consideration in addition to the contour extraction algorithms.

Phen Lan Lin et al. [8] have demonstrated the point reliability measuring methods that are used for contour alignment, and the matching is performed by Hausdorff distance-based contour matching. This algorithm is more suitable only for bitewing dental images. Vijayakumari et al. [9] explained the concept of rapidly linked fast connected component-based contour shape extraction and then Mahalanobis distance is used to suit it. Fast connected component labeling fails if the dental image is occluded. Thresholding-based contour feature extraction and Hierarchial matching for both AM and PM dental images are explained by Omaima Nomir et al. [10]. This algorithm is suited only for bitewing radiograph images.

Minh N. Do et al. [11] proved that the concept of non-separable contourlet filter bank gives multi-resolution and multi-direction smooth contours. In this paper, they compared that contourlet transform gives better multi-resolution and multi-direction than the wavelet filters. In order to improve the multi-directional information from the filter bank, Truong T. Nguyen [12] proposed a uniform and non-uniform directional filter bank. The non-uniform directional filter bank gives high magnitude coefficients in the sub-bands of the filter bank that corresponds to extract more geometrical features like edges and textures of an image, and it has better perceptual quality to improve the edges. Kavitha et al. [13] proposed the local binary pattern (LBP) of texture-based classification for hyperspectral images. The modified algorithm for center-symmetric local binary pattern (CS-LBP) for object tracking images has been demonstrated by Hanane Rami et al. [14]. The wavelet transform-based co-occurrence texture feature extraction was explained by Arivazhagan et al. [15] for automatic target detection. In case of derisory accessibility of dental X-rays, it may be important to examine family albums or photographs taken during certain functions for the identification of missing person [16]. In order to match both the AM and PM images, the Forensic Odontologist requires the whole teeth structure type. This work clearly explained the contourlet transform to extract the contour features of dental images. Contour features are not alone enough for victim identification. So this work includes an additional texture like features in order to improve the matching performance of both AM and PM images. This work is organized as follows. The first section will explain bilateral filtering as pre-processing technique. In the second section, contour-based shape extraction is done, and the third section will explain the texture-based shape extraction. Observing shape matching of both contour and texture is explained in fourth section. Figure 1 displays the pipeline for the proposed approach.

Pre-processing:

In this work, input AM/PM images are pre-processed by bilateral filter. It performs smoothening operation of the image, while preserving the edges in non-linear combinations of nearby image pixel values. The gray level distance between neighborhood and center pixel values are given by

$$D_s = \sqrt{\left|s^2(x_1, y_1) - s^2(x, y)\right|} \tag{1}$$

The sub-kernel function is expressed as

$$K_s = \exp\left(-\frac{1}{2}\left(\frac{D_s}{W_s}\right)^2\right) \tag{2}$$

where $s(x_1, y_1)$ is the intensity values at location (x_1, y_1) and the center pixel value $s(x, y)$. W_s are the distribution function of K_s. The spatial distance function can be expressed by using Euclidean distance, and it is given by

$$ED_s = \sqrt{s(x_1, x)^2 + s(y_1 - y)^2} \tag{3}$$

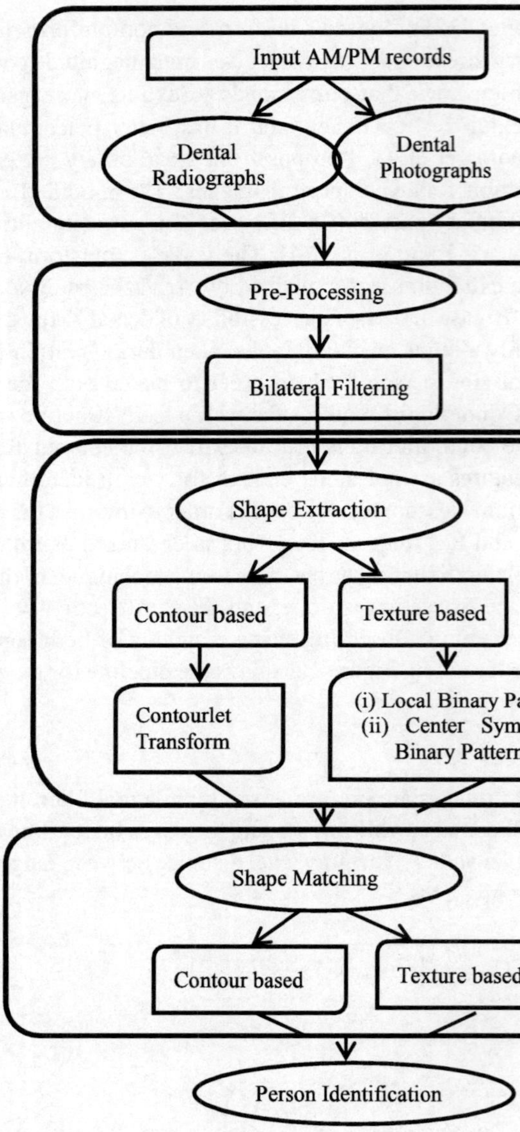

Fig. 1 Pipeline of proposed methodology

$$K_{EDs} = \exp\left(-\frac{1}{2}\left(\frac{ED_s}{W_{EDS}}\right)^2\right) \tag{4}$$

where K_{EDs} is the sub-kernel function of spatial distance function and W_{EDs} are the distribution function of K_{EDs}.

Kernel function for bilateral filtering is obtained by multiplying two sub-kernel functions of K_s and K_{EDs}

$$K_b = K_s K_{EDs} \tag{5}$$

The estimated pixel $\hat{g}(x, y)$ for the filtering kernel K_b is expressed as

$$\hat{g}(x, y) = \frac{\sum_{s=-a}^{a} \sum_{t=-b}^{b} K_b(s, t) s(x + s, y + t)}{\sum_{s=-a}^{a} \sum_{t=-b}^{b} K_b(s, t)} \tag{6}$$

2 Shape Extraction

Shape extraction is the major part to extract the shape details of an image. It can be performed by two approaches. One method is contourlet transform-based contour extraction and the other approach is texture-based feature extraction.

Contour Extraction:

Contour extraction is performed by using contourlet transform. It contains double filter bank structure; specifically for Laplacian pyramid with directional filter bank. It gives a pyramidal directional filter bank structure to extract the exact contour of an image. Contourlet transform gives the multi-resolution and directional decomposition for images, as it allows for a different number of directions for each scale [17]. The main advantage of the double filter bank is to acquire continuity in the tooth contour.

Texture Feature Extraction:

Texture plays an important role in human vision, and it is important in image segmentation and classification. The texture pattern for individual tooth may vary depending upon the image taken from different environment and color of the tooth. In order to extract the individual tooth texture information contour information is also added additionally. In this work, two texture analysesare carried, and it is compared with contour shape extraction.

(i) **Local Binary Pattern:**

Local binary pattern (LBP) is one of the texture operators which label the pixels by thresholding each neighborhood pixels (P_n) with center pixel (P_c) and the result obtained as a binary number. This binary number can be multiplied with the weighted mask [18]. Replace center pixel value with the addition of neighborhood values. The binary values of thresholding function can be obtained as

$$B(P_n - P_c) = \begin{cases} 0, & \text{if } P_n - P_c \geq 0 \\ 1, & P_n - P_c < 0 \end{cases} \tag{7}$$

The LBP can be defined as

$$LBP = \sum_{p=0}^{p-1} B(P_n - P_c)2^p \tag{8}$$

After applying LBP of an image, co-occurrence matrix is applied to extract some of the texture features.

(ii) **Center-Symmetric Local Binary pattern (CS-LBP):**

Similar concept is used for center-symmetric local binary pattern. In CS-LBP, the thresholding value is not compared with the center pixel. The discrepancy is determined by symmetrically measuring the opposite pixel with respect to the middle pixel and the result is obtained with a binary number. Then, the same procedure is applied for CS-LBP to obtain the resultant image. Then, co-occurrence matrix is calculated for final CS-LBP image to extract some of the texture features. The expression for CS-LBP is given as

$$CS(p) = \begin{cases} 1, & \text{if } p \geq 0 \\ 0, & p < 0 \end{cases} \tag{9}$$

The CS-LBP can be defined as

$$CS - LBP = \sum_{p=0}^{(p/2)-1} CS(p)2^p \tag{10}$$

3 Shape Matching

As with shape extraction, the next desirable element is shape matching as well. Contour shape matching is performed by Euclidean and Mahalanobis distance and then features are considered for texture-based matching.

Contour-based matching:

The distance metrics are better metrics to match the query image (PM) with database image (AM). Metric choices are necessary to determine the distance between query points and data points. By finding, the similarity between query and database image is more significant for better matching of images. Euclidean distance (ED) measure is one of the essential distance measures which are used to find the distance between query and database images. The Euclidean distance (ED) between two points $S(S_1, S_2, ...etc)\ and\ T(T_1, T_2, ...etc)$ is calculated by

$$ED = \sum_{p=1}^{n} (S_p - T_p)^2 \tag{11}$$

where S_p and T_p are the contour points of both query (PM) and database image (AM). The distance between AM and PM is small then it will be the better matching. The Mahalanobis distance (MD) can be observed by

$$MD_i^{Mahalanobis} = \sqrt{D_i} = \sqrt{(X - \mu_i)^T \Sigma_i^{-1} (X - \mu_i)} \tag{12}$$

where Σ_i^{-1} is the inverse of covariance matrix. Mahalanobis distance is also known as weighted Euclidean distance in which the weight is determined using the covariance matrix [20].

In addition to this, the additional measure for matching is optimized for sum of absolute distance (OSAD), and it is calculated by

$$OSAD = \sum_i \sum_j \frac{|I_1(i, j) - I_2(i, j)|}{\max(I_1(i, j), I_2(i, j))} \tag{13}$$

where I_1 and I_2 are two different images. And another measure, Average Difference (AD) is used to find the difference between two images as,

$$AD = \frac{\sum_{i,j} \sum_{i,j} I_1(i, j) - I_2(i, j)}{MN} \tag{14}$$

where I_1 and I_2 are two different images, M and N are the size of image.

Texture based Matching:

Texture is an important image attitude and is a useful measure for matching and retrieving images. After taking LBP, CS-LBP of an image, the matching performance evaluation can be done by using confusion matrix.

Performance Evaluation:

In order to evaluate the performance of texture analysis confusion matrix is used for both radiographic and photographic images. The performance measures can be evaluated by precision, recall, f-measure, false positive rate, and true negative rate.

$$precision\ (P) = \frac{TP}{TP + FP} \tag{15}$$

$$\text{Rec}all\ (R) = \frac{TP}{TP + FN} \tag{16}$$

$$F - measure = 2 * \frac{P * R}{P + R} \tag{17}$$

$$False\, positive\, rate\, (FPR) = \frac{FP}{FP + TN} \tag{18}$$

$$True\, negative\ rate\, (TNR) = \frac{TN}{FP + TN} \tag{19}$$

where TP, TN, FP, and FN refer true positive, true negative, false positive, and false negative. Precision is the correctness measure, recall is the completeness measure, F-measure is the measure of matching quality, and the false positive and true negative is complementary.

Results and Discussion

This work is evaluated for contour shape extraction and texture feature extraction of both radiographic and photographic images. For radiographs, a database from Digital X-ray center Madurai is used, which includes bitewing, periapical, and panoramic images. The photographic images are captured with high-resolution camera. The developed algorithm can be tested for 155 database images. Among 155 dental images, 90 images are radiographic images and the remaining 65 images are photographic images. The sample radiographic and photographic images are shown in Fig. 2.

The radiographic image comprises bitewing, periapical, and panoramic images. Such that the bitewing image contains 8–10 teeth, periapical image has 3–4 teeth, and the panoramic view has its whole 32 teeth. In this work, contourlet transform can be used for three levels of Laplacian pyramid and then directional filter bank (DFB) is used. Each level of Laplacian pyramid contains $REDUCE\ and\ EXPAND$ function. In order to avoid discontinuity present in the contour, DFB is used. The pyramidal result of radiographic and photographic image is shown in Fig. 2c. From this Fig. 2c, $REDUCE$ of function the image size can be reduced to half. It can be taken as first the row size of an image can be reduced and then column size of the image reduced. Similarly, for $EXPAND$ function, the size of image can be increased row-wise and then column-wise. After the pyramidal result, in order to get smooth

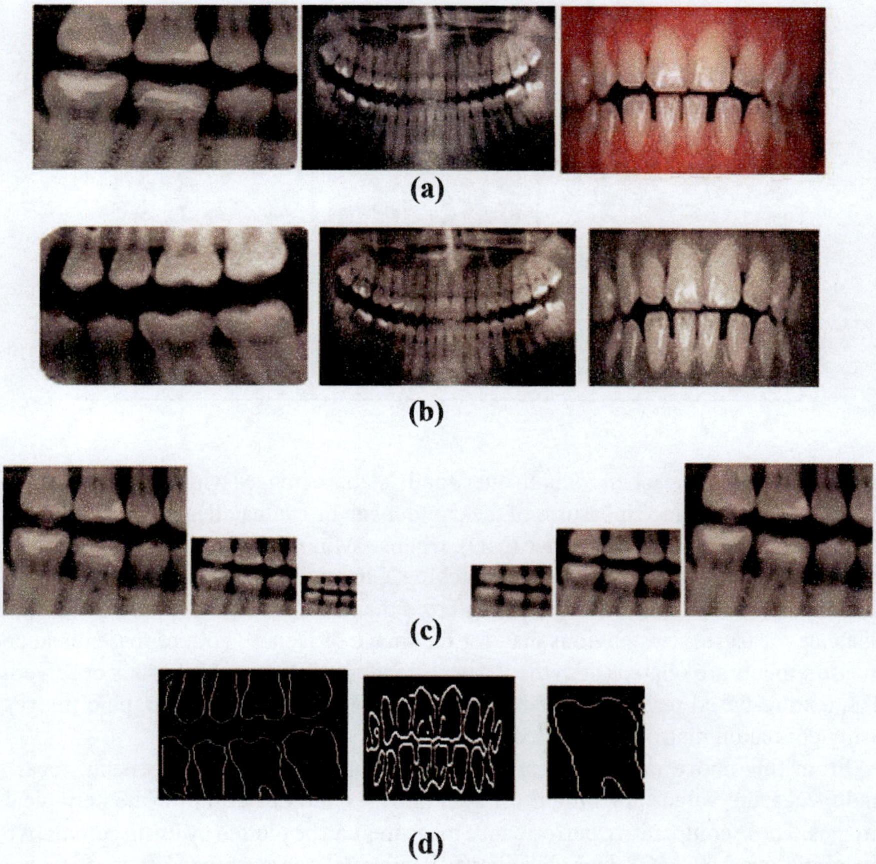

Fig. 2 **a** Sample Radiographic & Photographic images, **b** Pre-processed output, **c** Pyramidal result, **d** Contourlet transform of dental images

contour and directional edges, DFB is used. The directional information is captured as horizontal, vertical, and for the 45-degree direction. Many levels of decomposition are also possible for contourlet transform. In this work, 3 levels of decomposition are used and which leads to 8 directional sub-bands of the image. Searching time is also very less when compared to other existing techniques for contour extraction of an image. The result of contourlet transform is shown in Fig. 2d. The comparison of different texture features is tabulated in Table 1

From this table, it is noticeable that LBP, CS-LBP texture characteristics can differ from radiographic and photographic images. The individual tooth-like molar, premolar, and incisor can vary their texture characteristics. The distance-based performance evaluation is calculated for contour shape extraction and texture-based features are split into smaller blocks using bounding boxes. The size of the box can vary for different scale level images. Key points are extracted for each box and then find the Euclidean distance between center key points to the neighborhood points.

Table 1 Comparison of different Texture Features of Dental images

Input images	LBP	CS-LBP
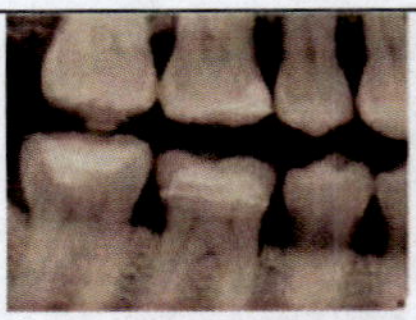	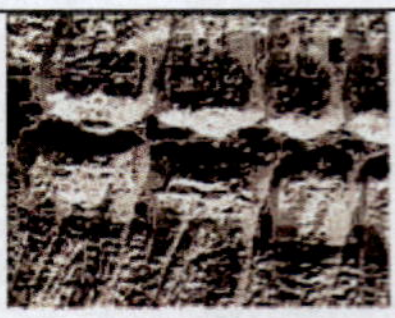	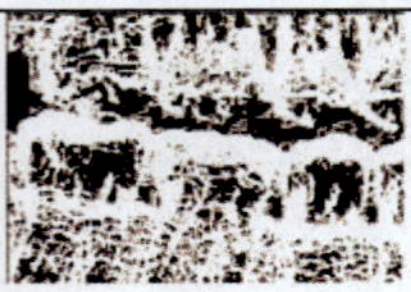
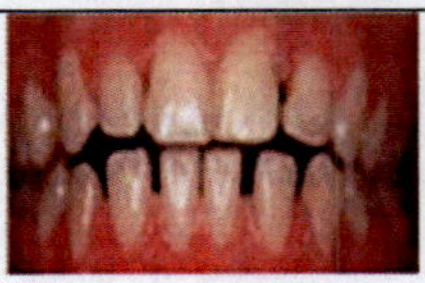	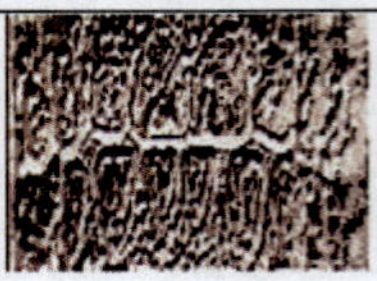	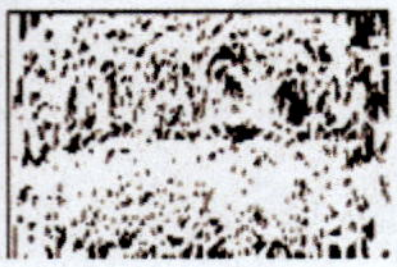

Smaller distances are taken for both query and database images which gives the better matching. Performance measures of the contour can be evaluated by using Euclidean distance and Mahalanobis distance (MD), whereas Minimum distance between query and database contour gives the better matching. Contour-based matching can be tabulated in Table 2. From this table, it is inferred that Dent 9 has lesser values for both distance measures. It is obvious that, for the image of Dent 9, both of these distance measurements are related (i.e.,) the lowest Euclidean distance of 0.1243 is observed. The texture-based performance analysis of radiographic and photographic images using confusion matrix is tabulated in Table 3.

From this above table, it clearly explains that the values of precision, recall, and F-measure values are similar for some images and deviation for the periapical images. These comparative performance measures can be plotted by using cumulative matching curve (CMC), and it is shown in Fig. 3. It is calculated by getting the percentage of top-ranking values and the percentage of test hit-rate. From this Fig, it is observed that radiographic contour of 79.3% images are within top-I position and 99.5% of images are retrieved within top-12 position. Similarly for photographic contour, 67.6% of images are observed for top-I retrieval and 95.3% of images are retrieved for top-12 position. It is inferred that the contour plot gives better matching

Table 2 Distance measures

Sample images taken		Euclidean distance (ED)	% of Similarity	Mahalanobis distance (MD)	% of Similarity
PM	AM				
Dent 15	Dent 9	0.1243	99.87	2.3651	97.63
	Dent 11	1.5625	98.43	4.6253	95.37
	Dent 42	1.7365	98.26	3.4562	96.54
	Dent 28	0.7452	99.25	3.0491	96.95
	Dent 2	3.4843	96.51	9.4517	90.54

Table 3 Performance measures of radiographic and photographic images

Sample images taken	Confusion matrix				Precision (%)	Recall (%)	F-measure (%)	FPR (%)	TNR (%)
	TP	TN	FP	FN					
Bitewing	95	83	08	10	92.23	90.47	91.34	0.08	0.91
Periapical	23	17	03	01	88.46	95.83	92	0.15	0.85
Maxila	115	92	12	11	90.55	91.26	90.9	0.12	0.88
Mandible	107	98	11	07	90.67	93.85	92.23	0.1	0.89
Photographic	31	22	13	12	70.45	72.09	71.26	0.37	0.62

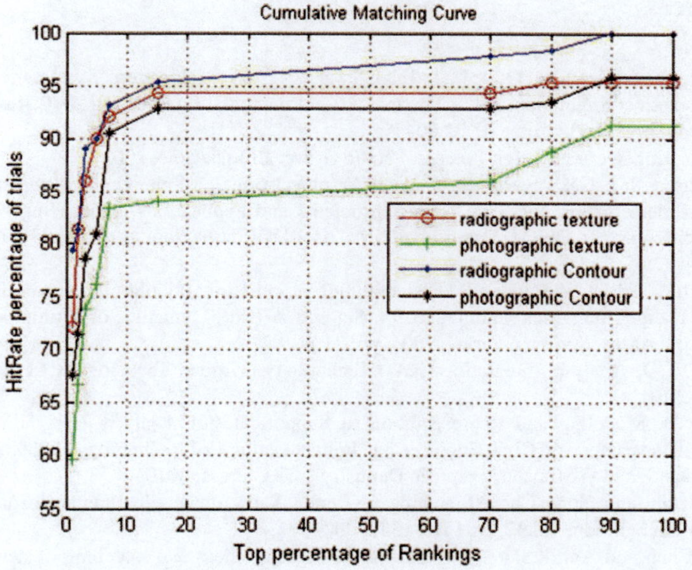

Fig. 3 Comparative performance analysis for texture and contour

compared to texture analysis plot. Since the texture seems to be varying for different illumination condition, it differs for same tooth itself.

4 Conclusion

In mass disaster situations, conventional means of biometrics are helpless other than dental pattern. Hence, the need for automated analysis of person identification using dental images has necessitated the development of elegant shape extraction and matching algorithms which can assist the forensic odontologist in person identification. This paper has explained the contour shape extraction and different texture analysis for radiographic and photographic images. While comparing the performance measures of both contour and texture radiographic and photographic analysis, contour gives better matching for person identification.

Acknowledgements The authors would like to thank the management of Aditya College of Engineering, Andhra Pradesh and also extend our thanks to Department of ECE, Kamaraj College of Engineering & Technology, Virudhunagar, Tamilnadu, for providing all the facilities to carry out this work.

References

1. I. Janajreh, S. Syed, R. Qudaih, I. Talab.Solar Assisted Gasification: Systematic Analysis and Numerical Simulation. Int. J. Thermal & Environmental Engineering 2010;1:81-90,DOI: https://doi.org/10.5383/ijtee.01.02.004
2. Haik, Y: Engineering Design Process. Pacific Grove: Brooks/Cole, 2003
3. Toukourou NM, Gakwaya B, Yazdani JJ. An object-oriented finite element implementation of large deformation frictional contact problems and applications. Proceedings of the 1st MIT conference on CFSM. Cambridge, MA, 2001.DOI: https://doi.org/10.5383/mitcfsm.010005084
4. OcceelliV, Tadrict W, Raddev H. Disintegration of cylindrical liquid columns in liquid-fluid systems: direct numerical simulation. In: Schmitt A (Ed), Dynamics of Multiphase Flows. across Interfaces. Springer-Verlag, 2006, pp. 21-60
5. Peky GK. X-Analysis Integration (XAI) Technology. Virginia Tech Report EL002-2000A, March 2010
6. Kumar D. Modeling and Representation to Support Design-Analysis Integration. Master Thesis, Department of Civil Engineering, Indian Institute of Technology; 2009. [1] David R.Senn and Paul G Stimson,"Forensic Dentistry", CRC Press, 2010
7. Anil K.Jain and Hong Chen,"Matching of Dental X-ray images for human identification", Pattern Recognition, vol.37, pp.1519-1532, 2004
8. Hong Chen and Anil.K.Jain," Dental biometrics: Alignment and matching of dental radiographs", IEEE Transactions Pattern Analysis Machine Intelligence, vol.27, Issue.8, pp.1319-1326, 2005
9. Samir Shah, Ayman Abaza, Arun Ross and Hany Ammar," Automatic tooth segmentation using Active Contour without edges", IEEE Biometrics Symposium, 2006
10. Banumathi.A, Vijayakumari.B, Geetha.A, Shanmugavadivelu.N and Raju.S,"Performance Analysis of various techniques applied in Human Identification using Dental images", Journal of Medical Systems, vol.31, No.3, pp.210-218, 2007
11. O.Nomir and Mohamed Abdel- Mottaleb," Human Identification from dental X-ray images based on shape and appearance of the teeth", IEEE Transactions on Information and Security, vol.2, No.2, pp.188-197, 2007
12. Hofer.M and Marana.AN," Dental Biometrics: Human Identification based on Dental work information", IEEE Brazilian Symposium on Computer graphics and Image Processing, pp.1530-1834, 2007.
13. PhenLan Lin, Yan Hao Lai, Po Whei Huang," Dental biometrics: Human Identification based on teeth and dental works in bitewing radiographs", Pattern Recognition, vol.45, pp.934-946, 2012
14. Vijayakumari pushparaj, UlaganathanGurunathan and BanumathiArumugam,"An Effective shape extraction algorithm using contour information and Matching by Mahalanobis distance", J Digital Imaging, June 2012

15. OmaimaNomir and Mohamed Abdel Mottaleb,"Hierarchical Contour matching for dental radiographs", Pattern Recognition, vol.41, pp.130-138, 2008
16. Minh. N.Do and Martin Vetterli,"The Contourlet Transform:An Efficient Directional Multiresolution Image Representation", IEEE Transactions on Image Processing,vol.14, issue.12, pp.2091-2106, 2005
17. Truong T. Nguyen and SoontornOraintara," Multi resolution Direction filter banks: Theory, Design and Applications", IEEE Transactions on Signal Processing, Vol.53, No.10, 2005
18. Anupa Maria Sabu,D.NarainPonraj and Poongodi,"Textural features based breast cancer detection: A Survey", Journal of Emerging Treads in computing and Information Sciences, vol.3,No.9,pp.1329-1334, 2012
19. Weszka, JS Dyer and Rosenfeld," A comparative study of texture measures for terrain classification", IEEE Transactions on systems, man and cybernetics, vol.6, No.4, pp.269-285, 1976
20. Younis.K,Karim.M,Hardie.R,Loomis.J,Rogers.S and Desimio.M,"Cluster merging based on weighted Mahalanobis distance with application in digital mammograph",IEEE conference of Aerospace and Electronics, 1998

An Efficient Low Latency Router Architecture for Mesh-Based NoC

Aruru Sai Kumar and T. V. K. Hanumantha Rao

Abstract NoC is a growing technology where interconnected patterns are developed in the state of multiprocessors. Due to the complicated routing links, many issues prevail regarding traffic congestion and latency which leads to the poor performance of a network. In this research work, Virtual router architecture is introduced which yields low latency resulting in improving the performance of a network. The proposed VIP-based VC architecture for a 4×4 mesh NoC has experimented for 128-bit wide system targeting up to 250 MHz using Xmulator. The experimental outcome exhibits a low latency that requires 500–600 cycles on an average with respect to other router architecture. This outperforms 33% of low latency when compared to the Wormhole router architecture.

Keywords Network on chip · Virtual point-to-point connection (VIP) · Router architecture · Latency and performance

1 Introduction

Network on chip is a peculiar approach in MpSoC applications [1]. The three fundamental appliances of Standard NoC such as Routers (for storing and forwarding), IP cores (Intellectual property), and Network Interface (NI) are illustrated in Fig. 1 [2].

Nowadays, interconnection technology is a limiting factor for NoC designs. A flexible and desirable approach is On-chip interconnection network rather than traditional global wiring that provides an On-chip communication [3]. Router plays a significant role in NoC. In fact, one of the major issues in NoC is implementing a

A. Sai Kumar (✉) · T. V. K. Hanumantha Rao
Department of Electronics and Communication Engineering, National Institute of Technology, Warangal, Telangana, India
e-mail: asaikumar.nitw@gmail.com
URL: http://www.nitw.ac.in

T. V. K. Hanumantha Rao
e-mail: tvkhrao75@nitw.ac.in

T. Laxminidhi et al. (eds.), *Advances in Communications, Signal Processing, and VLSI*, Lecture Notes in Electrical Engineering 722,
https://doi.org/10.1007/978-981-33-4058-9_21

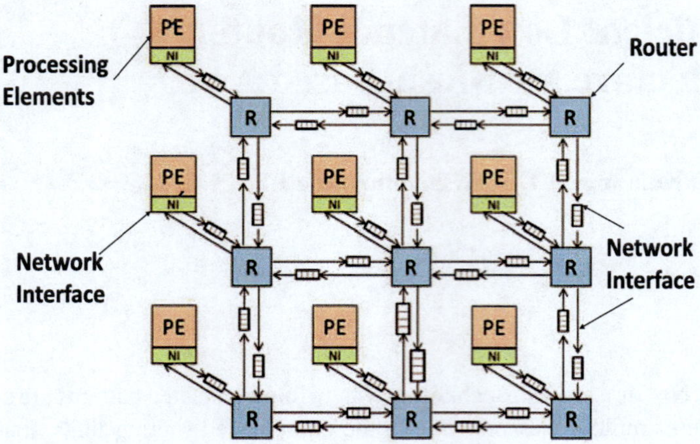

Fig. 1 3 × 3 2D NoC MESH

vital router architecture. Micro-architecture of a router should be robust, the design must be able to prevent the deadlock and congestion issues. Generally, the router contains five input and output ports such as South, North, East, West, and Local port. Each local port is linked to one PE of each router and the remaining ports linked to neighboring routers. In case, the output channel is busy then every input port of router stores the packet in the buffers. Each input port has been associated with a single buffer. These buffers are used in both Wormhole (WH) router or Virtual Channel (VC) routers. Wormhole routing is a versatile technique, composed of a single queue system, in this method packets will be transmitted into one after another. Due to the single queue buffer method, it gives a limited performance in terms of transmitting the packets and avoiding congestions. Virtual channel router is associated with multiple buffers located at every input port of the router. It is able to increase the performance in terms of latency reduction and congestion avoidance. NoC performance progressively improves, while reducing latency and power consumption [4].

The sections in this research paper are organized such as Sect. 2 discussing about related work of the router architectures, Sect. 3 discusses about Existing architecture of router, Sect. 4 states the Proposed VC router architecture, Sect. 5 discusses about Result analysis and Sect. 6 provides the Conclusion of our proposed work.

2 Related Work

This section addresses the different techniques used to identify complications in NoC. The router has many challenges, such as traffic control, avoiding deadlock and routing packets to the proper location. By addressing all these constraints and overcoming issues such as the area overhead and congestion problems in which different algorithms are introduced to improve efficiency.

Tran and Baas [5] proposed RoShaQ router architecture where input packets are allowed for bypassing the shared queues, resulting in the zero-load low latency. A comparison of the energy efficiency and performance of routers is also performed for embedded applications. Li et al. [6] introduced RoB-Router techniques for scheduling the packets to the input buffers as switch allocation is the critical problem in the traditional routers of NoC systems. This design allows the packets that are not located at the head of the VC in advance to the head packets. This results in improving the performance and throughput. Kavaldjiev et al. [7] implemented a simplified dynamic arbitration virtual channel router architecture for the On-chip network. This improves the efficiency in throughput and area for two-dimensional NoC.

Modarressi et al. [8] proposed packet-switched router resulting in the improvement of performance and energy metrics in NoC. VIP Connections are present on each virtual channel where two schemes are proposed for implementing these VIP connections. Su et al. [9] provided a dynamic router with allocation mechanisms such as inter-port and intra-port that operates on controlling the traffic for application-specific NoC systems. This results in increasing the buffer utilization, optimizing delay and throughput. Sai Kumar and Hanumantha Rao [10] implemented an effective core mapping named BMAP by considering the NoC architectures. This technique provides the suitable topology for a respective architecture which outperforms with a minimum objective function. Xu et al. [11] has proposed two new mechanisms for virtual channel router, namely, fixed and dynamic VC assignment for VC allocation for reducing the blockage of head of line which increases the throughput by 41% and reduces the latency by 66%. Liu et al. [12] proposed EnPSR technique where information of the output channel is provided on the basis of the virtual channel (VC) states that are available in neighboring routers. This improves performance and area overhead.

3 Wormhole Router Architecture

Router is the heart of NoC, the basic components of the router are crossbar network, buffers, and scheduling or router computation block and switching. Let us consider an existing router architecture named as Wormhole, i.e., illustrated in Fig. 2.

The wormhole router consists of the router computation block which schedules the input packets and buffered through FIFOs [13]. Input and output states provide the information to the RC (Router Computation). Switch allocator receives the information from the router computation and vice versa, crossbar allocation permits the inputs and outputs based on the availability of the bandwidth. The crossbar network receives the necessary schedule information by the router computation block and switch allocator. Switch allocator works based on the credit inputs, which decides the schedule to the crossbar network. The credit-out decides the buffer bandwidth whether the FIFO is free or not. The router architecture designed with respect to the

Fig. 2 Wormhole router design

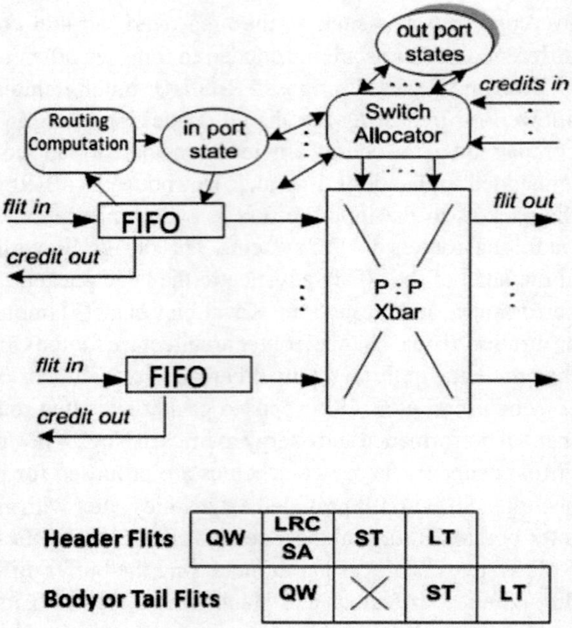

pipeline method. Typically, the wormhole router is designed with four stages such as queue write, look-ahead routing computation, switch traversal, and link traversal.

It requires one clock cycle for execution in every stage. In the first stage of router operation, each input data is allowing to write into the FIFO buffer represented as queue write (QW). Look-ahead routing computation schedules the input data based on the switch allocator in the second stage. Switch allocator decides the input and output states of a neighboring router. Switch Traversal allows transferring the data from the FIFO buffer via the crossbar network. The communication between the neighboring routers is established using the link traversal (LT).

4 VIP-Based Virtual Channel (VC) Router Architecture

As wormhole router architecture is a single input buffering mechanism, it has a greater challenge in traffic congestion which leads to the deadlocks. So to overcome this problem, we proposed a router architecture where VC allocator is used for transmitting the packets present in the queue to the switch allocator [14]. The VIP-based VC router can solve this head of the line blocking issue as represented in Fig. 3.

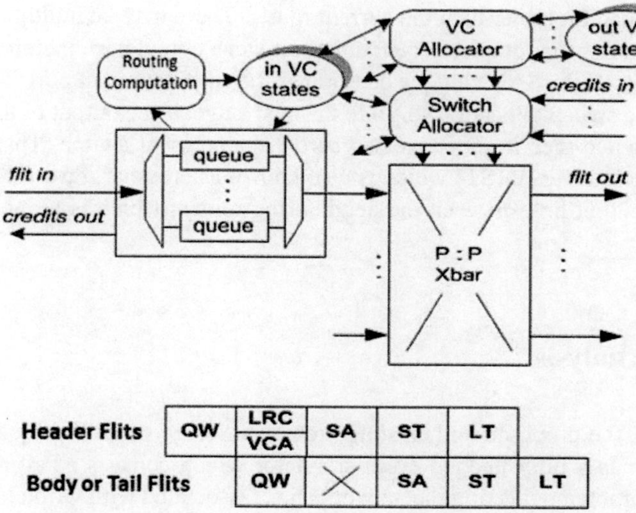

Fig. 3 VIP-based virtual channel router design

The VIP-based VC Router Architecture helps in resolving the traffic congestion and latency issues of a network. The proposed router architecture aimed to provide low latency with a performance network and also reduces traffic congestion. This router architecture contains a volatile storage system that buffers the packets [15].

The VIP-based VC Router Architecture contains an input buffer which comprises multiple parallel queues where each queue is represented as VC. The router computation block processes based on the input buffered from the queues. VC Allocator acts as a bridge between the input, output VC states, and the switch allocator. In wormhole router architecture, the packets are transferred one after the other, suppose if any of the packets are blocked, i.e., present at the head of the queue, then all the packets present behind it will also block its execution. This is overcome in VIP-based VC Router Architecture, which enables the packets from the distinct queues for bypassing each other to progress to the crossbar stage rather than being blocked which is present at the head of the queue by a packet.

The information that is required for scheduling, i.e., retrieved from router computation block and switch allocator is received to the crossbar switch.

This VIP-based VC Router Architecture is designed using the pipelining method which consists of five stages, namely, queue Write (QW), look-ahead routing computation (LRC), switch traversal (ST), switch allocator (SA), and link traversal (LT). This requires one clock cycle for the execution of each stage. Initially, the flits that are arrived at the input port are written to its respective buffer queues. This phase is termed as the queue write (QW) which is also known as the buffer write. The look ahead routing computation stage is defined as the packet that is present in the queue (assuming that there are no packets in front of the queue) starts initiating the output port to its next corresponding router on the basis of the destination information present in its head flit rather than the current router. There may be multiple packets present in the various input queues containing the same output port, therefore, in the current router itself the SA technique determines the specific output port. This stage is termed as the switch allocation (SA). In the next stage, if the output of the switch allocator is granted then it gets traversed across the crossbar switch. This stage is termed as switch traversal (ST) which is also known as crossbar traversal. Later, it produces the output link between the neighboring routers which is termed as link traversal (LT).

5 Result Analysis

In this section, the proposed and existing architectures are simulated by Xmulator [16]. Xmulator is a fully fledged event simulator which consists of all necessary components for interconnecting the networks, i.e., associated with Orion library for calculating the latency of the network.

Also, to construct the VIP or wormhole architecture requires sum of the components such as a router, routing computation block and control networks which are emulated in this environment. Simulation results of this VIP-based VC router architecture are performed for a 128-bit wide system. However, the size of the process is 70 nm and operating frequency is targeting up to 250 MHz, respectively. The following experiments are evaluated for different network configurations such as router structures and traffic loads.

Here, the results are reported for the following configurations and workloads: 4×4 Wormhole NoC and VIP-based VC routers. Wormhole switching is designed with a single flit buffer for the input VCs (including those for packet-switched network and are assigned to VIP paths).

Experiments show that VIP-based architecture in a 4×4 mesh-based NoC takes about 500–600 cycles, on average. The experimental results for VC router architecture are distinguished in Fig. 5 when compared with the wormhole-based architecture as shown in Fig. 4, it takes about 800–900 cycles. The horizontal axis represented in the above experimental figures contains the traffic generation rate (message/node/cycle) through which an IP core generates and inserts messages into the network. The simulation outcomes of our proposed router architecture were more favorable than wormhole router architecture. Finally, The VC router shows better outcomes in terms of latency for improving the performance of a network.

6 Conclusion

In this research work, we proposed the VIP-based VC router architecture for 4×4 mesh NoC which controls the traffic congestion and reduces the latency, resulting in high performance. This proposed architecture has experimented for a 128-bit wide

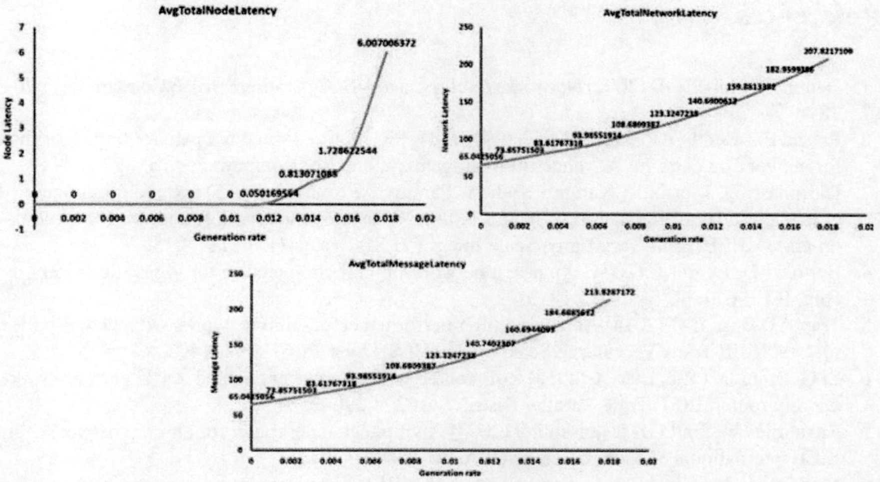

Fig. 4 Various latencies versus generation rate of wormhole router design

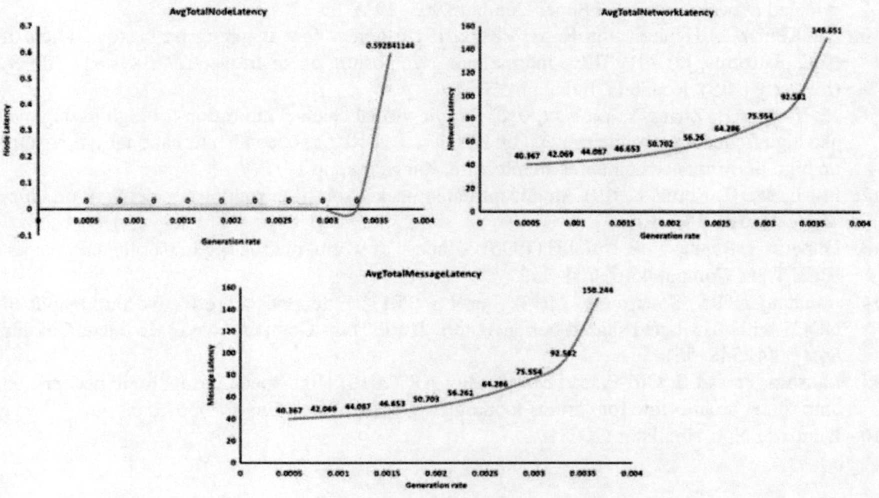

Fig. 5 Various latencies versus generation rate of VC router design

system, i.e., targeting up to 250 MHz to a size of 70 nm process. The simulation outcomes of our proposed algorithm were more favorable than wormhole routing design, which outperforms on an average of 500–600 cycles. In future, we plan to extend the work in two directions. Firstly, considering various topologies for our architecture. Second, finding out the shortest path routing with low latency.

References

1. Benini L, Micheli GD (2002) Networks on chips: a new SoC paradigm. IEEE Comput 35(1):70–78
2. Prasad EL, Reddy AR, Prasad MNG (2018) MWPR: minimal weighted path routing algorithm for network on chip. In: Advances in intelligent systems and computing, pp 15–22
3. DiTomaso D, Morris R, Karanth Kodi A, Sarathy A, Louri A (2013) Extending the energy efficiency and performance with channel buffers, crossbars, and topology analysis for network-on-chips. IEEE Trans Very Large Scale Integr (VLSI) Syst 21(11):2141–2154
4. Bertozzi D, Benini L (2004) Xpipes: a network-on-chip architecture for gigascale system on chip. IEEE Circuits Syst 4(2):18–31
5. Tran AT, Baas BM (2014) Achieving high-performance on-chip networks with shared-buffer routers. IEEE Trans Very Large Scale Integr (VLSI) Syst 22(6):1391–1403
6. Li C, Dong D, Lu Z, Liao X (2018) RoB-router: a reorder buffer enabled low latency network-on-chip router. IEEE Trans Parallel Distrib Syst 29:2090–2104
7. Kavaldjiev N, Smit GJM, Jansen PG (2004) A virtual channel router for on-chip networks. In: IEEE international SOC conference, USA, pp 289–293
8. Modarressi M, Tavakkol A, Sarbazi-Azad H (2010) Virtual point-to-point connections for NoCs. IEEE Trans Comput Aided Des Integr Circuits Syst 29(6):855–868
9. Su N, Gu H, Wang K, Yu X, Zhang B (2018) A highly efficient dynamic router for application-oriented network on chip. J Super Comput 2905–2915
10. Sai Kumar A, Hanumantha Rao TVK (2019) Efficient core mapping on customization of NoC platforms. In: 2019 IEEE international symposium on smart electronic systems (iSES) (formerly iNiS), Rourkela, India, pp 57–62
11. Xu Y, Zhao B, Zhang Y, Yang J (2010) Simple virtual channel allocation for high throughput and high frequency on-chip routers. In: HPCA-16, 2010 the sixteenth international symposium on high-performance computer architecture, Bangalore, pp 1–11
12. Liu L, Ma R, Zhu Z (2019) An encapsulated packet-selection routing for network on chip. Microelectron J 96–105
13. Felperin S, Raghavan P, Upfal E (1996) A theory of wormhole routing in parallel computers. IEEE Trans Comput 45(6):704–713
14. Ramanujam RS, Soteriou V, Lin B, Peh LS (2011) Extending the effective throughput of NOCS with distributed shared-buffer routers. IEEE Trans Comput Aided Des Integr Circuits Syst 30(4):548–561
15. Lakshmi Prasad E, Giri Prasad MN, Reddy AR (2018) High-speed virtual logic network on chip router architecture for various topologies. Comput Electr Eng 67:536–550
16. Xmulator NoC simulator (2008)

High-Performance Image Visual Feature Detection and Matching

N. Sambamurthy and M. Kamaraju

Abstract The fundamental task in high-speed and high-accuracy CCTV video surveillance system is attention-based Image visual point detection and matching. The image viewpoints are different at different instants of time, so that the image recognition and matching systems are having large computational complexity. The designed work is to present an ROI-based SIFT Image detection and BRIEF matching algorithms with a processing speed of 44 ns, and it meets the real-time requirement. The performance of the designed system is improved with a hybrid sorting-based median filter. These algorithms are able to establish accurate feature extraction and matching at 30fps for 640 Pixel video. This can be implemented using MATLAB and Xilinx tools.

Keywords Image features · Computer vision · Visual detection · ROI · SIFT · BRIEF

1 Introduction

Nowadays, computer vision has significantly improved in the field of image classification and object detection, although with very high computational cost [1]. To realize the potential benefits, from these advances in real-world settings, different methodologies must be viable on low-power embedded systems, like automotive vehicles, smartphones, wearable devices, and robots. Generally, visual matching features, image indexing, and object recognition are used embedded visual detection, extraction, and matching algorithms. Real-time performance and high throughput is a critical demand for most of the vision-based embedded system applications,

N. Sambamurthy (✉) · M. Kamaraju
Department of ECE, Gudlavalleru Engineering College, Gudlavalleru, India
e-mail: sambanaga009@gmail.com

M. Kamaraju
e-mail: profmkr@gmail.com

T. Laxminidhi et al. (eds.), *Advances in Communications, Signal Processing, and VLSI*, Lecture Notes in Electrical Engineering 722,
https://doi.org/10.1007/978-981-33-4058-9_22

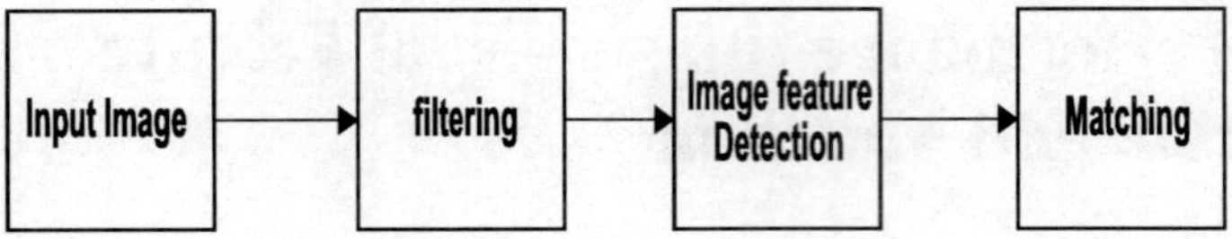

Fig. 1 Key steps in image visual detection and matching

which require the memory-efficient and energy-efficient visual point detection and matching of the visual extreme key points. The use of image feature extraction and matching in video surveillance applications are considered as highly demanding for low-power image recognition applications [2]. The feature detection and matching of the attention-controlled image features are computationally intensive, because their implementations involves hardware also. There are several algorithms to detect image features such as Speedup Robust Feature (SURF) [3–5] and Fast Accelerated Segment Test (FAST) [6]. SIFT and SURF designs are the most efficient methods to detect the image features from the video. BRIEF and BRISK techniques are the binary descriptors. Compared to vector descriptors like SIFT and SURF binary descriptors are very-high-speed processing algorithms. The main component of the embedded visual feature detection and matching system is shown in Fig. 1.

1.1 Median Filtering

The median filtering is used in the preprocessing stage and post-processing stage of the keypoint detection and matching [7, 8]. The median filter smoothens the impulse noisy pixels and easily find out the visual feature points in the given input image. The proposed median filter execution time increases with combined parallel and pipelining techniques.

1.2 Feature Detection

Image feature points are used to compute the abstraction of image viewpoints pixel by pixel. It is used for detecting an interest point in a given image by comparing each pixel in a given window with a given threshold value.

1.3 Feature Descriptor

Image descriptors describe the information about the features of the visual information.

1.4 Image Matching

The image feature matching automatically determines the geometric transformation between a pair of images. Image matching is the point-wise comparison of given two image visual points [4, 9].

2 Existing Work

The image visual Detection and matching algorithms are depends on intensity based and information based. The intensity based algorithm is comparing the entire pixel with finding absolute differences between overlapping pixels. Feature based algorithm is find out the visual correspondence between image point pairs. It compares all point features in one current image with previous image.

The Lowe et al. [1, 3] designed a software based visual point detection and matching methods shown very good results [10]. However power consumption and cost of the hardware increased. Cartney et. al. [11] investigated an image recognized algorithm with image extraction and matching system with image resolution of 640 × 480, and 0.8 s matching speed. the design requires optimization of matching module. Software based visual detection matching system with resolution of 640 × 480 images and speed about 32 ms. however hardware implementation takes maximum time [10, 12]. Bonato and Marques [6] designed NIOS II soft-core processor based feature descriptor with a speed of 11.7 ms. Zhong and Kang [13] demonstrated a SIFT algorithm based Image descriptor system with a image resolution of 320 × 256 images. Generally any single image have so many visual points, it is still far from satisfactory for the real-time performance and memory storage capabilities.

The timing and space complexity of existed computer vision algorithms are increased due to their processing version of software and hardware implementations. The GPU is also one of the most alternative solutions for the image recognition applications. The implementation of image features and matching with software model which increases the cost. So it does most of the implementation using Filed Programmable gate Array (FPGA). The FPGA based hardware is suitable for real-time image processing applications, and it gives better performance and optimal power. Still, the research is going onto search for the right detection and matching techniques. Image filtering plays a critical role in image visual detection and matching. Because for an appropriate feature or perfect match, a proper description of Image filters are needed first. For this, the median filter is designed and the pre-processing and post processing stages to be described in Sect. 3.2. Smart video surveillance and smart vision systems are having the problem with data transmission errors. The median filter eliminates data transmission errors caused by the noise in the received images. The region of selected image technique is helpful for FPGA based visual detection points. Because of hardware implementation, memory resources and hardware utilization are reduced in the FPGA.

3 Proposed System

The design consists of keypoint detection using a modified RoI-based SIFT algorithm and modified FSM-based BRIEF matching system as shown in Fig. 2. The pre- and post-processing stages of ROI-based keypoint detection module uses the median filter to smoothen the keypoint descriptors. The FSM-based brief descriptor is used to find out matching points by Hamming distance algorithm [8, 14]. The median filtering algorithm [7, 8] computational complexity is less compared to other filters. where each operation needs a single clock cycle and each 3 × 3 window performs sorting in nine clock cycles and ROI-based SIFT algorithm uses six clock cycles for every visual point detection. Detected key points are stored in the BRAM and hamming distance matching is performed between the current pixel and previous pixel in the consecutive frame. The design flow consists of the following visual detection and matching process is depicted in Fig. 3. The image key pointers and matching points are filtered with a designed median filter so that it improves the performance of the system compared to Gaussian filter. The proposed filter smoothens the detected points.

3.1 Difference of Median Filter Module

In Fig. 4, the process of SIFT algorithm [2] based feature point module consists of difference between median and key point module. The difference of the median filter is the subtraction of two denoising image filters with the two consecutive scales. The process of the difference of median filter is given in the following Eq. (1).

$$(x, y; \Psi) = M(x, y : k) - M(x, y; \Psi) \tag{1}$$

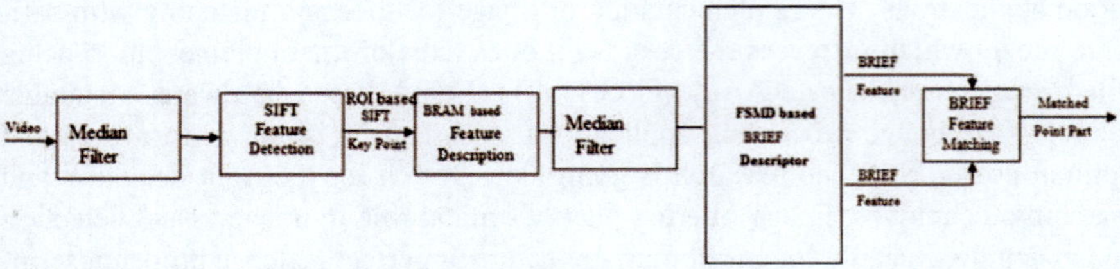

Fig. 2 The process diagram of the designed system

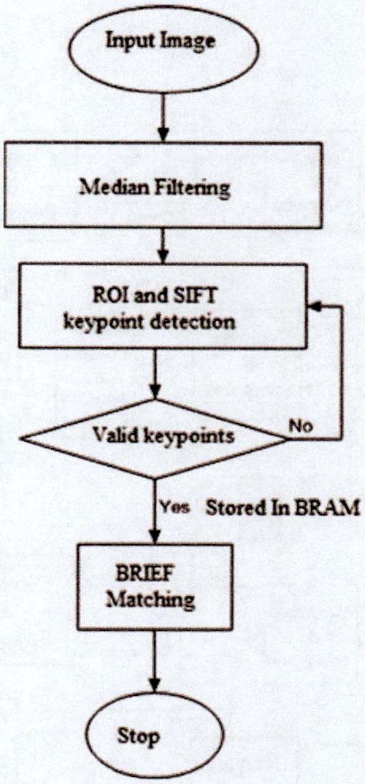

Fig. 3 Visual detection and matching system

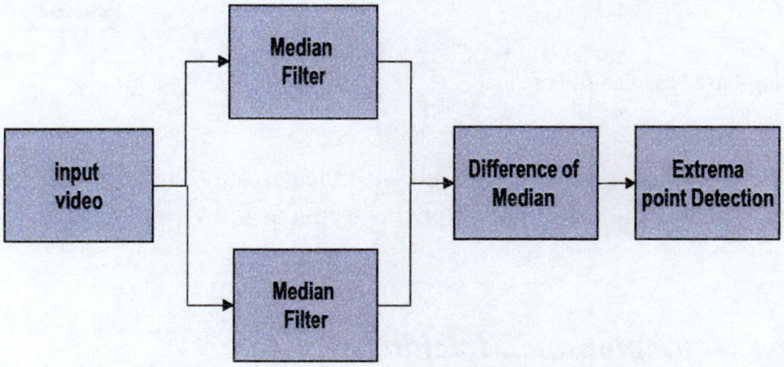

Fig. 4 Difference of median filter module

3.2 *Median Filter*

In Fig. 5, the hybrid sorting network consists of compare and swap unit is designed. and maintains a unit delay. The total delay of 9 samples ($\times 1$–$\times 8$) is 8 clock cycles. For each completion of window size, the parallel output contains sorted data and it is stored in BRAM consecutive memory locations and to produce the filtered output. The CAS unit is the 8-bit size and it handles 2 pixels simultaneously. Filtered window

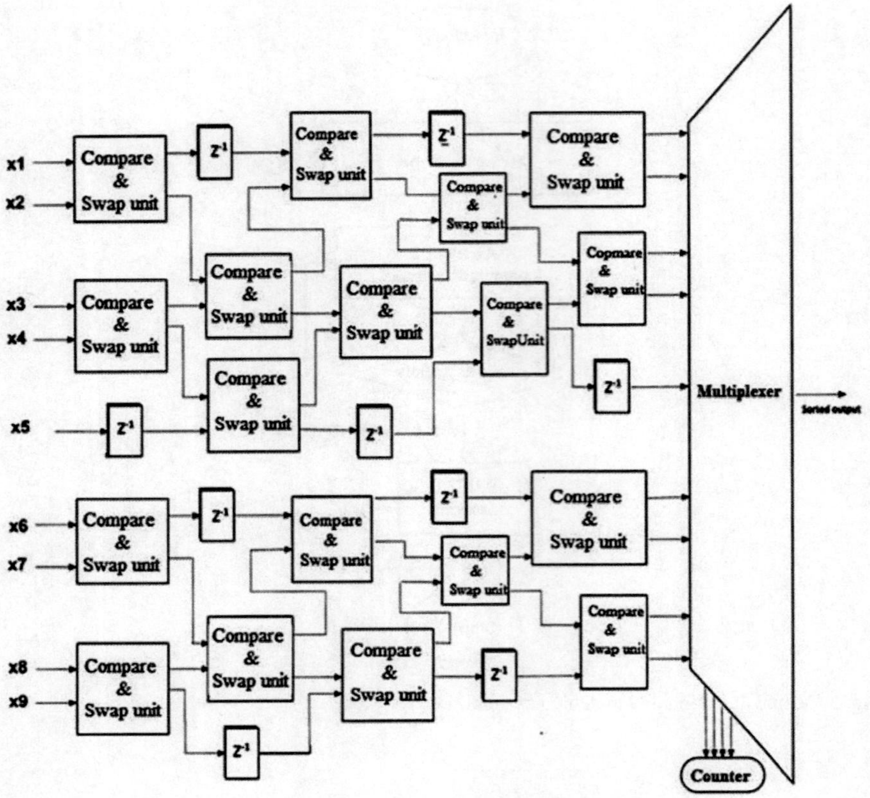

Fig. 5 Sorting-based median filter

pixel value is maxima and it is compared with the surrounding 26 pixels in 3, 3 × 3 neighborhood scales [3, 15]. Finally, local extrema as the visual point.

3.3 *Brief Descriptor and Matching*

The descriptor and matching system [10, 12] is very sensitive to noise to eliminate low contrast and strong points using Hessian matrix and set the threshold value as shown in Eq. (2).

$$\beta(\mathrm{p}; \mathrm{u}, \mathrm{v}) := \{1; \quad \text{if } \mathrm{I}(\mathrm{p}, \mathrm{u}) < \mathrm{I}(\mathrm{p}, \mathrm{v}) \\ 0; \quad \text{otherwise.}\} \tag{2}$$

The BRIEF descriptor is performed by comparison between two filtered image visual point pairs U and V as shown in Fig. 6.

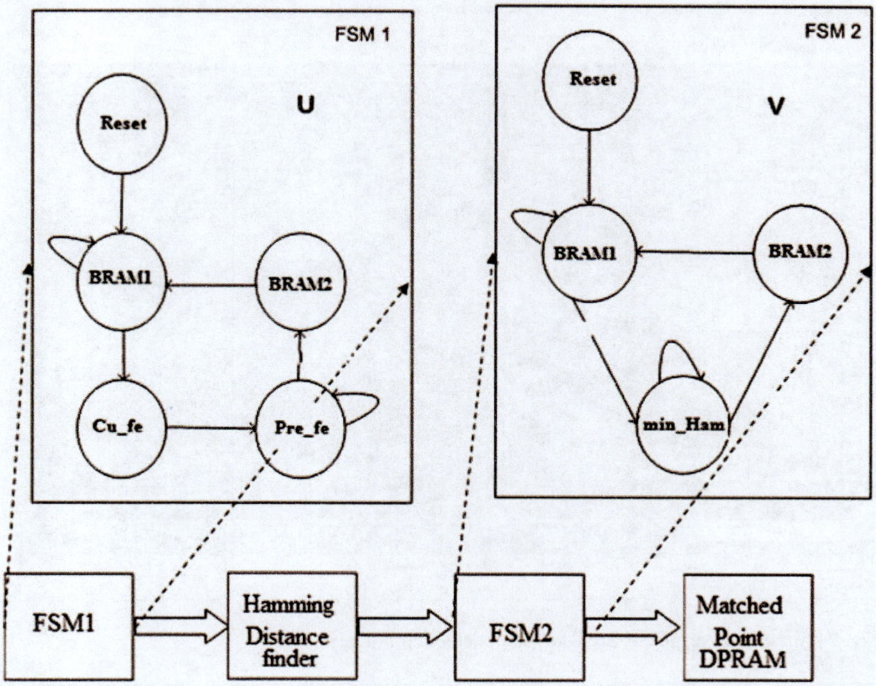

Fig. 6 BRIEF matching system

The brief algorithm is a 256-dimensional bit string that is used to find a point pair that has the minimum Hamming distance [9]. By using this algorithm, the accuracy of point pairs increases. If any noise is present in the designed hardware-based median filter, it is eliminated in noise pre- and post-processing stages.

4 Experimental Results

In order to verify the designed visual detection and matching system, here we give the software and hardware results in three aspects. First, we assess the designed system on real-time recorded videos and check the resource utilization in VIRTEX FPGA. Simulate the designed system using Modelsim software as shown in Figs. 7 and 8. The matching results are given in Fig. 9a-i. It is synthesized and design utilization profile is done using Xilinx software tools. We compare the designed system with the existing system timing and device utilization profiles are shown in Tables 1 and 2. System with the existing system timing and device utilization profiles are shown in Tables 1 and 2.

Figure 7 shows that the SIFT algorithm based key points are detected with different scales and it is displayed in the simulation results.

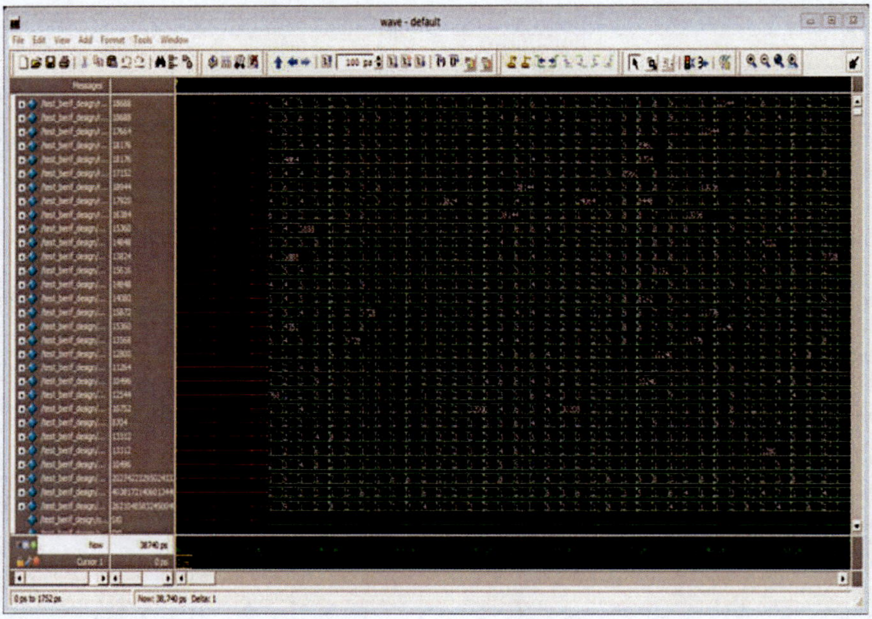

Fig. 7 Simulation results of SIFT keypoint detection module

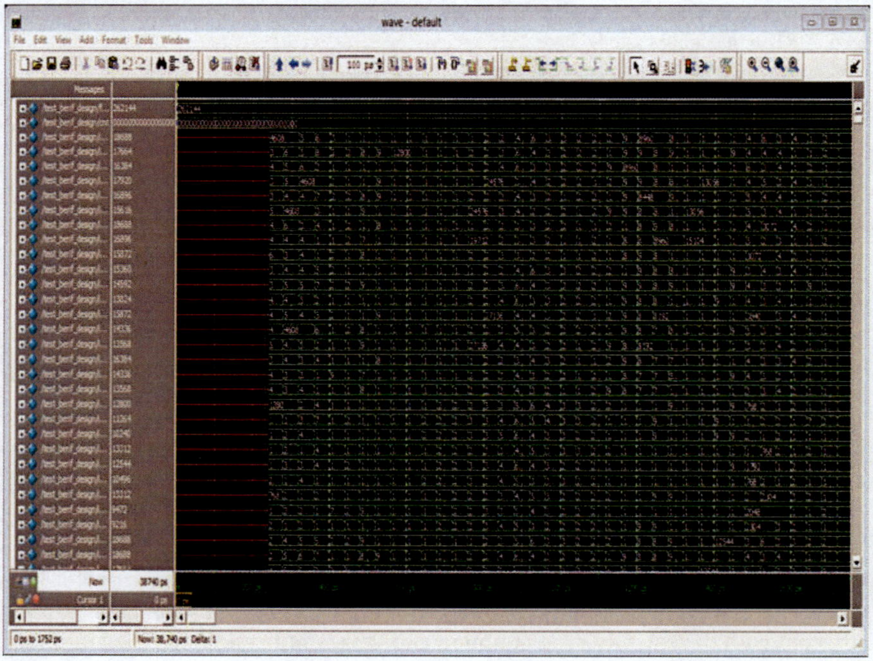

Fig. 8 Simulation results of BRIEF descriptor and matching

Table 1 Comparative analysis

Parameter	Designed system		Existed system [15]	
	Resolution	Speed	Resolution	Speed
SIFT detector	640X480	2.8 ns	640X480	3.3 ns
BRIEF descriptor	640X480	3.2 ns	640X480	3.4 ns
Brief matching	640X480	38 ns	640X480	33.1 **ms**

Table 2 Device utilization profile using FPGA

	LUT's	FF'S	Slices	DSP48E	BRAM
Designed system					
SIFT detector	15,694	10,469	4654	2	2
BRIEF descriptor	452	29	29	2	2
BRIEF matching	653	395	242	0	2
Existing system [2, 12]					
SIFT detector	13,982	9836	5694	52	–
BRIEF descriptor	689	35	34	–	–
BRIEF matching	946	491	350	–	–

The brief module used Hamming distance algorithm for correct matching of descriptors and their binary values are shown in Fig. 8.

The matched points are distributed over all images as shown in Figs. 9, 10, 11, 12, 13, 14, 15, 16, 17 and 18. The designed system is tested with MATLAB on a 640×480 pixel video. The detected features are 52,487. Among which 19,786 are matched with a processing speed of 44 ns.

In Table 1, the designed system utilizes high processing Speed compared to the existing system. By using the proposed median filter in SIFT detector, the performance is highly improved compared to the existing system.

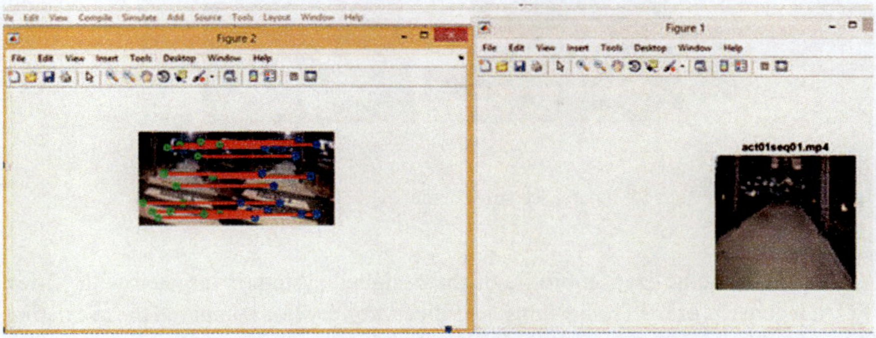

Fig. 9 Visual point detection and matching results with different natural scenarios

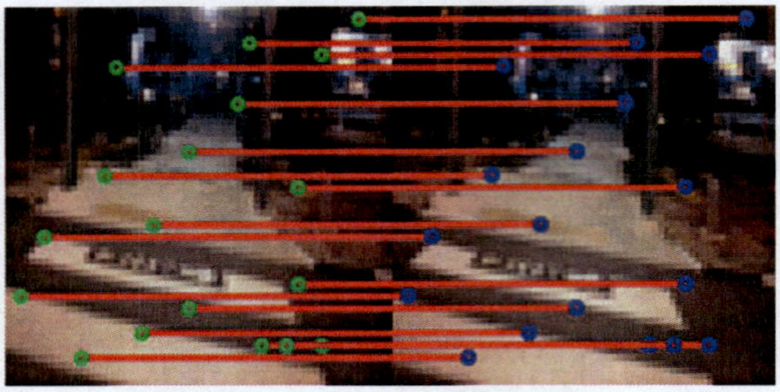

Fig. 10 Matching results for natural scenario-1 video

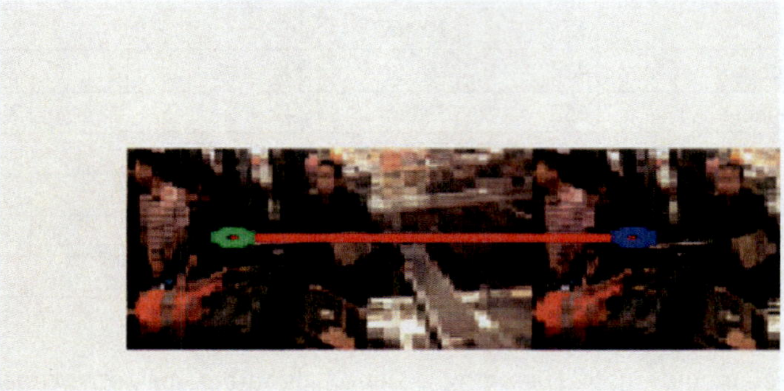

Fig. 11 Matching results for natural scenario-2 video

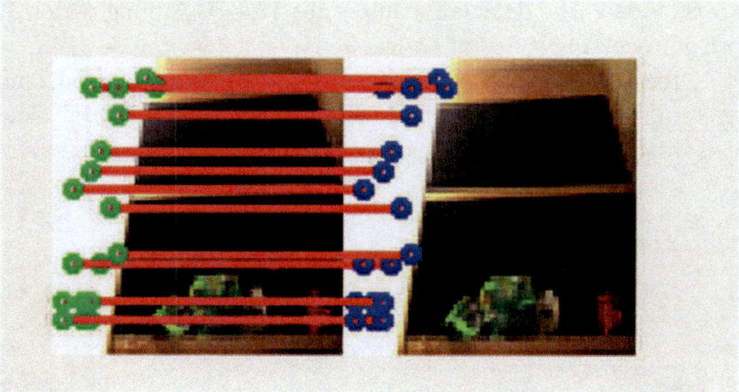

Fig. 12 Matching results for natural scenario-3 video

In Table 2, the utilization profile of the designed system is targeted with Virtex FPGA. It consists of LUTs, flip-flops, and slices are very less compared to the existing system [2, 12].

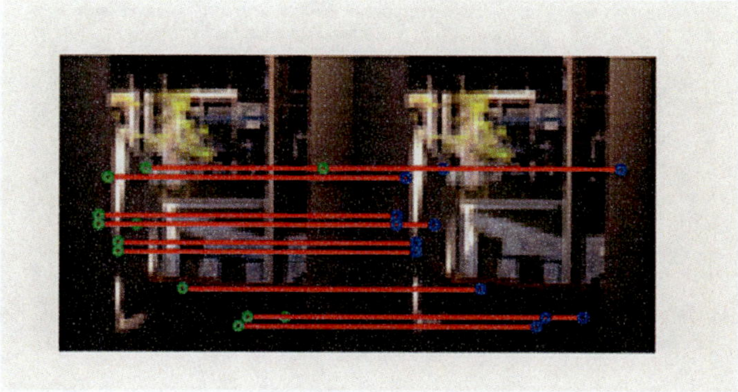

Fig. 13 Matching results for natural scenario-4 video

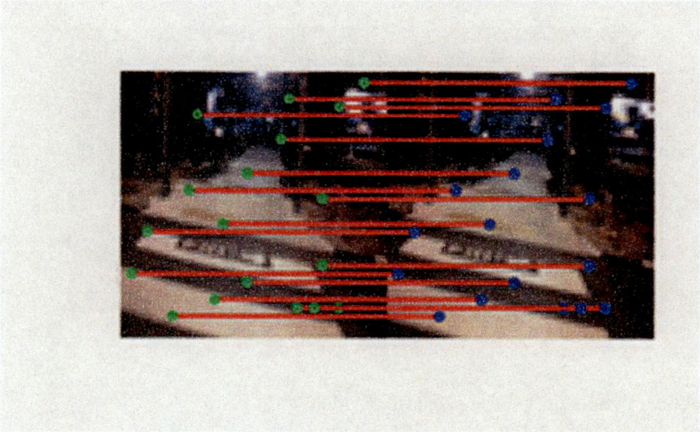

Fig. 14 Matching results for natural scenario-5 video

Fig. 15 Matching results for natural scenario-6 video

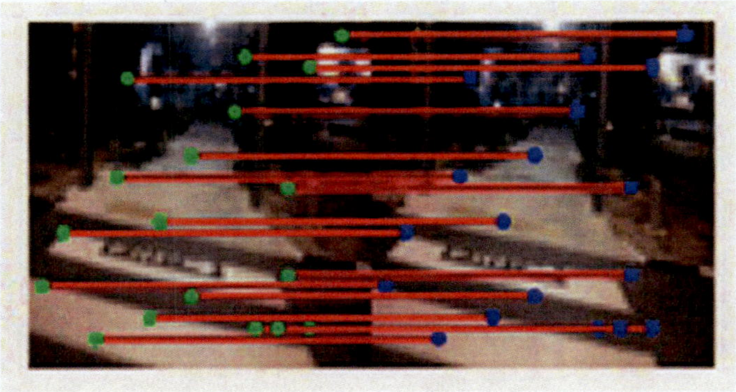

Fig. 16 Matching results for natural scenario-7 video

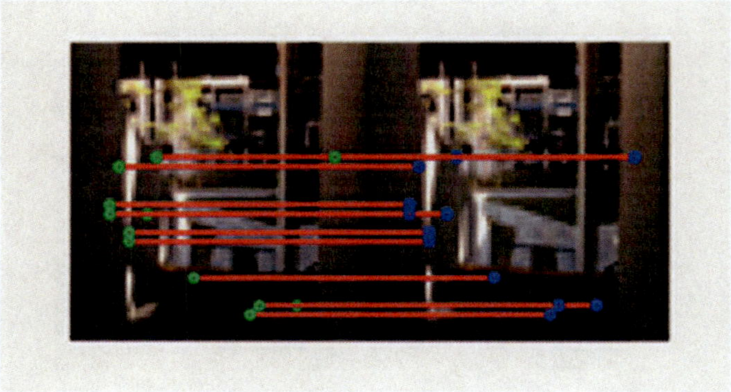

Fig. 17 Matching results for natural scenario-8 video

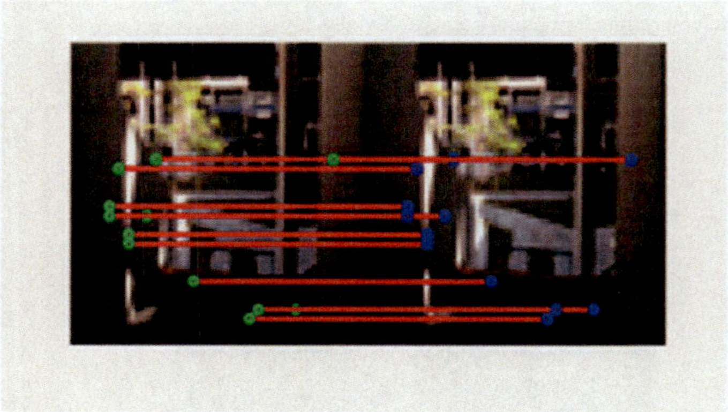

Fig. 18 Matching results for natural scenario-9 video

5 Conclusion and Future Scope

A search for the perfect detection and matching technique is still going on. Moreover, this work presents the design of visual detection using RoI-based SIFT and BRIEF algorithm with a processing speed of 44 ns. The high accuracy is possible with RoI-based Median filter of 640 × 480 pixel video in MATLAB. High accuracy is possible with the hybrid sorting-network-based median filter. The targeted FPGA utilization profile shows very less resources for the designed system compared to the existing system. In the future, SoC-based feature extraction and matching can be implemented with higher accuracy and speed.

References

1. Lowe DG (2004) Distinctive image features from SIFT based key points detection. Int J Comput Vis Appl 60(2):91–110
2. Mizuno K, Noguchi (2011) A low-power real-time scale and rotation invariant feature detection (SRIFD) based descriptor generation engine for full-HDTV video recognition. IEICE Trans Electron System 94(4):448–457
3. Chen SJ, Zheng SZ (2018) An improved image matching method based on SURF algorithm. Int J Photogramm Remote Sens 3:179–184
4. Mikolajczyk K, Schmid C (2005) A performance and evaluation of BRIEF local descriptors. IEEE Trans Pattern Anal Mach Intell 27(10):1615–1630
5. Bay H, Less A, Tutelary T, Gool LV (2008) Speeded-up robust features (surf). Comput Vis Image Underst 110(3):346–359
6. Bonato V, Marques E (2008) A parallel hardware architecture for scale and rotation invariant feature detection (SIFT). IEEE Trans Circuits Syst Video Technol 18(12):1703–1712
7. Thirusangu T, Priya TL et al (2017) Some studies on detection and filtering algorithms for the removal of random valued impulse noise. IET Image Process 11(11):953–963
8. Saranya B, Ramya A, Murugan D (2018) Adaptive fuzzy based nonlinear filter for de speckling ultrasound images. I-Manager's J Comput Sci 5:1–3
9. Skea D, Barrodale I (1993) A control point matching algorithm. Int J Pattern Recogn 26:269–276
10. Ethan R, Rabaud V, Bradski G (2011) Orient BRIEF (ORB): and efficient alternative to SIFT or SURF. In: IEEE ınternational conference on computer vision
11. McCartney MI, Malkani M (2009) Image registration for sequence of visual images captured by unmanned aerial vehicle. In: Proceedings of IEEE symposium computer ıntelligent multimedia signal vision process, March–April 2009, pp 91–97
12. Sinha N, Michael J, Pollefeys M (2006) GPU-based video feature tracking and matching. In: Proceedings workshop edge computer using new commodity architecture, vol 278, pp 695–699
13. Zhong S, Kang L (2013) A real-time embedded architecture for scale-invariant feature points detection (SIFT). IEEE Trans Circuits Arch 59(1):16–29
14. Calonder M, Lepetit V (2012) BRIEF algorithm based computing a local binary descriptor using very fast algorithm. IEEE J Pattern Anal Mach Intell Syst 34(7):1281–1298
15. Huang FC, Chen YC (2012) High performance SIFT hardware accelerator for real-time image feature extraction. IEEE J Circuits Syst Video Technol 22(3):340–351
16. Zhong S, Wang J, Yan L, Kang L, Cao Z (2013) A real-time embedded architecture for SIFT. J Syst Archit 59(1):16–29

Systolic-Architecture-Based Matrix Multiplications and Its Realization for Multi-Sensor Bias Estimation Algorithms

B. Gopala Swamy, U. Sripati Acharya, P. Srihari, and B. Pardhasaradhi

Abstract The accelerators are gaining predominant attention in the HW/SW designs and embedded designs due to the less power consumption and parallel data processing capabilities compared to standard microprocessors and FPGA's. In this paper, MSSKF (Multi-sensor Schmidt–Kalman filter)-based coupled bias estimation problem is considered for single target multiple sensors case. Here MSSKF augments the state vector and bias vector for bias estimation, results in computationally expensive as the dimensions of the state and sensors increases. Hence to address the computational complexity, digital signal processing (DSP) architectures are proposed and accelerated the algorithm to meet the real-time constraints. In the MSSKF algorithm, the overload of the algorithm is due to state covariance prediction and innovation covariance prediction. To realize the state covariance and innovation covariance, a folded DSP architecture and parallel processing based folded DSP architecture are proposed, respectively. The matrix multiplications are addressed with systolic arrays to gain the advantage of latency and parallel processing. Moreover, MSSKF using systolic array architectures simulated and synthesized in Vivado 2018.1 using Verilog and implemented on FPGA-Zynq-7000 board. The performance of the systolic-based accelerator realization was compared with normal matrix multiplication.

B. Gopala Swamy (✉) · U. Sripati Acharya · P. Srihari · B. Pardhasaradhi
National Institute of Technology Karnataka, Surathkal 575025, India
e-mail: gopalswamybgs@gmail.com; bodduboingopalaswamy.183vl001@nitk.edu.in

U. Sripati Acharya
e-mail: sripati.acharya1@gmail.com

P. Srihari
e-mail: srihari@nitk.ac.in

B. Pardhasaradhi
e-mail: bethi.pardhasaradhi@gmail.com

T. Laxminidhi et al. (eds.), *Advances in Communications, Signal Processing, and VLSI*,
Lecture Notes in Electrical Engineering 722,
https://doi.org/10.1007/978-981-33-4058-9_23

1 Introduction

The main purpose of bias estimation is to correct the bias and to estimate the state of the target. The bias correction is very important in sensor networks, multi-target multi-sensor target tracking [1], localization problem with multiple sensors, drones based surveillance, autonomous vehicle interactions, and collision avoidance. Sensors are affected with different types of biases like residual bias, measurement bias, offset, scaled bias, clock bias, etc. [2]. In the literature, bias estimation algorithms are classified into two types as coupled and decoupled. In coupled algorithms, the state vector is coupled with bias vector to predict and update the state of the target [3], whereas in decoupled algorithms the state vector is not coupled with bias vectors and bias estimation is done by measurement differences. However, decoupled algorithms are not accurate in comparison to coupled algorithms. Multi-radar-Multi-target bias estimation presented in. The Multi-sensor Schmidt–Kalman Filter (MSSKF) is a commonly used coupled bias estimator which requires high-performance systems to compute higher dimensional matrix multiplications involved in algorithm steps.

More number of sensors installation, sensor interaction, and decision-making are increasing in autonomous driving systems and robotics. Whereas, these applications are mostly realized in microprocessor-based hardware architecture and hence limited by power consumption and latency of the platform used. As an alternative solution to this problem, efficient embedded systems should be designed and make use of DSP architecture minimization techniques to provide high computational power and minimum hardware. Here the modern FPGA's are equipped with a large number of reconfigurable fabrics, configurable slices, and digital signal processor blocks that can be used for floating-point operation. Computationally expensive algorithms can be parallelized an FPGAs with floating-point precision compared to standard PC and GPU platforms. FPGAs have gained rapid growth over the past decade because they are useful for a wide range of applications includes digital signal processing, bioinformatics, device controllers, software-defined radio, random logic, ASIC prototyping, medical imaging, computer hardware emulation, integrating multiple SPLDs, voice recognition, cryptography, filtering and communication encoding, and many more [4].

MSSKF is the precise approach for sensor bias estimation by augmenting the target state with a bias of the sensor [3]. By augmenting the target state with sensor bias, the dimensions of state vector increases, and the calculation of filter gain, state prediction, and state and covariance update are computationally infeasible. As the dimensions of the state vector increases, the computational complexity increases which result in time delay, high power consumption. To overcome the computational complexity systolic array architecture is one of the choices. Systolic architectures are similar to pipelining to implement on FPGA platforms [5]. Systolic array architecture is regular in geometry and can be scalable to different dimensions and we can achieve a high degree of pipelining. The importance and applications of the systolic arrays are mentioned in [6]. Systolic arrays are very efficient to solve simultaneous equations. By using systolic arrays, we can solve a linear system of equations with less latency, accuracy, and cost-effective. The memory is also not large enough in FPGAs to process the large dimension matrices and also there are many I/O constraints. To overcome those issues, parallel computation elements are required known as processing elements (PE) [7]. Each PE has storage and is connected to neighboring PEs. These architectures are compatible with the VLSI design layout. Different approaches for systolic implementations of Kalman filter discussed in [8–11]. FPGA-based implementations of signal processing systems are discussed in [12]. Different matrix multiplication architectures implementation in FPGA are mentioned in [13, 14]. Tracking algorithms' implementation on different FPGAs with different models are compared and also sinusoidal signal tracking example is discussed [15].

The paper is organized as follows. Section 2 discusses problem formulation regarding the complexity of MSSKF and need of acceleration. Section 3 presents the proposed DSP architectures for acceleration. Section 3.1 explains about systolic arrays. In Sects. 4 and 5, results and conclusion are presented, respectively.

2 Problem Formulation

2.1 MSSKF

In the MSSKF approach, the state vector is augmented with measurement bias vector. We calculate the filter gain for augmented vector and then update the target state with filter gain. Let us consider M = 2 sensors and measurements from each sensors coming at same time stamp (synchronous) [1, 3]. The augmented state with bias is given by

$$x = \begin{bmatrix} x \\ b^i \\ b^j \end{bmatrix} \tag{1}$$

where i and j are the measurement provided sensor indexes. The state equation of augmented state is given by [3]

$$x_k = F_{k-1}x_{k-1} + v_{k-1} \tag{2}$$

where F is state transition matrix and v is the process noise (Fig. 1).

$$F_{k-1} = \begin{bmatrix} F_{k-1} & 0 & 0 \\ 0 & I_{k-1} & 0 \\ 0 & 0 & I_{k-1} \end{bmatrix} \tag{3}$$

normally F_{k-1} is 4×4 matrix for 2D coordinate (C = 2) position and velocity when two sensor (M = 2) bias estimations for position and velocity are stacked to state transition matrix which then results in 8×8. For M sensors and C coordinate case, F matrix is $(2C + 2M) \times (2C + 2M)$. v_{k-1} is the process noise

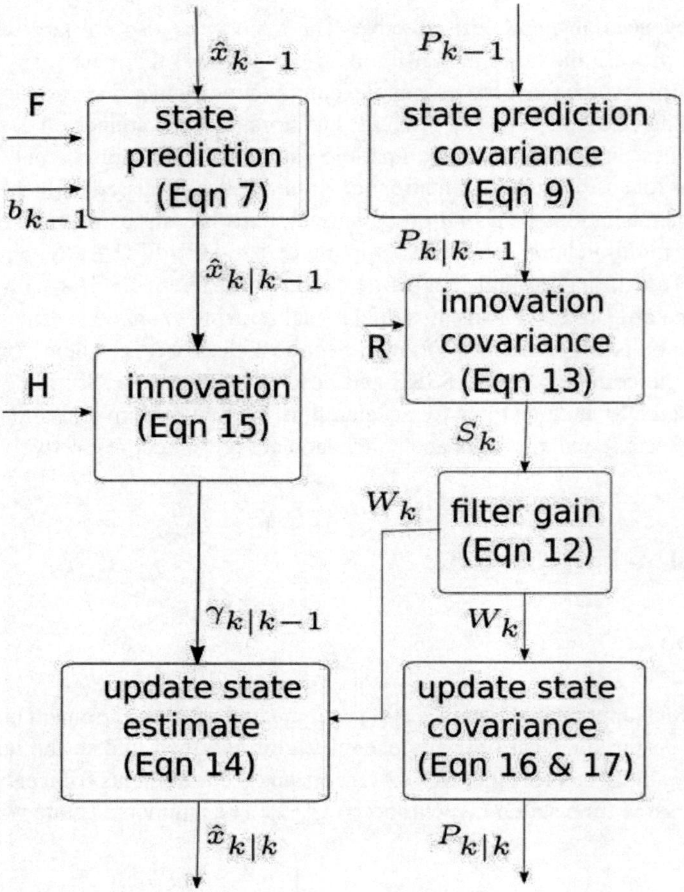

Fig. 1 Schmidt–Kalman filter flow diagram for bias estimation

$$v_{k-1} = \begin{bmatrix} v_{k-1} \\ 0 \\ 0 \end{bmatrix} \tag{4}$$

The biases are assumed to be constant here. The measurement equation is given by

$$z_k^i = H_k^i x_k + w_k^i \tag{5}$$

where H^i is the ith sensor measurement transition matrix, which is stacked with both state and bias measurement transition matrix and w is the measurement noise.

$$H_k^i = \begin{bmatrix} H_x^i \ H_b^i \ 0 \end{bmatrix}_k \tag{6}$$

The state prediction for stacked state is given by

$$\hat{\mathbf{x}}_{k|k-1} = F_{k-1}\hat{\mathbf{x}}_{k-1|k-1} + b_{k-1} \quad (7)$$

The stacked state prediction covariance is represented by

$$P_{k|k-1} = \begin{bmatrix} \{P_{xx}\}_{k|k-1} & \{P_{xb^i}\}_{k|k-1} & \{P_{xb^j}\}_{k/k-1} \\ \{P_{xb^i}\}'_{k|k-1} & \{P_{b^ib^i}\}_{k/k-1} & 0 \\ \{P_{xb^j}\}'_{k|k-1} & 0 & \{P_{b^jb^j}\}_{k|k-1} \end{bmatrix} \quad (8)$$

The covariance matrix size is (2C + 2M) × (2C + 2M) for generalized case. The prediction covariance of x_k given by

$$P_{k|k-1} = F_{k-1}\{P\}_{k-1|k-1}F'_{k-1} + Q_{k-1} \quad (9)$$

where Q_{k-1} is the process noise covariance. The prediction covariance involves higher order dimension matrix multiplication.

The optimal filter gain for updating x_k is given by

$$\left\{W^i\right\}_k^{OPT} = P_{k|k-1}\left\{H_k^i\right\}'\left\{S_k^i\right\}^{-1} \quad (10)$$

$$\left\{W^i\right\}_k^{OPT} = \begin{bmatrix} W_x^i \\ W_{b^i}^i \\ W_{b^j}^i \end{bmatrix}_k \quad (11)$$

The idea of the SKF is to use only $\left\{W_x^i\right\}_k$ Which involves with cross-covariance of the state and bias [3].

$$\left\{W_x^i\right\}_k = \left[\{P_{xx}\}_{k|k-1}\left\{H_x^i\right\}'_k + \{P_{xb^i}\}_{k|k-1}\left\{H_b^i\right\}'_k\right]\left\{S^i\right\}_k^{-1} \quad (12)$$

where the innovation covariance is

$$\begin{aligned} S_k^i = {} & \left\{H_x^i\right\}_k\{P_{xx}\}_{k|k-1}\left\{H_x^i\right\}'_k \\ & + \left\{H_x^i\right\}_k\{P_{xb^i}\}_{k|k-1}\left\{H_b^i\right\}'_k \\ & + \left\{H_b^i\right\}_k\{P_{b^ix}\}_{k|k-1}\left\{H_x^i\right\}'_k \\ & + \left\{H_b^i\right\}_k\{P_{b^ib^i}\}_{k|k-1}\left\{H_b^i\right\}'_k + R_k^i \end{aligned} \quad (13)$$

where R^i is the measurement noise covariance of the sensor i. The innovation covariance, covariance of the state vector, and cross-covariance of the state and bias vector are given in [3]. Here it is assumed that the bias from each sensor is uncorrelated. The state update is given by

$$\hat{\mathbf{x}}_{k|k} = \hat{\mathbf{x}}_{k|k-1} + \left\{W_x^i\right\}_k \gamma_k^i \tag{14}$$

where the innovation corresponding to z_k^i is

$$\gamma_k^i = z_k^i - \left\{H_x^i\right\}_k \hat{\mathbf{x}}_{k|k-1} \tag{15}$$

The updated covariance [3] and cross-covariance of the state is given by

$$\begin{aligned}\{P_{xb^i}\}_{k|k} = {} & \left[I_{n_x} - \left\{W_x^i\right\}_k \left\{H_x^i\right\}_k\right]\left\{P_{xb^i}^i\right\}_{k|k-1} \\ & - \left\{W_x^i\right\}_k \left\{H_b^i\right\}_k P_{b^i} P_{b^i}\end{aligned} \tag{16}$$

$$\{P_{xb^j}\}_{k|k} = \left[I_{n_x} - \left\{W_x^i\right\}_k \left\{H_x^i\right\}_k\right]\left\{P_{xb^j}^i\right\}_{k|k-1} \tag{17}$$

As we are stacking the biases of sensor to the state, it results in increase of the dimensions of the state vector and remaining matrices and hence leads to high latency and complexity. The complexity of the algorithm is presented in Table 1.

Table 1 Generalized augmented matrix dimensions for MSSKF (C = number of coordinates and M = number of sensors)

Symbol	Dimension	Description
x	(2C + 2M) × 1	Target state vector
F	(2C + 2M) × (2C + 2M)	State transition matrix
$\hat{x}$	(2C + 2M) × 1	State prediction matrix
z	(2C) × 1	Measurement vector
H	(2C) × (2C + 2M)	Measurement transition matrix
P	(2C + 2M) × (2C + 2M)	Prediction covariance
S	(2C) × (2C)	Innovation covariance
v	(2C + 2M) × 1	Process noise vector
w	(2C) × 1	Measurement noise vector
Q	(2C + 2M) × (2C + 2M)	Process noise covariance
R	C × C	Measurement noise covariance
P_{xx}	2C × 2C	State-to-state cross-covariance
P_{xb^i}	(2C) × M	State to sensor 1 cross-covariance
P_{xb^j}	(2C) × M	State to sensor 2 cross-covariance
W_x^i	(2C) × C	Filter gain
H_x^i	2 × (2C)	Measurement transition matrix
H_b^i	2 × (M)	Bias measurement transition matrix

3 DSP Architecture Designs

3.1 Systolic Arrays

Matrix multiplication involves in a lengthy sequence of arithmetic computations. The memory also is not large enough in FPGAs to process the high-dimensional matrices and also there are many I/O constraints. To overcome those issues, parallel computations elements are required known as processing elements (PE). Each PE has storage and is connected to neighboring PEs. The systolic array also provides an ideal layout for VLSI implementation [14].

Different types of systolic architectures are mentioned and presented in [6]. Matrix multiplication for $C_{3\times3} = A_{3\times3} \times B_{3\times3}$ is shown below. Systolic architectures consist of processing element which acts as a MAC (multiply and accumulate) that means it will take two inputs and multiply those two values, it will add that value to previously stored value and pass those two inputs to adjacent processing elements.

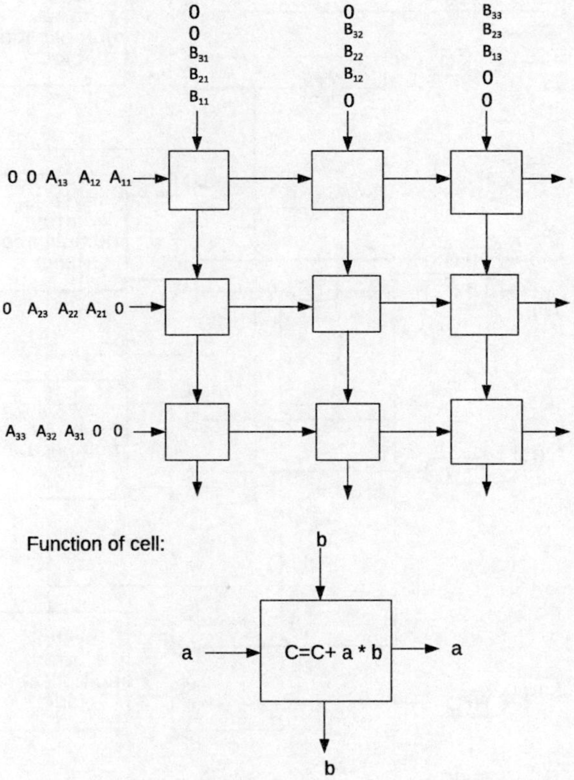

Fig. 2 3 × 3 matrix multiplication using systolic array architectures

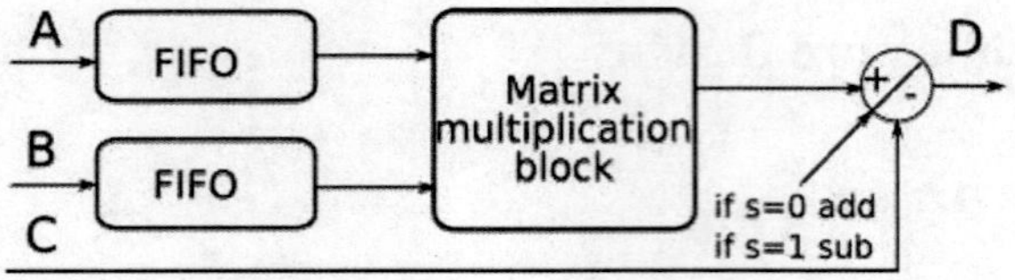

Fig. 3 Architecture 1

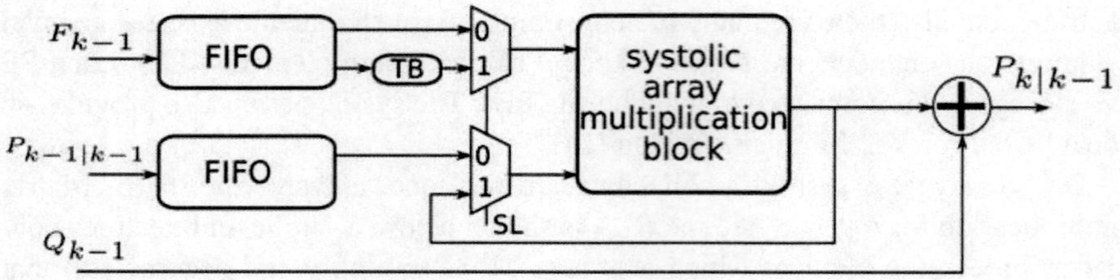

Fig. 4 Architecture 2

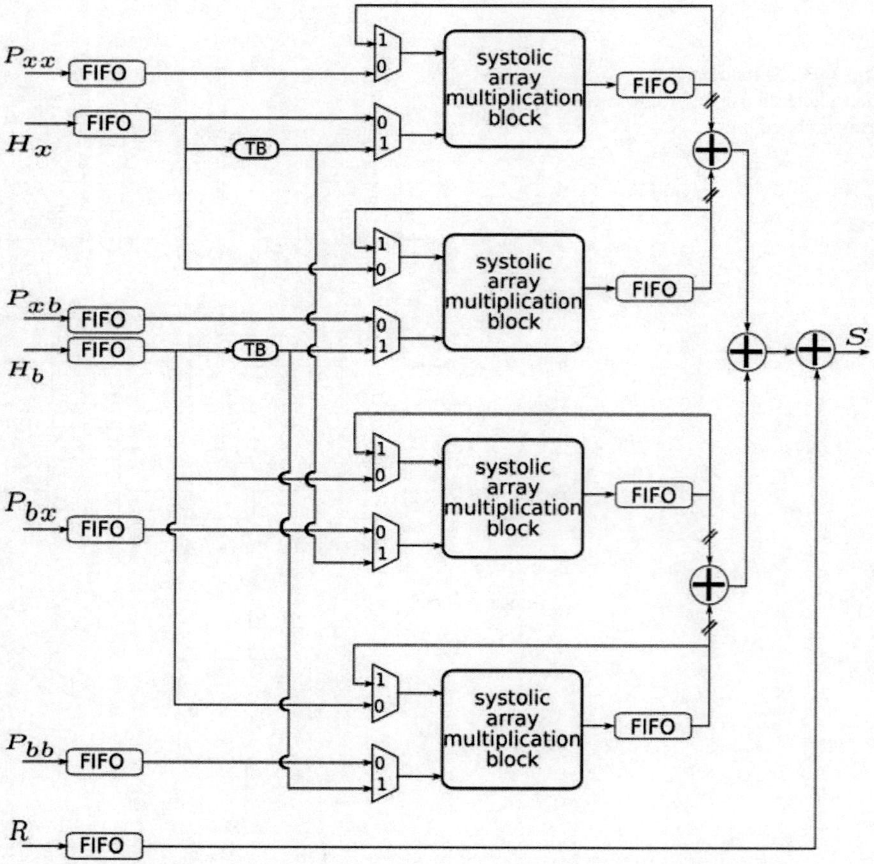

Fig. 5 Architecture 3

Table 2 Resource utilization table

Parameter	Architecture 1	Architecture 2	Architecture 3
LUT	337 (0.63%)	9467 (17.80%)	3975 (7.47%)
FF	295 (0.28%)	8387 (7.88%)	1612 (1.52%)
IO	79	75	63
BUFG	1 (3.13%)	1 (3.13%)	1 (3.13%)
DSP	–	64 (29.09%)	32 (14.55%)
$f_{\max}$ (MHZ)	176	68	74
Power (W)	0.13	0.115	0.136

3.2 *Proposed DSP Accelerators*

See Figs. 2, 3, 4 and 5.

4 Results and Discussions

In this paper, we consider C = 2 coordinates and M = 2 sensors. For those values, MSSKF algorithm has the following dimensions as shown in Table 2.

5 Conclusion

In this paper, we presented systolic-based matrix multiplication for synchronous sensor bias estimation and its hardware implementation in Zynq-7000 FPGA board. Tracking the application's latency is a very important parameter. By using systolic architectures, we can reduce the latency which leads to faster computation of the target state and bias estimation.

References

1. Taghavi E, Tharmarasa R, Kirubarajan T, Bar-Shalom Y (2013) Bias estimation for practical distributed multiradar-multitarget tracking systems. In: Proceedings of the 16th international conference on information fusion. IEEE, pp 1304–1311
2. Taghavi E (2016) Bias estimation and sensor registration for target tracking. PhD dissertation
3. Bar-Shalom Y, Willett PK, Tian X (2011) Tracking and data fusion. YBS Publishing Storrs, CT, USA

4. Sonawane D, Sutaone M, Malek I (2009) Resource efficient 64-bit floating point matrix multiplication algorithm using FPGA. In: 2009 IEEE region 10 conference, TENCON 2009. IEEE, pp 1–5
5. Bigdeli A, Biglari-Abhari M, Salcic Z, Lai YT (2006) A new pipelined systolic array-based architecture for matrix inversion in fpgas with kalman filter case study. EURASIP J Appl Sig Process 2006:7575
6. Kung H-T (1982) Why systolic architectures? IEEE Comput 15(1):37–46
7. Parhi KK (2007) VLSI digital signal processing systems: design and implementation. Wiley
8. Yeh H-G (1986) Kalman filtering and systolic processors. In: IEEE international conference on acoustics, speech, and signal processing, ICASSP'86, vol 11. IEEE, pp 2139–2142
9. Chen Z, Sun Z, Wang W (2011) Design and implementation of Kalman filter
10. Gaston F, Irwin G (1990) Systolic Kalman filtering: an overview. In: IEE Proc D-Control Theory Appl (IET) 137(4):235–244
11. Sung T-Y, Hu Y-H (1986) VLSI implementation of real-time Kalman filter. In: IEEE international conference on acoustics, speech, and signal processing, ICASSP'86, vol 11. IEEE, pp 2223–2226
12. Woods R, McAllister J, Lightbody G, Yi Y (2017) FPGA-based implementation of signal processing systems. Wiley Online Library
13. Sonawane DN, Sutaone MS, Malek I (2009) Systolic architecture for integer point matrix multiplication using FPGA. In: 2009 4th IEEE conference on industrial electronics and applications. IEEE, pp 3822–3825
14. Qasim SM, Abbasi SA, Almashary B (2008) A proposed FPGA-based parallel architecture for matrix multiplication. In: 2008 IEEE Asia pacific conference on circuits and systems, APCCAS 2008. IEEE, pp 1763–1766
15. Belaabed A, Benbouchama C (2012) On the FPGA implementation of tracking algorithms. In: 2012 20th mediterranean conference on control and automation (MED). IEEE, pp 1456–1461

Fast Encoding Using X-Search Pattern and Coded Block Flag Fast Method

S. Karthik Sairam and P. Muralidhar

Abstract High Efficiency Video Coding (HEVC) is the most promising video coding standard which helps in reducing the bit rate of the encoder by 50% compared to the H.264 Advanced Video Coding (AVC) standard. The drawback associated with the HEVC standard is the time required for encoding is very high. This problem is mainly due to the motion estimation process. In this paper, to reduce the motion estimation time, a novel X-Search pattern is proposed. Along with the search pattern the fast encoder settings are used which helps in reducing the encoder time of encoder by 43.82% compared to the TZ (Test Zone) algorithm with diamond search pattern. The proposed work also reduces the bit rate by 0.42% with small decrease in video quality which is a negligible value.

Keywords X-Search · Diamond search · Motion estimation · Encoding time · Coding unit · Bit rate

1 Introduction

In the day to day life, communication plays a key role in transmitting the data from one place to another place. The growing demand for data particularly high-quality video content leads to the requirement of high bandwidth. As the quality of the video increases, the size of the video automatically increases. To effectively utilize the bandwidth, the video should be compressed before transmitting. So HEVC [1] came into existence which compresses the high-quality videos without losing more quality. HEVC is a hybrid video coding standard which involves breaking of CTU into Coding Units (CU), motion estimation, intra prediction, transform and quantization, filtering and entropy coding. To reduce the motion estimation time, several authors have proposed different search patterns like Three Step Search (TSS) [2], logarithmic

S. Karthik Sairam (✉) · P. Muralidhar
Department of Electronics and Communication Engineering, National Institute of Technology, Warangal, Telangana, India
e-mail: karthik_sai_ram@yahoo.com

T. Laxminidhi et al. (eds.), *Advances in Communications, Signal Processing, and VLSI*, Lecture Notes in Electrical Engineering 722,
https://doi.org/10.1007/978-981-33-4058-9_24

search [3], improvements in TSS [4–6], one-dimensional full search [7], hexagonal search [8], etc. Liquan et al. [9] used a method which skips specific depths that results in decrease of encoding time by 26%. Ismail et al. [10] presented a method which selects the best PU using depth zero of residual quad tree to reduce encoding time by 14.50%. Kim et al. [11] have presented a method which uses only one reference list motion information out of two reference lists when the two lists are having same information during the bidirectional process. Lee et al. [12] have given a method which detects skip mode early. Huang et al. [13] have used the correlation between the current block and co-located block of current and reference frame to reduce the encoding time.

In this paper, a novel X-Search pattern is proposed to reduce the motion estimation time. Along with the search pattern, Coded Block Flag (CBF) Fast Method (CFM) and fast encoder settings are used which helps in decreasing the encoding time of encoder. The rest of the paper is organized as follows. Section 2 describes the related work, Sect. 3 explains the proposed work, experimental results are discussed in Sect. 4 and Sect. 5 concludes the work.

2 Related Work

In the block-based video coding standard, the Coding Tree Unit (CTU) is breakdown into CUs. The CU can have size from 8×8 to 64×64. Then the CUs will be further divided based on the complexity of texture. Maximum four depths are allowed in the HEVC, i.e., depth 0, 1, 2, and 3 which is shown in Fig. 1. The best CU among all the CUs will be determined based on the Rate Distortion (RD) cost. The RD cost will be calculated by using Eq. (1).

$$J = D + \lambda R \tag{1}$$

where
D = SAD between original and reference blocks
λ = Lagrangian multiplier
R = Number of bits need to be transmitted.

The CU with low RD cost will be considered as the best CU. The advantage of determining the best CU is for the CU with not having minimum RD cost, the checking of RD cost for the sub-splitted CUs can be skipped out. After determining the best CU, the best motion compensation prediction mode has to be determined. This can be determined based on the Motion Vector Difference (MVD) and Coded Block Flag (CBF) information. During the inter prediction process, motion estimation has to be done. To perform fast motion estimation, a fast motion vector search pattern is required. Several search patterns have been used like diamond search pattern.

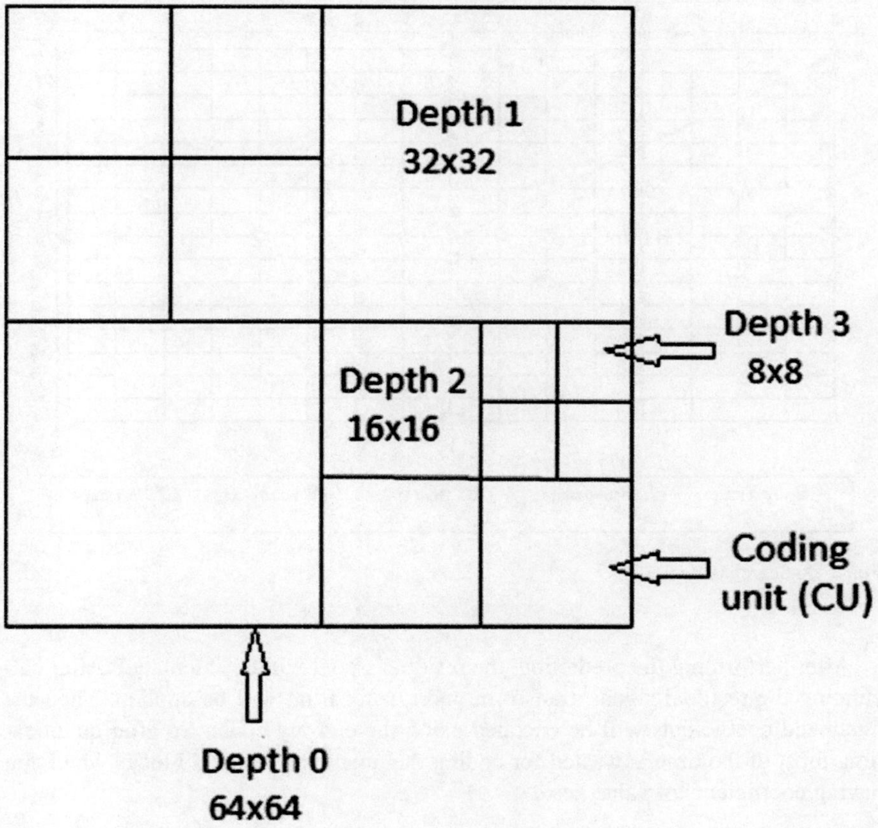

Fig. 1 CTU partitioning

During the motion estimation process, the Sum of Absolute Difference (SAD) value will be calculated. The SAD value will be calculated by using Eq. (2).

$$SAD = \sum_{i,j} \left| S_A(i, j) - S_B(i, j) \right| \tag{2}$$

where
$S_A(i, j) = (i, j)_{th}$ sample in current frame block
$S_B(i, j) = (i, j)_{th}$ sample in reference frame block.

Here both the blocks are of the same size.

In HEVC, during the calculation of the SAD value for large blocks, more number of computations are required. For example, to calculate the SAD value for 16×16 block, 256 differences and 255 additions are required which will consume a lot of time.

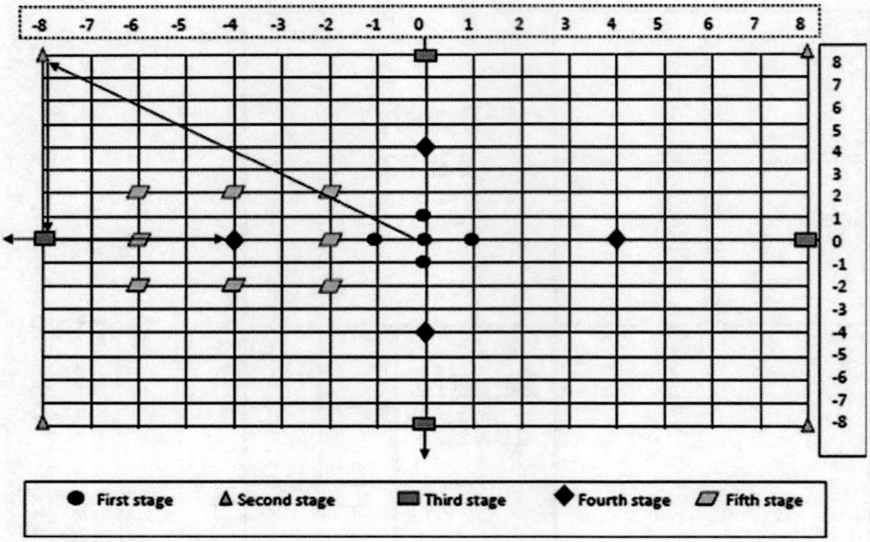

Fig. 2 X-Search pattern

After performing the prediction, the residual signal will be calculated. After calculating the residual signal, transform and quantization will be applied. Then the quantized coefficients will be encoded using the entropy coder. So after quantization, most of the time is wasted for coding the quantized residual blocks which are having coefficients of value zero.

3 Proposed Work

The proposed work includes finding of motion vector using X-Search pattern, Coded Block Flag (CBF) Fast Method (CFM) and Fast encoder settings. Each of them will be discussed below.

3.1 X-Search Pattern

In this paper, X-Search pattern is used to find the motion vector fastly during the motion estimation process. The searching process consists of five stages and each of them will be discussed below. Figure 2 represents the X-search pattern with the example directions to determine the best motion vector.

step 1: Initially search the four points around the origin. If the origin is having the minimum Block Distortion (BD) value then the search process stops and origin (minimum RD cost search point) is considered as the Best Motion Vector (BMV). Otherwise move to step 2.

step 2: Now consider the corner points at a distance of search range like second stage which was shown in Fig. 2. Find the BD value for the search points and compare with the RD cost of BMV of step 1. If the BD value of BMV of step 1 is still minimum then the search process stops, otherwise move to step 3.

step 3: Now consider the two nearby search points which are represented by '■' close to the BMV of step 2 and find the minimum BD value search point.

step 4: Now reduce the step 3 distance to half and consider the search point which is represented by 'rhombus' symbol. Then again calculate the BD value and find the minimum BD search point.

step 5: Finally apply eight search points around the BMV of step 4. Find the minimum BD search point. If the BMV of step 5 is at centre, the search process stops, otherwise step 5 repeats.

The number of search points required for the X-Search pattern is given in Eq. (3).

$$\text{Number of search points} = 9 + (m * n) + 8 \tag{3}$$

where
$m = 3$ or 5
$n =$ number of iterations.

3.2 *Coded Block Flag (CBF) Fast Method (CFM)*

After the quantization process, the status of the CBF will be checked. The flag CBF $= 0$ represents all the coefficients of residual block after quantization are zeros and hence no coding of coefficients is required. Otherwise, the coefficients coding is required.

3.3 *Fast Encoder Settings*

During the calculation of SAD values for the blocks which are having more than 8×8 size, large numbers of computations are required. This can be reduced by choosing only even number of rows from the blocks which are having size greater than 8×8. The example of frame conversion using fast encoder settings is shown in Fig. 3.

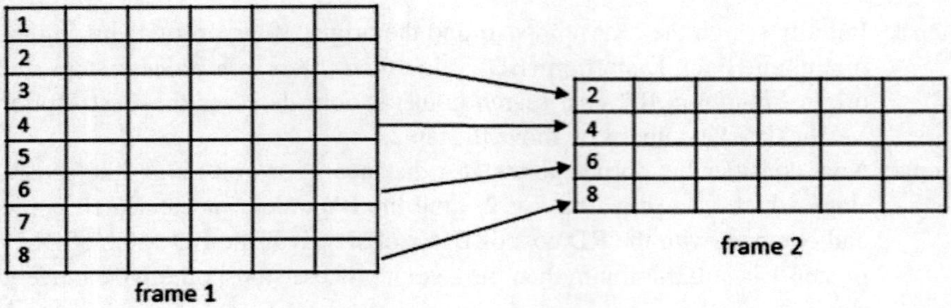

Fig. 3 Original frame to even row frame conversion

By using the fast encoder settings, for 16 × 16 frame, the number of SAD computations required are only 128 differences and 127 additions.

4 Experimental Results

In this paper, the proposed work is evaluated using HM 16.5 [14] HEVC reference software. During the evaluation random access main configuration is used and the performance is analyzed using bit rate, YPSNR, and encoding time parameters. The change in YPSNR (Δ YPSNR) gives the difference between the original and proposed YPSNR values. The change in bit rate (Δ BR) and encoding time (Δ t) represents the percentage decrease of bit rate and encoding time due to proposed algorithm compared to the original algorithm. The change in bit rate, encoding time, and YPSNR are calculated by using Eqs. (4), (5), and (6).

$$\Delta\ \text{Bitrate}\ (\%) = \frac{\text{Bitrate}_{\text{orig}} - \text{Bitrate}_{\text{prop}}}{\text{Bitrate}_{\text{orig}}} \times 100 \tag{4}$$

$$\Delta\, t\ \text{saved}(\%) = \frac{T_{\text{orig}} - T_{\text{prop}}}{T_{\text{orig}}} \times 100 \tag{5}$$

$$\Delta\ \text{YPSNR (dB)} = \text{YPSNR}_{\text{orig}} - \text{YPSNR}_{\text{prop}} \tag{6}$$

The experimental conditions required for evaluation of proposed work are shown in Table 1. In Table 2, 'orig' represents the "TZ with Diamond search pattern (TZD)" and 'prop' represents the "proposed work".

Table 2 shows the comparison results of TZD [15] and proposed work under random access configuration. Different video sequences like BQSquare, BasketballPass, and RaceHorses are used. For the BQSquare video sequence, maximum of 51.46% encoding time was saved. Similarly, for RaceHorses and BasketballPass, maximum of 39.73 and 52.73% of encoding time was saved due to the proposed work.

Table 1 Experimental conditions

Configuration	Encoder random access main
Input	Racehorses, BQSquare and BasketballPass sequences
QP	32 and 39
Max CU size	64 × 64
Max CU depth	4
Search range	64
GOP size	8
Number of frames to be encoded	50

Table 2 Comparison results of TZD and proposed method

Input	QP	YPSNR orig	YPSNR prop	Δ YPSNR (dB)	BR orig	BR prop	Δ BR (%)	Enc time orig	Enc time prop	Δ t (%)
BQSquare	32	32.18	32.13	0.05	357.09	357.03	0.01	2173.92	1249.73	42.51
	39	28.05	28.01	0.04	156.56	155.89	0.42	1780.34	864.02	51.46
RaceHorses	32	31.84	31.69	0.15	581.23	577.95	0.56	3107.89	2192.99	29.43
	39	28.05	27.91	0.14	223.31	221.92	0.62	2457.06	1480.80	39.73
BasketballPass	32	34.77	34.67	0.10	222.76	221.12	0.73	1890.57	1000.25	47.09
	39	30.87	30.79	0.08	92.94	92.76	0.19	1682.58	795.24	52.73
Average				0.09			0.42			43.82

Figure 4a, b shows the bar graphs which compares the encoding time between TZD and proposed method. The graphs show that the encoding time was decreased by large value for basketballPass sequence. On an average, the proposed work has saved the encoding time by 43.82% and also the bit rate was decreased by 0.42% with small degradation in quality.

5 Conclusions

HEVC is a standard which can provide more compression by maintaining good video quality. Even though the efficiency is increased, the time required for encoding also increased. So in this paper, a novel X-Search pattern along with CFM and fast encoder settings are used to decrease the encoding time of encoder. The proposed work has saved encoding time and bit rate by 43.82 and 0.42% compared to TZ algorithm with diamond search pattern.

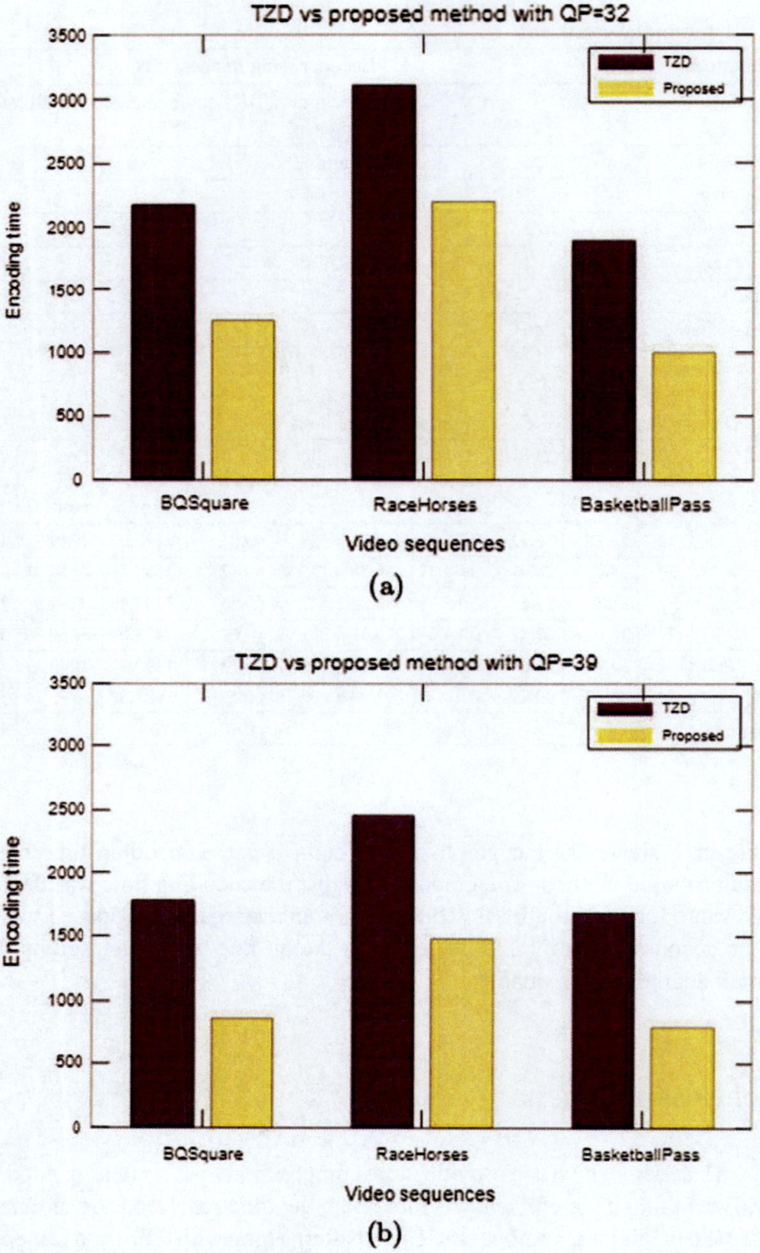

Fig. 4 TZD versus Proposed method with **a** QP = 32, **b** QP = 39

References

1. Ohm J, Sullivan GJ, Schwarz H, Tan TK, Wiegand T (2012) Comparison of the coding efficiency of video coding standards including high efficiency video coding (HEVC). IEEE Trans Circuits Syst Video Technol 22(12):1669–1684
2. Koga T et al (1981) Motion compensated interframe coding for video conferencing. In: Proceedings of networks telecommunications conference, pp G5.3.1–G5.3.5
3. Jain JR, Jain (AK 1981) Displacement measurement and its application in interframe image coding. IEEE Trans Commun COMM-29, pp 1799–1808
4. Chun KW, Ra JB (1994) An improved block matching algorithm based on successive refinement of motion vector candidates. Signal Process Image Commun 6:115–122
5. Lee LW, Wang JF, Lee JY, Shie JD (1993) Dynamic search-window adjustment and interlaced search for block matching algorithm. IEEE Trans Circuits Syst Video Technol 3:85–87
6. Li R, Zeng B, Liou ML (1994) A new three-step search algorithm for block motion estimation. IEEE Trans Circuits Syst Video Technol 4:438–442
7. Chen MJ, Chen LG, Chiueh TD (1994) One dimensional full search motion estimation algorithm for video coding. IEEE Trans Circuits Syst Video Technol 4:504–509
8. Hamosfakidis A, Paker Y (2002) A novel hexagonal search algorithm for fast block matching motion estimation. EURASIP J Adv Signal Process 595–600
9. Liquan S, Zhi S, Xinpeng Z, Wenqiang Z, Zhaoyang Z (2013) Correspondence: an effective CU size decision method for HEVC encoders. IEEE Trans Multimed 15(2):465–470
10. Ismail M, Ma J, Sim D (2014) Full depth RQT after PU decision for fast encoding of HEVC. In: IEEE international symposium on consumer electronics
11. Kim K-Y, Kim H-Y, Choi J-S, Park G-H (2014) MC complexity reduction for generalized P and B pictures in HEVC. IEEE Trans Circuits Syst Video Technol 24(10)
12. Lee H, Shim HJ, Park Y, Jeon B (2015) Early skip mode decision for HEVC encoder with emphasis on coding quality. IEEE Trans Broadcast 61(3):388–397
13. Huang X, An P, Zhang Q (2017) Efficient AMP decision and search range adjustment algorithm for HEVC. EURASIP J Image Video Process 75
14. Software reference test model HM16.5. https://hevc.hhi.fraunhofer.de/
15. Randa K, Nejmeddine B, Fatma B, Fatma S, Mohamed A et al (2016) Fast motion estimation for HEVC video coding. In: 2016 international image processing, applications and systems (IPAS), Hammamet, Tunisia

Energy Efficient and Accurate Hybrid CSA-CIA Adders

Gayathri Devi Tatiknoda, Kalyani Thummala, Aruna Kumari Neelam, and Sarada Musala

Abstract An adder can be treated as a fundamental component to perform arithmetic operations. A large number of operations can be performed by using adders such as additions, subtractions, multiplications, and divisions. In this paper two structures for hybrid CSA-CIA adder were proposed. This paper gives a comparative study of existing hybrid CSA-CIA adder and proposed hybrid CSA-CIA adders. The existing CSA-CIA hybrid adder does not work for all the combinations of inputs. The proposed designs work good for all input combinations. Also the proposed hybrid CSA-CIA adder2 has less energy and delay values compared to existing CSA-CIA hybrid adder. The code is written in Verilog hardware description language (HDL) and the simulations done by using Cadence Nclaunch tool. The layout reports are generated using Cadence Encounter tool.

Keywords Carry save adder (CSA) · Carry increment adder (CIA) · Hybrid adder · Area · Power · Delay · Verilog

G. D. Tatiknoda · K. Thummala · A. K. Neelam · S. Musala (✉)
Department of Electronics and Communication Engineering, Vignan's Foundation for Science, Technology & Research, Vadlamudi 522213, India
e-mail: sarada.marasu@gmail.com

G. D. Tatiknoda
e-mail: gayathridevit29@gmail.com

K. Thummala
e-mail: kalyanithummalaece@gmail.com

A. K. Neelam
e-mail: arunavignan4@gmail.com

T. Laxminidhi et al. (eds.), *Advances in Communications, Signal Processing, and VLSI*, Lecture Notes in Electrical Engineering 722,
https://doi.org/10.1007/978-981-33-4058-9_25

1 Introduction

In digital design, adders are very important components. Adders are used not only for arithmetic operations but also in applications like designing of filters, multiplexing, digital image processing, communication, etc. The overall circuit performance depends on individual gates of the circuit. Using small number of gates for design can increase the performance in terms of delay, area, and power. To get high speed the critical path should be as minimum as possible. Similarly to get low power less number of gates has to be used at circuit level without compromising the accuracy of the circuit. The demand for low power vlsi is increasing rapidly in mobile communication to decrease power consumption so that portability will become simple. The speed of adder plays a crucial role in deciding the overall delay of the circuit.

Adders [1–9] are extensively used in different types of very large-scale integrated circuits like central Processing Unit (CPU), Arithmetic Logic Unit (ALU), etc. In complex computations ALU and FPU play an important role. These units use adder as their basic component. By using logic gates like AND, XOR, and OR gates adder circuit is designed. To achieve high speed the propagation delay of the adder has to be decreased. So, optimizing the adder, in terms of speed, area, or power dissipation will give improved and more reliable computational unit and will lead to a better circuit performance.

To perform addition on multiple bits cascading of FAs and HAs is done. Some adders that perform n-bit addition are Ripple Carry Adder (RCA), Carry Increment Adder (CIA), Carry Look Ahead Adder (CLA), Carry Save Adder (CSA), Carry Skip Adder (CSKA), Carry Select Adder (CSLA), etc., [10]. Among all the mentioned adders RCA is simple to implement and takes large delay. For n-bit addition RCA takes delay of n times the full adder. CLA calculates the all carrys before sum so that the delay can be reduced to a good extent. But as the number of bits increases area will increase and is complex to design. And it is having a disadvantage of irregular layout. CSLA computes all the possible carrys prior to sum. So the delay may be reduced.

Carry save adder is a type of carry propagates adder which computes addition of 3 numbers. CSA can operate on 3 numbers at a time and carry is propagated through different stages [11]. In multiplications carry save adder used to calculate partial products. In this architecture carry is stored at present stage and passed to next as addend. Hence the delay can be decreased.

Compared to other adders Carry increment adder gives better performance in terms of area and delay. Carry increment adder uses RCAs and increment circuitry. The increment circuitry consists of half adder chain so that the propagation delay is reduced. Regarding power dissipation and delay Carry Increment Adder (CIA) gives good performance.

CSA-CIA combines both the advantages of carry increment adder and carry save adder. Here there is an advantage of adding three inputs at a time as well as reduced delay because of CIA. Instead of using only CSA combining CSA and CIA yields good results for delay and PDP.

In this paper, section II gives the description of existing CSA-CIA hybrid adder and its demerits, section III describes the proposed adders and how the problem with existing adder was eliminated. Section IV gives results, corresponding RTLs, layouts, and conclusion.

2 Existing CSA-CIA Hybrid Adder

The existing hybrid CSA-CIA adder [10] uses both carry save adder and carry increment adder. The architecture is shown in Fig. 1. The advantage of CSA is the ability to do operation on three inputs at a time. By using carry increment adder least propagation delay is achieved since it has increment block. So, in hybrid adder the addition is performed on three inputs.

At first stage the adders take inputs and all the full adders operate in parallel and produce sum and carry which then gets passed to the next stage with left shifting [10]. This adder splits the operands into 4-bit block. Every 4-bit block produces the sum and carry at the same time and the output carry from the first stage will transfer to the increment circuit.

Carry propagation problem with the above-proposed adder:

Example 1
For the inputs A[7:0] = 00,001,111(15),
B[7:0] = 00,001,111(15) and C[7:0] = 00,001,111(15).

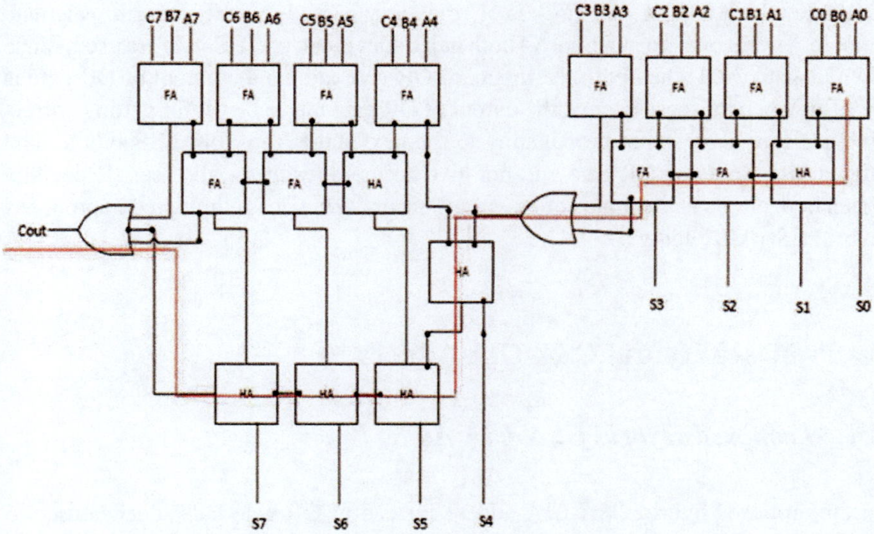

Fig. 1 Architecture of existing CSA-CIA adder[10]

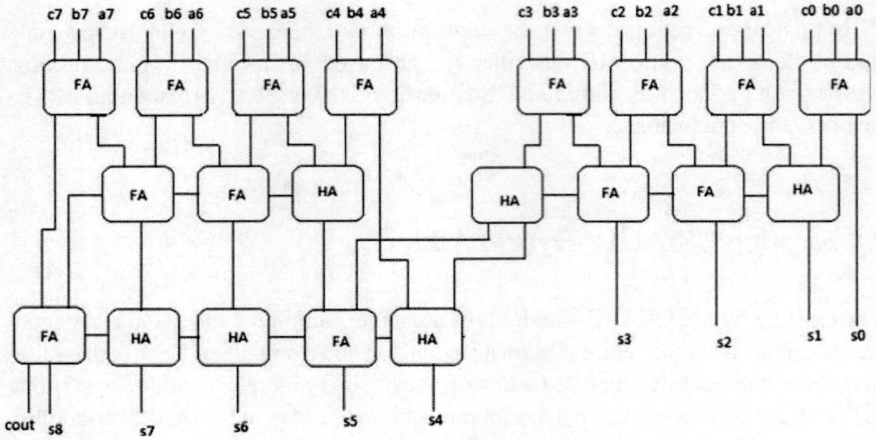

Fig. 2 Proposed hybrid CSA-CIA adder1

The expected sum is s[7:0] = 00,101,101(45).
But the actual sum is s[7:0] = 00,011,101(29).
Figure 4 shows this result.

Example 2
Similarly for the inputs A[7:0] = 00,101,100(44), B[7:0] = 00,111,100 (60) and C[7:0] = 00,111,110(62) the expected sum is s[7:0] = 010,100,110 (166). But the actual sum is s[7:0] = 010,010,110(150).

The operation of Fig. 1 can describe the above-said problem as follows. In the structure first stage contains CSA and second stage contains CIA. When output of OR gate which is after 4bit CSA is '1' the carry passed to next stages to calculate remaining sum bits. To generate S4 both half adder block and OR gate are used. Since the OR gate used when both the inputs of OR gate are 1 s the output of OR gate is '1'. But when this case occurs, the output of OR gate has to be '0' for getting correct output. Here carry is not propagating to the next stage from four LSB bits to next higher bits. So using OR gate will not give accurate result for all cases (Especially when both are 1 s at the inputs of or gate). This drawback was eliminated in proposed hybrid CSA-CIA adders.

3 Proposed Hybrid CSA-CIA Adders

3.1 Proposed Hybrid CSA-CIA Adder1

In the proposed hybrid CSA-CIA adder1 instead of OR gate half adder is used to generate S4. When both the inputs of this half adder are logic 1 's then the generated carry passed to next stages. The operation is same as existing adder, but here the carry can be passed to next stages so that it gives correct operation. (Fig. 2 shows the proposed hybrid CSA-CIA adder1).

This adder gives correct output for given combinations of inputs. Also it works well for all combinations.

3.2 *Proposed Hybrid CSA-CIA Adder 2*

To reduce the delay of Proposed hybrid CSA-CIA adder1, the proposed hybrid CSA-CIA adder2 is implemented.

Figure 3 shows the architecture of the proposed hybrid CSA-CIA adder2. Here instead of using half adder the full adder was used to generate S4 bit and one more full adder at the end to generate S8 and cout. By using full adder the problem that was present in existing hybrid CSA-CIA adder is eliminated.

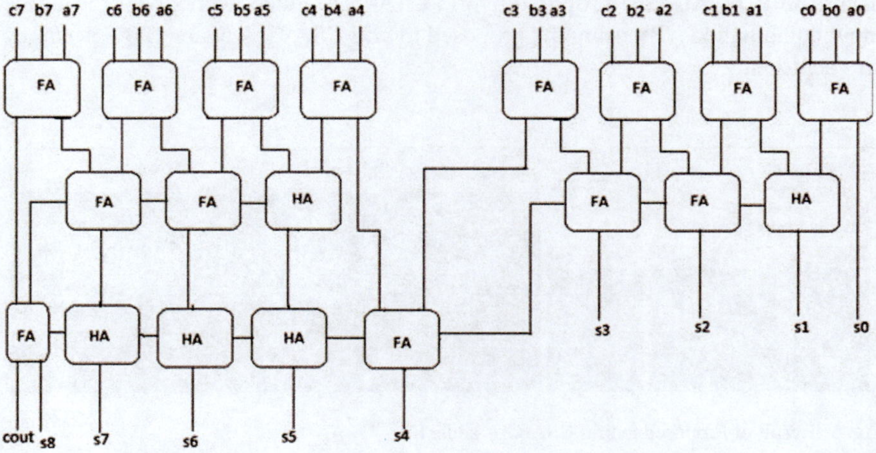

Fig. 3 Proposed hybrid CSA-CIA adder 2

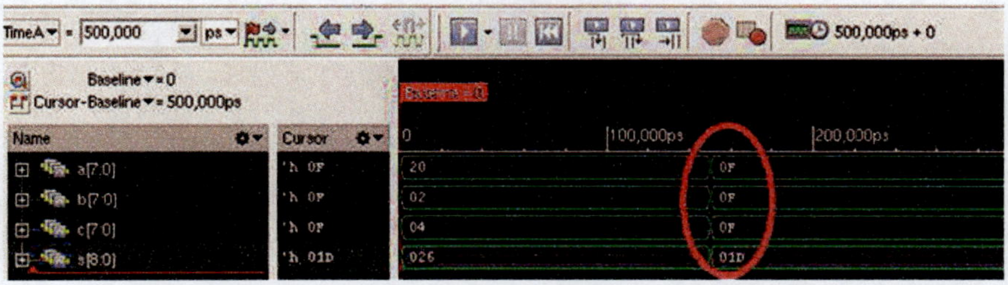

Fig. 4 Result of Existing CSA-CIA adder

Advantages of proposed hybrid CSA-CIA adder2

The proposed adder2 works for all input combinations. It also operates just like CSA and delay reduced as compared to the proposed hybrid CSA-CIA adder1. It gives better performance over proposed hybrid CSA-CIA adder1 in terms of delay and PDP.

4 Simulation Results

The simulation done by using Cadence Nclaunch tool. Figures 4, 5, and 6 show the transient responses of existing hybrid CSA-CIA adder, proposed hybrid CSA-CIA adder1, and proposed hybrid CSA-CIA adder2, respectively.

From Fig. 4 it can be observed that the output is not correct for the inputs a = 'h0F, b = 'h0F, c = 'h0F. However the problem was eliminated in Figs. 5 and 6.

The results of delay, area, and power are tabulated in Table 1, which are obtained from synthesis report. The proposed CSA-CIA hybrid adder2 got least values of delay and PDP. Also the proposed hybrid CSA-CIA adders works well for all the input combinations. By using the proposed hybrid CSA-CIA adder 100% accuracy was achieved.

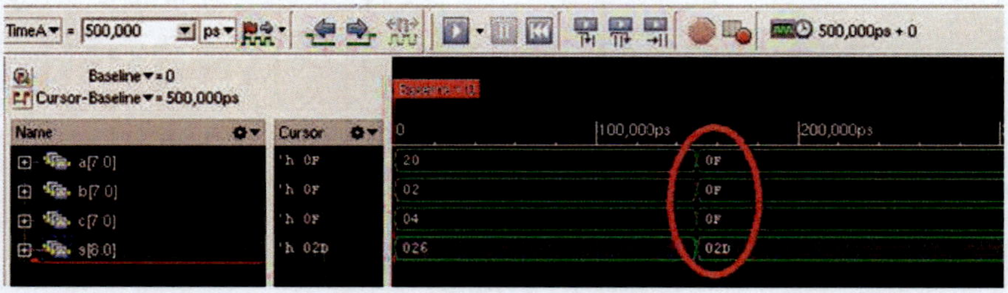

Fig. 5 Result of Proposed hybrid CSA-CIA adder1

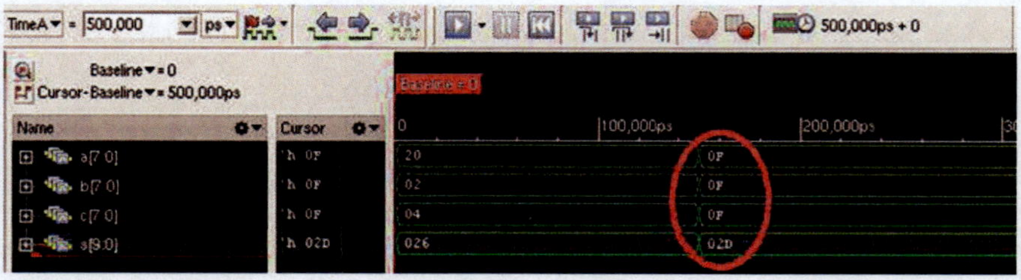

Fig. 6 Result of Proposed hybrid CSA-CIA adder 2

Table 1 Comparison between CSA, existing and proposed hybrid CSA-CIA adders

Hybrid adder	CSA	Existing CSA-CIA	Proposed CSA- CIA 1	Proposed CSA-CIA2
Delay(ns)	2.109	1.739	1.898	1.687
Area (μm^2)	299.372	319.412	348.174	336.064
Leakage power(μW)	1.357	1.631	1.607	1.600
Dynamic power(μW)	13.019	14.085	14.094	14.056
Total power(μW)	14.376	15.716	14.701	15.656
PDP	30.319	27.330	29.231	26.411
Working for all input combinations	Yes	No	Yes	Yes

Fig. 7 RTL of Existing hybrid CSA-CIA adder

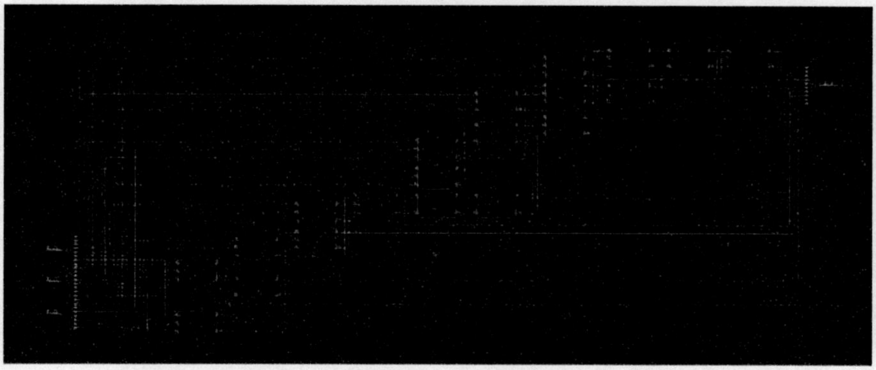

Fig. 8 RTL of Proposed hybrid CSA-CIA adder1

Figures 7, 8, and 9 show the RTLs(register-transfer level) diagram of existing hybrid CSA-CIA adder, proposed hybrid CSA-CIA adder1, and proposed hybrid CSA-CIA adder2, respectively.

The layouts are generated using Cadence encounter tool. The Figs. 10, 11, and 12 show the layouts of existing hybrid CSA-CIA adder, proposed hybrid CSA-CIA adder1, and proposed hybrid CSA-CIA adder2, respectively.

From design summary report, various areas values are tabulated in Table 2. The comparison done using parameters like Standard cell area, area of the core, area of the chip, and total length of wire.

The problem with existing CSA-CIA adder is eliminated in the proposed hybrid CSA-CIA adders. The proposed hybrid CSA-CIA adder2 got least values of delay and PDP.

Fig. 9 RTL of Proposed hybrid CSA-CIA adder2

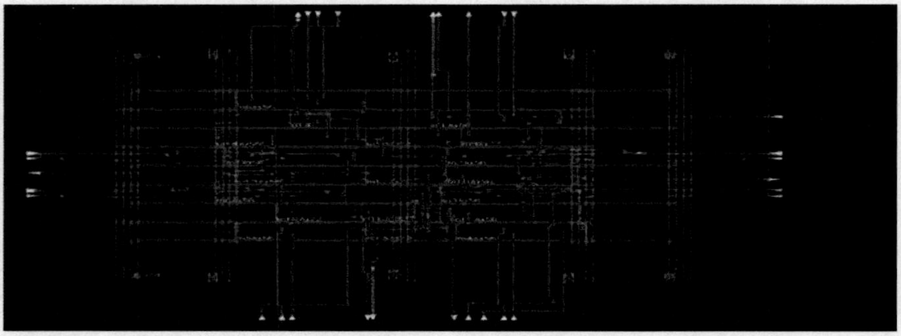

Fig. 10 Layout of Existing hybrid CSA-CIA adder

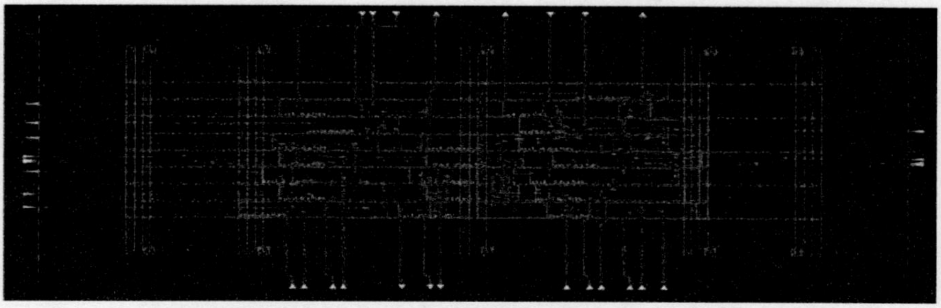

Fig. 11 Layout of Proposed hybrid CSA-CIA adder1

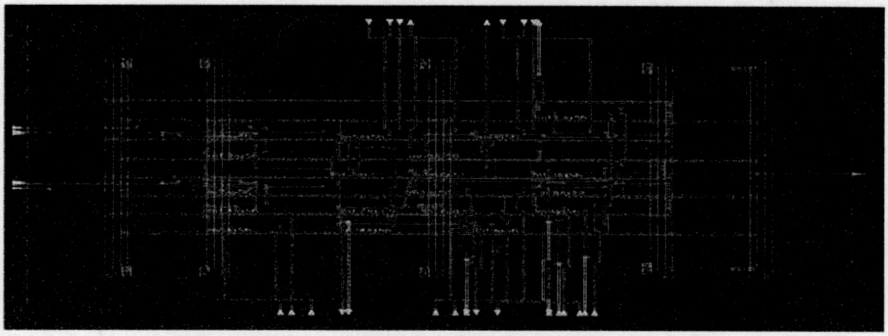

Fig. 12 Layout of Proposed hybrid CSA-CIA adder2

Table 2 Comparison of existing and proposed hybrid CSA-CIA adders

Parameter	Existing hybrid CSA-CIA	Proposed hybrid CSA-CIA adder1	Proposed hybrid CSA-CIA adder2
Total standard cell area (μm^2)	319.412	348.174	336.064
Core area (μm^2)	456.332	497.466	480.136
Chip area (μm^2)	1735.943	1817.067	1796.591
Total wire length (μm)	666.2850	748.4500	708.2700
Average wire length/net (μm)	10.2505	10.8471	11.0667

5 Conclusion

The carry propagation problem with existing CSA-CIA adder is eliminated by using proposed hybrid CSA-CIA adders. The proposed hybrid adders work for all the combinations of inputs. Also the proposed hybrid CSA-CIA adder2 got least values of delay and PDP.

The simulation, synthesis, and layout results are presented. So for accurate and fast operations the proposed adders can be used.

References

1. Hardy B, BR759875, "A Study of The Advancement of CMOS ALU & Full Adder Circuit Design For Modern Design", Orlando, FL 32816–2362. https://doi.org/10.1080/10655140290011122
2. Dubey N, Akashe S (2014) Implementation Of an arithmetic logic using area efficient carry look-ahead adder. Int J VLSI Des Commun Syst (VLSICS) 5(6):29. https://doi.org/10.5121/vlsic.2014.5604
3. Girdher A, Devi P, Singh B (2010) improved carry select adder with reduced area and low power consumption. Int J Comput Appl 3(4). https://doi.org/10.5120/723-1016
4. Salivahanan S (2012) Digital circuit and design. Fourth edition
5. Themozhi G, Thenmozhi V (1992) Propagation delay based comparison of parallel adders. J Theoret Appl Inform Technol ISSN: 1992–8645, E-ISSN: 1817–3195
6. Saradindu P, Banerjee A , Maji B., Dr. Mukhopadhyay AK (2012) Power and delay comparison in between different types of full adder circuits. Int J Adv Res Electric, Electron Instrument Eng 1(3). ISSN 2278 – 8875
7. Jacob A ppt, Department of Electrical and Computer Engineering. The University of Texas Austin. Design of adder
8. Sood L, Kaur J (2015) Comparison between various types of adder topologies. IJCST 6(1). ISSN: 09768491 (Online) | ISSN: 2229–4333 (Print), 62 International
9. Shirakol S (2014) Conference Paper. https://doi.org/10.13140/2.1.3303.9045. Design and Implementation of 16-bit Carry Skip Adder using Efficient Low Power High Performance Full Adders. https://www.researchgate.net/publication/268632532.
10. Sarkar1 S, Sarkar2 S, Mehedi J (2018) Modified CSA-CIA for Reducing propagation delay. 2018 international conference on computer communication and informatics (ICCCI -2018), Jan 04–06, 2018, Coimbatore, India https://doi.org/10.1109/ICCCI.2018.8441482
11. Sarkar S, Mehedi J (2017) Design of hybrid (CSA-CSkA) adder for improvement of propagation delay. 2017 third international conference on research in computational intelligence and communication networks (ICRCICN), Kolkata, 2017, pp 332–336. https://doi.org/10.1109/ICRCICN.2017.8234530

Machine Learning Oriented Dynamic Cost Factors-Based Routing in Communication Networks

Rohit Misra and Rahul Jashvantbhai Pandya

Abstract Increasing Internet usage leads us to the increased number of users in communication networks. This results in high data rate loading in the networks and increases the blocking. Traditional routing algorithms work mostly on the shortest path-based routing and result in higher congestion and sub-optimal network solution. Moreover, they do not consider multiple cost factors as a hybrid cost factor in finding efficient network routes. Therefore, in the current paper, we present a multiple cost factor-based routing algorithm to improve the network performance and reduce the cost. Our algorithm forms a composite cost factor considering multiple individual parameters, such as the probability of error, throughput, latency, and bandwidth. In addition, the dynamic behavior of the network leads us to the higher complexity and sub-optimal routing solution if the parameters and network situation are not continuously tracked. In order to overcome this, we apply the supervised monitoring and machine learning method in our routing to identify the optimal solution dynamically.

Keywords Dynamic routing · Machine learning · Composite cost · Probability of error · Throughput · Latency · Bandwidth

1 Introduction

As the number of users in communication networks is increasing, the traffic load on the existing networks is also growing. Therefore, traditional and static routing leads to sub-optimal network solutions. To support the next generation of communication networks, we need dynamic routing algorithms focusing on the automatic adaption based on machine learning. In the recent year following efforts have been carried out by the research community to improve the network routing.

R. Misra · R. J. Pandya (✉)
Department of Electronics and Communication Engineering, National Institute of Technology, Warangal, India
e-mail: rpandya@nitw.ac.in

R. Misra
e-mail: rohit931646@student.nitw.ac.in

T. Laxminidhi et al. (eds.), *Advances in Communications, Signal Processing, and VLSI*, Lecture Notes in Electrical Engineering 722,
https://doi.org/10.1007/978-981-33-4058-9_26

Q. Zhang et. al. presented the low latency routing algorithm considering the field location of the objects and network architecture connectivity [1]. E. Akin et. al. presented the detailed comparison of the several routing algorithms considering the static and dynamic link cost factors in software-defined networks [2]. K. Thangramya et. al. presented the energy-aware algorithm focusing on clustering and neuro-fuzzy method [3]. J. Wang et. al. presented the ant colony-based optimization approach for the routing in wireless sensor networks [4]. F. Shirazi et. al. have presented a detailed survey on the existing routing protocols for the communication networks [5]. A. N. Hassan et. al. presented distanced-based inter-vehicle connectivity aware routing for vehicle to vehicle communication networks [6]. G. Chen et. al. presented the optimal routing considering the trusted connectivity probability for multi-hop device to device communication networks [7]. J. Liu et. al. presented a wide survey summarizing the position oriented routing for the vehicle to vehicle ad hoc communication networks [8]. R. J. Pandya et. al. have demonstrated the application of least cost routing algorithm in finding the optimized network solution for the high-speed optical communication networks focusing on survivability, impairments, and power consumption [9–11].

2 Machine Learning-Based Dynamic Routing

In this section, we present our machine learning oriented dynamic cost factors-based routing algorithm. In order to find out the enhanced network solution, we considered a composite cost factor considering the multiple cost factors such as probability of error, throughput, latency, and bandwidth. Moreover, these parameters are continuously monitored and machine learning approach has been employed when identifying the optimal paths. We divide the network operation into two parts called network initialization and dynamic costing. Network initialization is essential to extract the initial network snapshot before beginning the actual network function. In order to achieve this, the test packets are sent across the entire network and machine learning oriented database is constructed. Based on this, the initial cost factors are considered. The network initialization phase is presented in Fig. 1.

Once the network has been initialized, the dynamic costing phase continuously learns the change in the network behavior and updates the cost parameters. Considering the multiple cost parameters, we can form a composite cost factor for efficient routing. This routing method significantly overcomes the imbalance caused by the single and a static cost factor-based routing. We considered the following cost parameters to form a composite cost factor: probability of error, throughput, latency, and bandwidth.

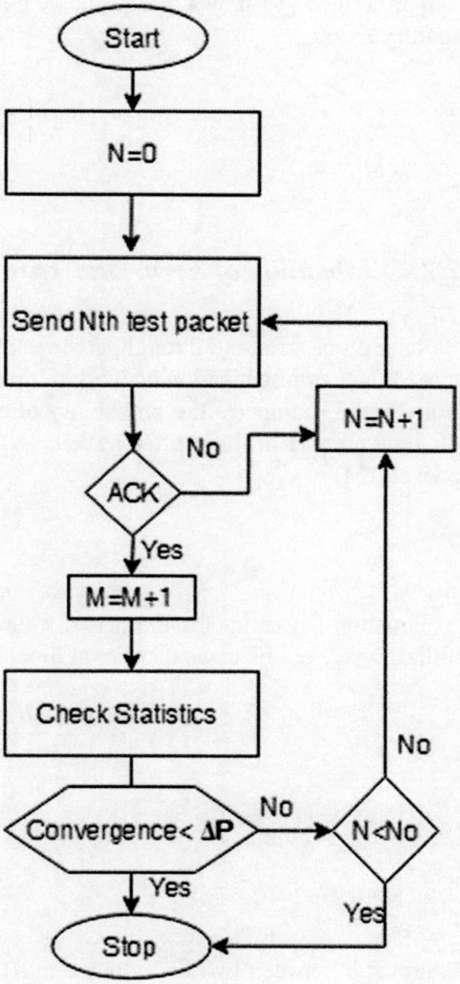

Fig. 1 Initialization Flow Chart

2.1 Latency

To achieve high speed and high data rate communication, latency is an essential network parameter to optimize. In the practical scenario, there is a threshold limit on the maximal allowable latency which is denoted by X_M. Beyond X_M the packets are dropped out of the network. As shown in Eq. (1), we collect the latencies after transmission of a certain number of packets (N_o). We find the mean of the data to obtain the average latency (X_{AVG}) of the link where X_k is the recorded latency of the k-th test packet.

$$X_{AVG} = \frac{1}{N_o}\sum_{k=0}^{N_o} X_k \tag{1}$$

Furthermore, as shown in Eq. (2), we normalize the latency, which is used in the routing action.

$$X' = \frac{X_{AVG}}{X_M} \tag{2}$$

2.2 Probability of Error and Throughput

Probability of error and throughput are considered as control performance parameters for any communication network to meet the desired Quality of Service (QoS). Equation (3) computes the probability of error based on the gathered database of the total number of transmitted packets (N_o) and received acknowledgment (ACK) packets (M).

$$P_e = 1 - \frac{M}{N_o} \tag{3}$$

Equation (4) computes the network throughput by computing the ratio of the utilized average link capacity to total link capacity.

$$T = 1 - \frac{Avg.\ utilized\ link\ capacity}{Total\ link\ caacity} \tag{4}$$

2.3 Bandwidth

Bandwidth between two nodes is the maximum amount of data per unit time that can be transmitted from one to another. The more the traffic on the link, the lesser is the available bandwidth. This factor is used in the composite costing of the link for dynamic routing of the traffic in the network. Equation (5) computes the bandwidth, where BW_M is the maximum available bandwidth of the link and BW' is the currently available bandwidth of the link.

$$BW = 1 - \frac{BW'}{BW_M} \tag{5}$$

Using the above individual cost factors, Eq. (6) computes the composite cost factor by vector forming as **P**.

$$P = \begin{bmatrix} X' \\ P_e \\ T \\ BW \end{bmatrix} \quad (6)$$

This composite cost factor is used in all the routing decisions. To define the cost function, we first introduce a coefficient matrix (**K**).

$$\mathbf{K} = [k_1 k_2 k_3 k_4] \quad (7)$$

The k_i values correspond to the coefficients of the different elements of the vector **P** in the composite cost function. Depending on the conditions in which the network is to be deployed, the coefficient matrix provides a means to adjust the priorities of the different matrix. Once **K** is fixed, it remains the same for the whole network unless changed manually. The composite cost of the link is defined as given in Eq. (7).

$$C = \boldsymbol{K}.\boldsymbol{P} \quad (8)$$

After the initialization, the routing information is sent to all the nodes and Dijkstra's algorithm is used to determine the minimum cost paths. Once the network is running, over usage of certain optimum paths leads to traffic imbalance and ultimately overloading of the optimum paths, thus, deteriorating the network's overall performance. To tackle this issue we use machine learning oriented dynamic routing. Continuous monitoring of the network parameters helps us to compute a dynamic cost C_d using the same algorithm as initialization. This effectively increases the cost of the overused links, thereby increasing the usage of the underused links.

3 Simulation and Results

The simulation was done using a C++ program. The value of the link parameters was increased for every link that was a part of the shortest path found using Dijkstra's algorithm to simulate the traffic dependent dynamic costing.

Pseudocode:

1. Consider the network topology as shown in Fig. 2.
2. Assign the cost factors (probability of error, throughput, latency, and bandwidth) for each link.
3. Add the cost parameters using the terrain factors as weights to get the composite cost matrix. (C = K.P).
4. Find the composite cost-based shortest paths using the Dijkstra's algorithm for the complete network.
5. Increase the cost of every link proportional to the traffic load on it.

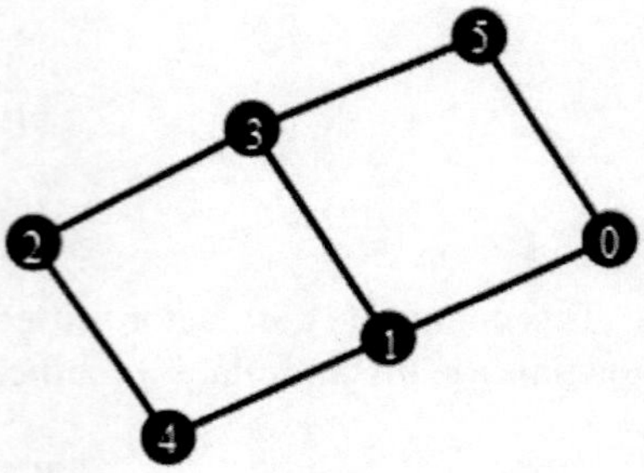

Fig. 2 Network topology

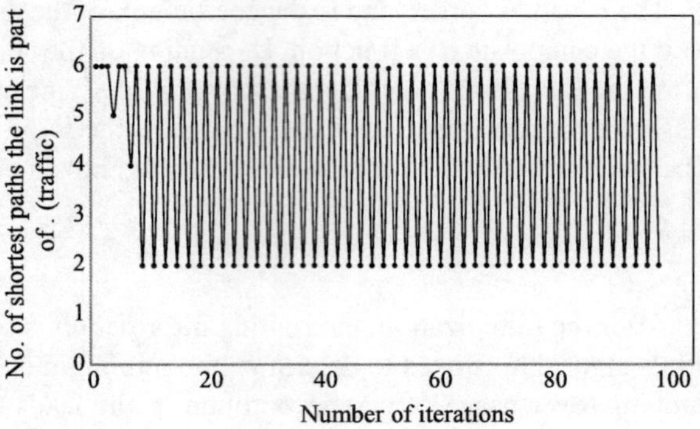

Fig. 3 Evolution of cost of link 0 → 1

6. Go back to Step 4 and repeat 100 times.

The links with the highest traffic have a higher chance of being the backbone of the network. Using the proposed algorithm we aim to stop the backbone from solidifying to a particular set of links. A decrement in the rate of change of link cost shows the effective reduction of traffic on the link. The number of shortest paths a given link is part of being a fair representation of the traffic present on the link.

Figure 3 shows the traffic status over a sample link (0 → 1). Initially, it is observed that link (0 → 1) is carrying very high traffic. Later, the dynamic costing-based routing employing a machine learning approach has segregated the traffic on other optimal paths and distributed the network load reducing the probability of error and increasing throughput.

Figures 4 and 5 show clearly that the proposed algorithm intelligently redistributes the traffic load by continuous monitoring of the link parameters over the iterations.

4 Conclusion

In this paper, we proposed a novel strategy of machine learning oriented dynamic costing-based routing algorithm for communication networks. The first objective was to incorporate multiple cost factors in the routing decisions so as to optimize

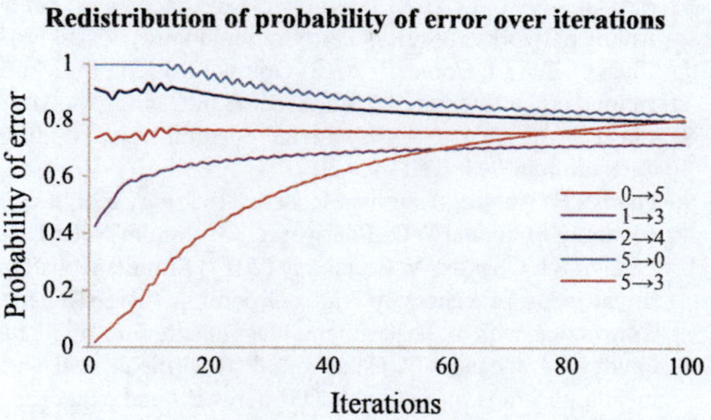

Fig. 4 Redistribution of the probability of error over iterations

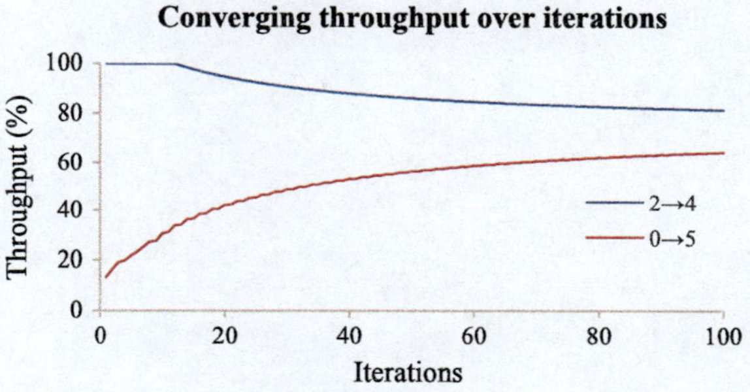

Fig. 5 Redistribution of throughput over iterations

the network. This was achieved by using a composite cost-based scheme which analyzes a database of various parameters and calculates a composite cost incorporating important parameters like the probability of error, throughput, latency, and bandwidth. The second objective was to intelligently and dynamically redistribute the traffic load in the network by continuous monitoring of the links to eliminate backbone formation in the network. As observed in the simulations, our objectives were met using the proposed novel routing algorithm.

References

1. Zhang Q, Jiang M, Feng Z, Li W, Zhang W, Pan M (2019) IoT enabled UAV: network architecture and routing algorithm. IEEE Int Things J 6(2):3727–3742
2. Akin E, Korkmaz T (2019) Comparison of routing algorithms with static and dynamic link cost in SDN. In: Proceedings of 16th IEEE annual consumer communications & networking conference (CCNC), pp 1–8, (2019)
3. Thangramya K et al (2019) Energy aware cluster and neuro-fuzzy based routing algorithm for wireless sensor networks in IoT. Comput Netw **151**:211–223
4. Wang J et al (2018) An improved Ant Colony Optimization-based approach with mobile sink for wireless sensor networks. J Super Comput 74(12):6633–6645
5. Shirazi F et al (2018) A survey on routing in anonymous communication protocols. ACM Comput Surv 51(3)

6. Hassan AN et al (2018) Inter vehicle distance based connectivity aware routing in vehicular Adhoc networks. Wireless Pers Commun 98(1):33–54 (2018)
7. Chen G, Tang J, Coon JP (2018) Optimal Routing for Multihop Social-Based D2D Communications in the Internet of Things. IEEE Int Things J 5(3):1880–1889
8. Liu J et al (2016) A survey on position-based routing for vehicular ad hoc networks. Telecommun Syst 62(1):15–30
9. Pandya RJ (in press) survivable virtual topology search with impairment awareness and power economy in optical WDM networks. J Commun Netw Distribut Syst
10. Pandya RJ, Chandra V, Chadha D (2014) Simultaneous Optimization of Power Economy and Impairment Awareness by Traffic Grooming, Mixed Regeneration, and All-Optical Wavelength Conversion with an Experimental Demonstration. IEEE J Lightw Technol 32(24):4166–4177
11. Pandya RJ, Chandra V, Chadha D (2014) Impairment-aware routing and wavelength assignment algorithms for optical WDM networks and experimental validation of impairment aware automatic light-path switching. Opt Switch Netw 11(A):16–28 (2014)

Design of Look-Up Table with Low Leakage Power Enhanced Stability 10 T SRAM Cell

R. Manoj Kumar and P. V. Sridevi

Abstract Leakage Power and cell stability are the most salient parameters that draw major attention while designing the SRAM cell. So, a new SRAM topology is proposed which reduces the Leakage power further and enhances the stability. Leakage Power is calculated and compared with existing SRAM cells and the Proposed Low Leakage Power Stable 10 T SRAM(P10T) cell. It is evaluated by making SRAM cell to operate in the idle mode of operation. There is a reduction of 29.068%, 23.23%, 39.30%, 22.58% leakage power when storing 1 in P10T SRAM cell compared with 6 T,8 T,8TG,9 T cells at TT Corner, respectively, at 0.9 V supply voltage. There is a reduction of 2.40%, 11.69%, 16.49%, 3.39% leakage power when storing 0 in P10T SRAM cell compared with 6 T, 8 T, 8 TG, 9 T at TT Corner, respectively, at 0.9 V supply voltage. Other design metrics related to the stability of the SRAM cell like HSNM, RSNM, and Write Margin are also calculated and compared. There is an increase of 19.40%, 18.71%, 19.40%, 18.71% in HSNM of P10T SRAM cell compared to 6 T, 8 T, 8 TG, and 9 T SRAM cell, respectively, at TT corner. There is an increase of 130.425%, 18.73%, 262.69%, 18.73% RSNM for the P10T SRAM cell compared to 6 T, 8 T, 8 TG, and 9 T SRAM cells, respectively, at TT corner. During Write 1, there is an increase of write margin by 29.03%, 28.61%,28.61% at TT corner at 0.9 V supply voltage. The impact of Process, Voltage and Temperature (PVT) variations on Leakage Power, HSNM, RSNM and Write Margin are evaluated. Monte Carlo Analysis with 2000 samples is performed on Leakage Power, which shows less variability for the P10T SRAM cell while storing 0. 4-input Look-Up table is implemented using 6 T and P10T SRAM cell. The LUT using P10T SRAM cell decreases the Leakage power by 8.58% than the 6 T SRAM cell. All SRAM cells are implemented in 45 nm CMOS Technology using Cadence Virtuoso.

Keywords Leakage power · PVT variations · Write margin · RSNM · LUT

R. M. Kumar (✉) · P. V. Sridevi
ECE Department, AUCE(A), Andhra University, Visakhapatnam, Andhra Pradesh, India
e-mail: manojkumar.rongali@gmail.com

P. V. Sridevi
e-mail: pvs6_5@yahoo.co.in

T. Laxminidhi et al. (eds.), *Advances in Communications, Signal Processing, and VLSI*, Lecture Notes in Electrical Engineering 722,
https://doi.org/10.1007/978-981-33-4058-9_27

1 Introduction

As there is rapid development in Very Large Scale Integration, the technology nodes are scaling at a faster pace which raises concern about the leakage power and the stability of the portable electronic devices [1]. SRAM is used as an embedded cache in these portable devices to improve the performance. Also, these portable devices use FPGA which use SRAM cell for storage purpose. One of the building blocks of FPGA is the Look-Up Table (LUT). LUT is just like a truth table. It gives a specific output for a particular input combination. This LUT uses SRAM for the storage purpose. So, reducing the leakage power of SRAM decreases the leakage power of LUT which further brings down the overall leakage power in the FPGA [2]. The on-chip computations in wireless sensor network also need SRAM cells due to their faster speed. Many times, majority of the SRAM cells will be present in standby mode. Due to which, its need of the hour to reduces the leakage currents in idle or standby mode which increases the battery life of the electronic portable devices like sensor networks, biomedical equipment, digital systems using FPGA as basis [3]. In addition to that, with the steep increase in Internet usage globally, the high leakage power in IoT Devices drive towards low leakage designs [4]. At present IoT has become more popular as it allows many electronics devices to connect and communicate with each other. The System-On chip block in IoT consists of many building blocks like on-chip memories which uses SRAM. Low Leakage is required in SRAM as it is used as a memory element in wearable devices. The Leakage current is responsible for consuming more than 40% of active energy in high-end processors [5, 6]. So, it is of a great value to reduce the leakage power which would enhance the battery life of the electronic devices.

Various approaches have been used earlier which try to improve one design metric at the expense of the other. Asymmetric Sizing [7] enhances the stability of SRAM cell but at the cost the increased leakage currents due to increased sizing of the transistors. The usage of high-Vth devices [8] for SRAM array reduce the leakage but it is more prone to process variations. Low leakage is obtained in [9] but due to the presence of the stacked transistor, there is more probability for the SRAM cell to move into metastable state. In [10], 4 T read port is used to reduce the leakage but at the cost of increased area. In [11], the threshold voltage is decreased using reverse short channel effect to improve the performance but with increased area and leakage currents.

Existing SRAM Topologies are dealt in Sect. 2. Proposed Low Leakage Stable 10 T SRAM Cell and its operation is explained in Sect. 3. Section 4 contains comparison of Leakage power, HSNM, RSNM, and WM for the Existing SRAM Cells and the P10T SRAM Cell with PVT variations. Section 5 details the 4-input LUT leakage power using 6 T and P10T SRAM cell. Section 6 concludes.

2 Existing SRAM Topologies

6 T SRAM cell is a conventional design and has been the standard in the industry for long which is shown in Fig. 1. MN1-PM1, MN2-PM2 are the two cross-coupled inverters forming the basis for holding the data at the storage nodes Q, QB using positive feedback. Two access transistors MN4, MN5 are used to carry out read and write operations using the Word Line (WL). In read mode of operation, both Bit Line (BL) and Bit Line Bar (BLB) are Precharged to V_{DD} and basing on the content of the cell, either of the Bit Lines is discharged by enabling WL. But as the read current flows into the storage nodes Q or QB, stability in read mode is very less which further declines with the reduction in the V_{DD}. To overwrite the content of the cell, the new Data and its complement is placed on Bit Lines by enabling WL. In the hold mode, WL is disabled and the SRAM cell holds the content. Cell Ratio (CR or β) is usually made between 1.2 and 3 related to read operation and Pull Up Ratio (PR) is usually made <1.8 related to write operation. CR is the ratio of the $W_{pull\text{-}down}$ to W_{access} transistor and PR is the ratio of the $W_{pull\text{-}up}$ to W_{access} transistor.

To increase the stability during the read operation in the conventional 6 T SRAM cell, the read current flowing into the storage nodes is isolated through a separate read port using 8 T SRAM Cell [12]. In Read mode, Read Bit Line (RBL) is Precharged to V_{DD} and Read Word Line (RWL) is made Logic High. Basing on the content of the cell, RBL either stays at V_{DD} or discharges. 8 TG SRAM Cell [13] has a similar structure of 6 T SRAM cell but with additional PMOS transistors used with NMOS access transistors but however, it suffers from high leakage and reduces RSNM. For the further decrease of Leakage power in 8 T SRAM Cell, 9 T SRAM cell [14] implements stacking of transistors in the read path. Other than the read port, as the structure is the same in remaining SRAM cells except for 6 T and 8 TG, they operate in an identical manner in the hold and write operations. The HSNM, RSNM, and WM determine the stability of the SRAM cell in Hold, Read, and Write modes, respectively.

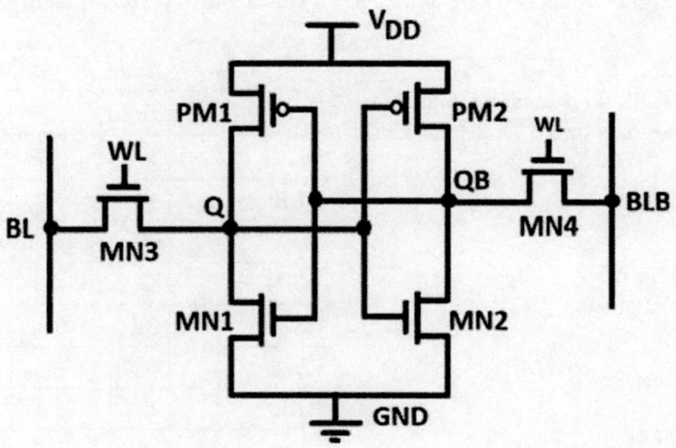

Fig. 1 Conventional 6 T SRAM Cell

3 Proposed Low Leakage Power Stable 10 T SRAM Cell(P10T)

Despite having the above SRAM Topologies, there is still a need for further reduction of Leakage Power and increase the stability in more than one mode of operation. So, a new 10 T SRAM Cell is proposed which would reduce the leakage power with enhanced stability for further use in the portable electronic devices. Figure 2. shows the Proposed Low Leakage Power Stable 10 T SRAM cell. A Separate read port is used to enhance the RSNM which would be equivalent to HSNM. The status of the control signals is shown in Table 1.

3.1 Hold Mode of Operation

In Hold mode, control signals BL, BLB, WLB, RBL are enabled and WL, CS, RWL are disabled. The absence of pull-down transistor in the right side of the inverter,

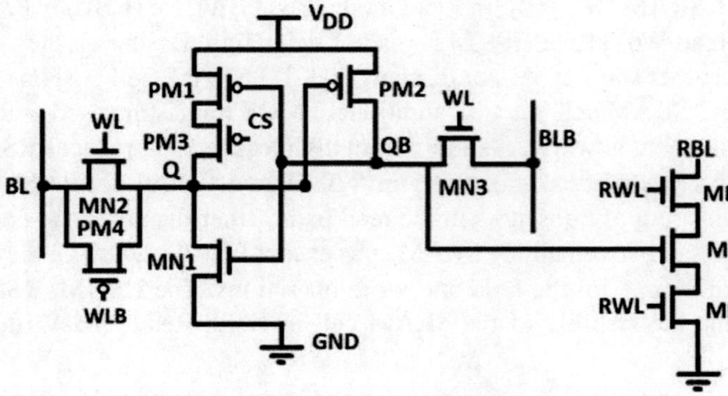

Fig. 2 Proposed Low Leakage Power Stable10T SRAM Cell(P10T)

Table 1 Status of Control Signals for P10T SRAM Cell

	Hold	Read	Write 1/0
BL	1	1	1/0
BLB	1	1	0/1
WL	0	0	1
WLB	1	1	0
CS	0	0	0/1
RWL	0	1	0
RBL	1	1	0

however, is not going to effect the storage values in hold mode as CS is kept at logic 0, PM1, PM3, and MN1 hold the data already written.

3.2 Read Mode of Operation

During the read mode of operation, WL, WLB are kept at logic 0 and 1, respectively. RWL is enabled which turns ON MN4, MN6, and RBL is precharged to V_{DD}. If the data stored is Logic High, i.e., $Q = V_{DD}$ and $QB = 0$, then MN5 is OFF which doesn't provide any discharging path for RBL. So RBL stays at V_{DD}. If the data stored is Logic Low, i.e., $Q = 0$, $QB = V_{DD}$, then MN5 is ON which provides a discharging path for RBL through MN4-MN5-MN6 to ground. The sense amplifier senses the values of RBL and gives the output accordingly in both of the above cases. The isolated read port provides independent read, write operations, and device sizing. The isolation of the cell current through Q, QB provides RSNM equivalent to HSNM.

3.3 Write Mode of Operation

During write operation, read port is disabled by making $RWL = 0$. WL, WLB are enabled which turn ON the access transistors MN2, MN3, and PM4. When the data stored is Low ($Q = 0$, $QB = V_{DD}$), To write 1 in the cell, CS is kept at 0, BL and BLB are kept at V_{DD} and 0, respectively. The storage node Q is charged using BL, VDD, and QB is discharged to 0 through MN3 which performs write 1 to happen. When the data stored is High ($Q = V_{DD}$, $QB = 0$), To write 0 in the cell, CS is kept at V_{DD}, BL and BLB are kept at 0 and V_{DD}, respectively. The storage node Q is discharged using MN2, PM4 to 0 and QB is charged to V_{DD} through MN3 which performs write 0 to happen. As CS turns OFF PM3, write margin is enhanced.

4 Results

The comparison of the SRAM cells is done based on Leakage Power, HSNM, RSNM and WM with Process, Voltage, and Temperature (PVT) variations in CMOS 45 nm technology using cadence virtuoso.

4.1 Leakage Power

Many SRAM cells stay idle in Hold mode which contributes a major part of power consumption. The product of Leakage current through power supply and V_{DD} gives the Leakage Power [15, 16]. WL and RWL are kept at Logic 0, WLB is kept at Logic 1 while calculating the leakage power. The impact of PVT variations has been studied. The Leakage power of the proposed P10T SRAM cell has been reduced because of the stacked PMOS transistor PM3 in the left inverter and absence of pull-down transistor in the right inverter.

The Leakage Power is compared for existing SRAM Cells and P10T SRAM cell which is shown in Fig. 3 at all process corners at 0.9 V supply voltage. There is a reduction of 29.068%, 23.23%, 39.30%, 22.58% leakage power when storing 1 in P10T SRAM cell compared with 6 T,8 T,8 TG,9 T cells at TT Corner, respectively, at 0.9 V supply voltage. There is a reduction of 2.40%, 11.69%, 16.49%, 3.39% leakage power when storing 0 in P10T SRAM cell compared with 6 T,8 T, 8 TG,9 T at TT Corner, respectively, at 0.9 V supply voltage. P10T SRAM cell consumes less

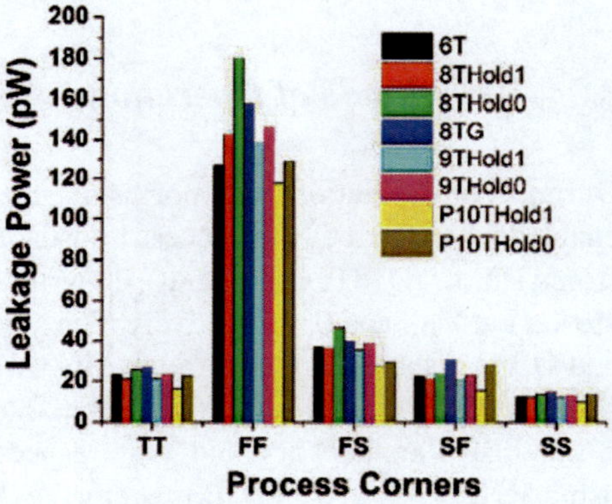

Fig. 3 Leakage Power Comparison for Existing SRAM cells with the P10T SRAM cell at all process corners at 0.9 V supply voltage

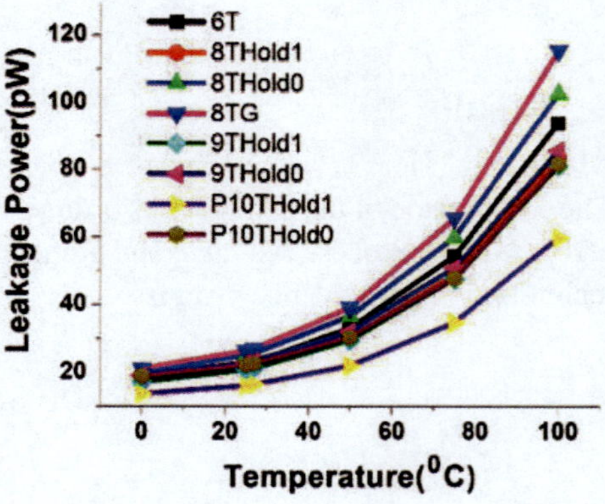

Fig. 4 Leakage Power Comparison for Existing SRAM cells with the P10T SRAM cell over wide range of temperatures at 0.9 V supply voltage

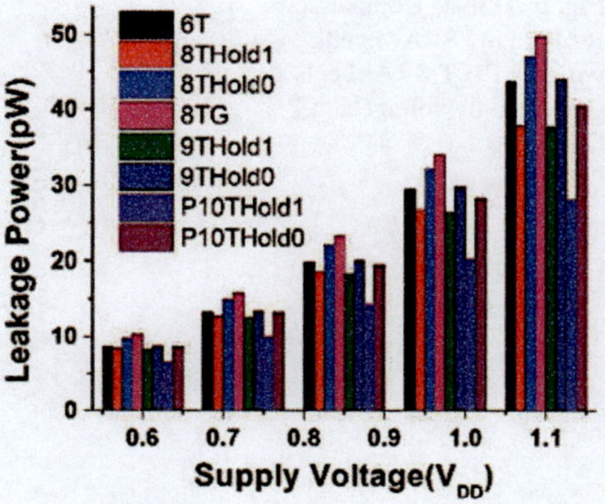

Fig. 5 Leakage Power Comparison for Existing SRAM cells with the P10T SRAM cell over different supply voltages

Table 2 Comparison of Means and Standard deviations for Leakage Power(pW) of different SRAM cells at 0.9 V supply voltage, TT Corner

	μ	σ
6 T	25.54	4.866
8 T Hold 0	24.11	4.73
8 TG	30.082	4.341
9 T Hold 0	25.95	5.117
P10T Hold 0	23.7892	1.98
P10T Hold 1	18.23	3.99

leakage power at different temperatures shown in Fig. 4 and over different supply voltages shown in Fig. 5.

Monte Carlo analysis has been done for the leakage power for 2000 samples with sigma = 6. Table 2 lists the respective means and standard deviations of different SRAM cells. P10T SRAM cell storing 0 has less variability compared to others.

4.2 Hold Stability

HSNM is used to describe the stability in Hold Mode of operation for SRAM. The voltage at one storage node is linearly increased and its effect on the other storage node is captured and vice versa. The largest side of the square in the smaller lobe of the curve gives HSNM. As the structure of 6 T, 8 T, 8 TG, and 9 T SRAM cells in hold mode is symmetrical, the HSNM curve forms the butterfly curve. But the P10T SRAM cell due to its asymmetric structure doesn't form a butterfly curve. So, there is an improved stability in hold mode compared to 6 T, 8 T, 8 TG SRAM cell as shown in Fig. 6. 8 T and 9 T have similar HSNM values due to the similar device sizing and structure in hold mode.

The effect of process corners on HSNM is shown in Fig. 7. There is an increase of 19.40%, 18.71%, 19.40%, 18.71% in HSNM of P10T SRAM cell compared to

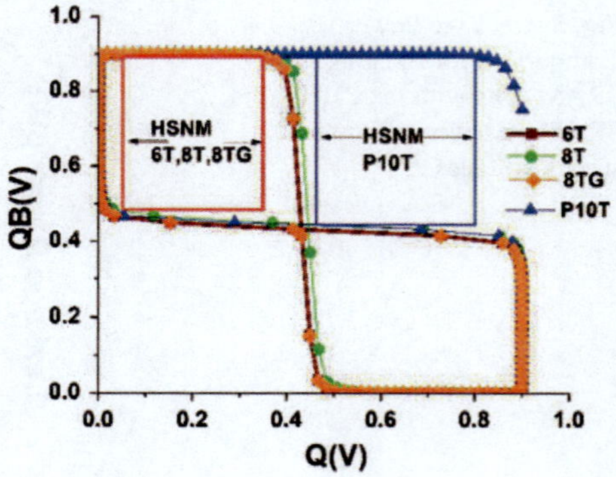

Fig. 6 HSNM Comparison for Existing SRAM cells with the P10T SRAM cell at 0.9 V supply voltage at TT Corner

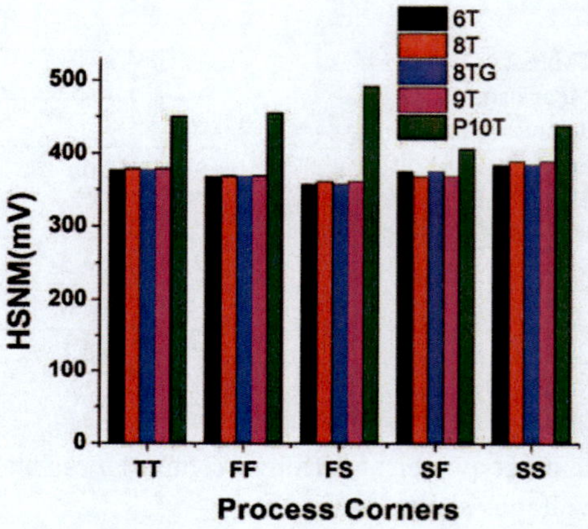

Fig. 7 HSNM Comparison for Existing SRAM cells with the P10T SRAM cell at all process corners at 0.9 V supply voltage

6 T, 8 T, 8 TG, and 9 T SRAM cell. So, P10T SRAM cell has higher stability in hold mode.

The effect of different voltages and different temperatures on HSNM is shown in Figs. 8 and 9, respectively. HSNM of the P10T SRAM cell is high compared to all SRAM cells at various voltages and at different temperatures.

4.3 Read Stability

Read stability is determined using the metric RSNM. It is calculated by using the butterfly curve formed by the inverter VTC's when SRAM cell is operating in the read mode. The Largest side of the square which can be inscribed in the butterfly curve of smaller lobe gives RSNM [17, 18].

Since there is a separate read port in P10T similar to 8 T and 9 T, its RSNM would be equivalent to HSNM. RSNM comparison of different SRAM cells is shown in Fig. 10. The RSNM at different process corners is shown in Fig. 11. There is an

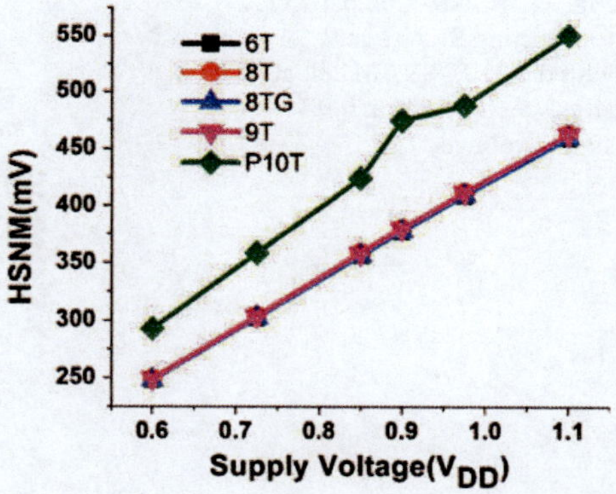

Fig. 8 HSNM Comparison for Existing SRAM cells with the P10T SRAM cell at different supply voltages

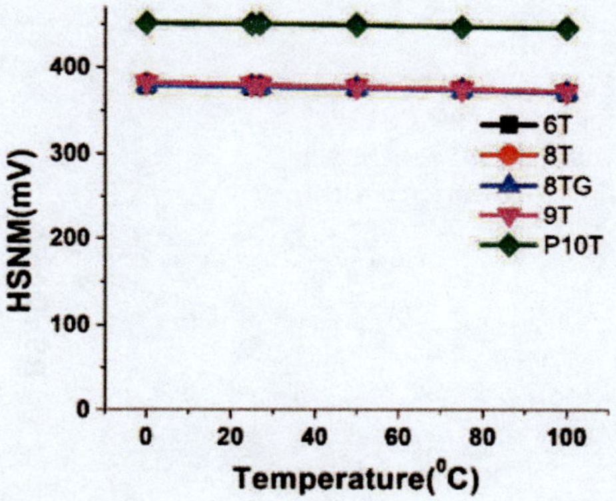

Fig. 9 HSNM Comparison for Existing SRAM cells with the P10T SRAM cell over wide range of temperatures at 0.9 V supply voltage

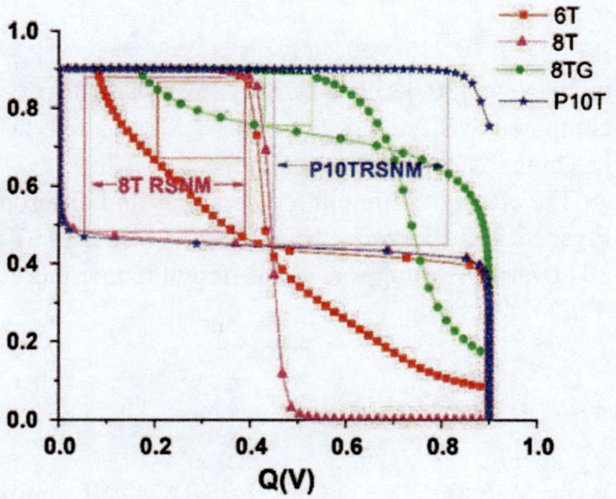

Fig. 10 RSNM Comparison for Existing SRAM cells with the P10T SRAM cell at 0.9 V supply voltage at TT Corner

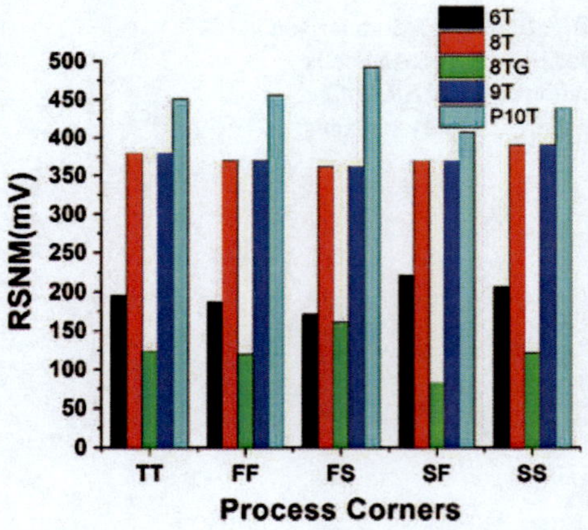

Fig. 11 RSNM Comparison for Existing SRAM cells with the P10T SRAM cell at all process corners at 0.9 V supply voltage

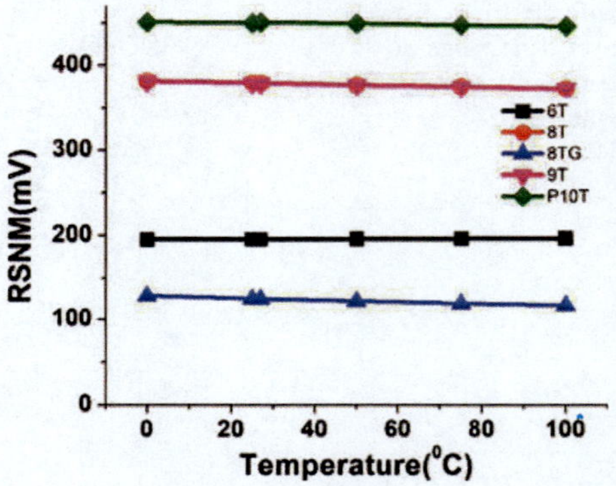

Fig. 12 RSNM Comparison for Existing SRAM cells with the P10T SRAM cell over wide temperature ranges

increase of 130.425%, 18.73%, 262.69%, 18.73% RSNM for the P10T SRAM cell compared to 6 T, 8 T, 8 TG, and 9 T SRAM cells at TT Corner. So, P10T SRAM cell has higher stability in read mode.

The effect of different temperatures and different voltages on RSNM is shown in Figs. 12 and 13, respectively. RSNM of the P10T SRAM cell is high compared to all at various voltages and at different temperatures.

4.4 Write Stability

Write Margin (WM) defines the SRAM cell stability in write mode of operation. It can be calculated by the word line sweep method [19]. In Write mode of operation, WL is swept from Low to High. The difference between the point where the storage nodes trip and V_{DD} gives the Write Margin. The Write 0 margin has been highest, as the CS = 1 turns off PMOS PM3 which disconnects the path from Q to V_{DD}. During write 0, there is an increase of 281.3%, 280.19%, 55.05%, 280.19% WM compared

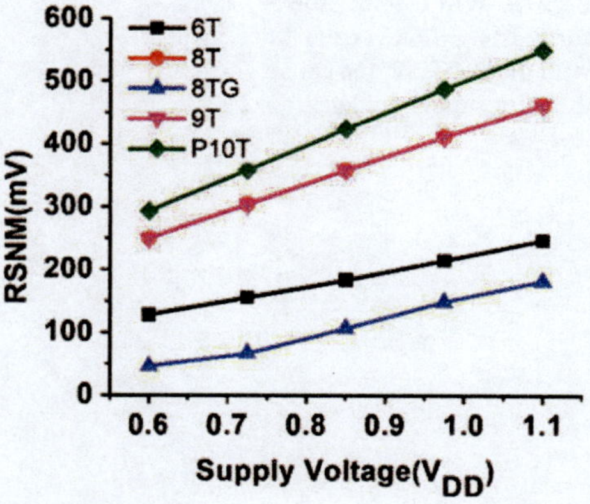

Fig. 13 RSNM Comparison for Existing SRAM cells with the P10T SRAM cell at different supply voltages

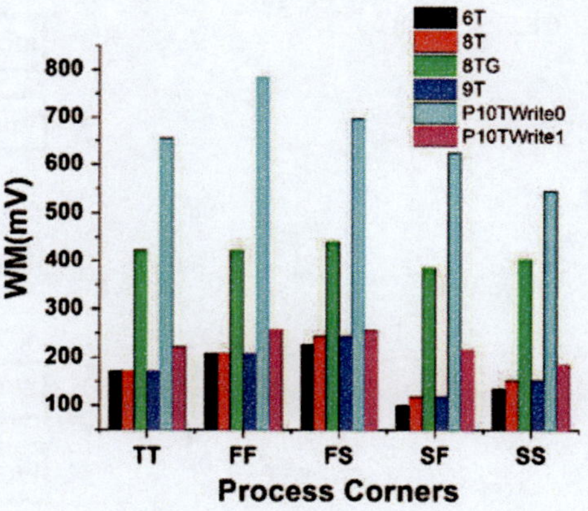

Fig. 14 WM Comparison for Existing different SRAM cells with the P10T SRAM cell at all process corners at 0.9 V supply voltage

to 6 T, 8 T, 8 TG, and 9 T SRAM cells, respectively, at 0.9 V supply voltage at TT corner shown in Fig. 14.

During Write 1, there is an increase of 29.03%, 28.61%,28.61% Write Margin at TT corner at 0.9 V supply voltage as shown in Fig. 14.

The Write 0 margin is more in the P10T SRAM cell over different supply voltages but Write 1 margin is less compared to 8 TG SRAM cell as shown in Fig. 15. However, write margin is more compared to 6 T,8 T, and 9 T SRAM cells.

5 Four Input LUT Using P10T and 6 T SRAM Cell

As SRAM cells form the basis for the Look-Up Tables (LUT's) in FPGA, it is necessary to have low leakage and stable SRAM which may otherwise flip the stored data in LUT and consume more leakage power. The P10T SRAM cell outperforms in terms of leakage and stability compared to other SRAM cells.

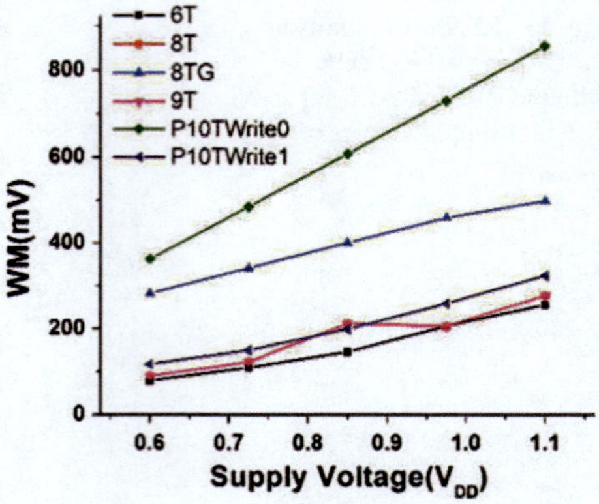

Fig. 15 WM Comparison for Existing SRAM cells with the P10T SRAM cell at different supply voltages

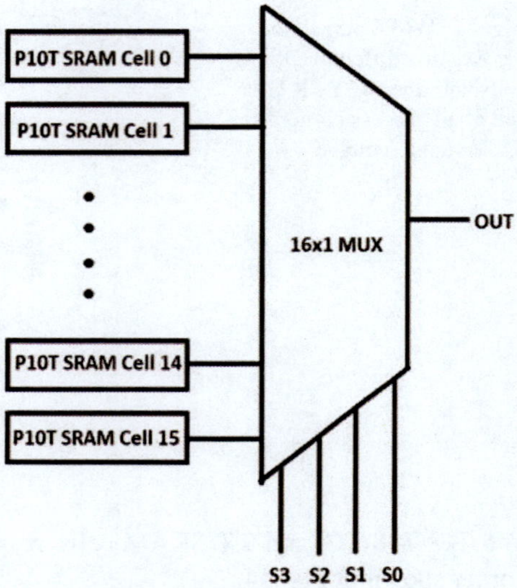

Fig. 16 4-input LUT using P10T SRAM cell

In order to further verify its leakage reduction in complex circuits, 4-input LUT is implemented as shown in Fig. 16 and leakage power is compared with the 6 T SRAM Cell. The LUT using P10T SRAM cell consumes 0.91 × leakage power as compared to the LUT using 6 T SRAM Cell at TT Corner at 0.9 V supply voltage. There is a reduction of 8.57% leakage power using P10T SRAM cell compared to 6 T SRAM cell at TT corner at 0.9 V supply voltage.

6 Conclusion

Leakage Power reduction is important in biomedical instruments, on-chip memories, wireless devices, LUTs in FPGA, etc. Cell stability is also important for reliable operation of these devices. Although the existing SRAM cells maintain superiority in cell stability in one of the modes of operation, they lag in leakage power reduction and stability in more than one mode of operation. The proposed P10T SRAM cell excels than the other SRAM cells in Less Leakage Power while enhancing the stability in the Hold, Read, and Write modes.

Acknowledgements The authors express sincere thanks to AUCE(A), Andhra University, Visakhapatnam for providing the facilities. Also, heartfelt thanks to University Grants Commission (MHRD) for granting the JRF fellowship for doing research.

References

1. Rad JS, Gauthas M, Hugey R (2014) Confronting the variability issues affecting the performance of next-generation SRAM design to optimize and predict the speed and yield. IEEE Access 2:577–601
2. Choi D, Choi K, Villasenor JD (2008) New non-volatile memory structures for FPGA architectures. IEEE Trans Very Large Scale Integr Syst 16(7):874–881
3. Singh P, Vishvakarma SK (2013) Device/circuit/architectural techniques for ultra-low power FPGA design. Microelec Solid State Electron 2(A):1–15
4. Harsh, N.P., Farah, B.Y., Benton, H.C.: Subthreshold SRAM: Challenges, Design Decisions, and solutions. IEEE 60th International Midwest Symposium on Circuits and Systems, Boston, USA (2017).
5. Kursun V, Friedman EG (2006) Multi-Voltage CMOS Circuit Design. John Wiley & Sons, New York, NY, USA
6. Roy K, Prasad SC (2009) Low-Power CMOS VLSI Circuit Design. John Wiley & Sons, New York, NY, USA
7. Nalam S, Calhoun BH (2011) 5T SRAM with asymmetric sizing for improved read stability. IEEE J Solid State Circ 46(10):2431–2442
8. Yahya FB, Harsh NP, James B, Arjit B, Benton HC (2016) A sub-threshold 8T SRAM macro with 12.29 nW/KB Standby power and 6.24 pJ/access for battery-less IoT SoC's. JLPEA, (2016).
9. Islam A, Hasan M (2012) Leakage characterization of 10T SRAM cell. IEEE Trans Electron Devices 59(3):631–638
10. Calhoun BH, Chandrakasan AP (2007) A 256-kb 65-nm sub-threshold SRAM design for ultra-low-voltage operation. IEEE J. Solid-State Circ 42(3):680–688
11. Kim TH et al (2009) A voltage scalable 0.26 V, 64 kb 8T SRAM with Vmin lowering techniques and deep sleep mode. IEEE J Solid-State Circ 44(6):1785–1795
12. Chang, Leland, Montoye, Robert K, Nakamura, Yutaka, Batson, Kevin A, Eickemeyer, Richard J, Dennard, Robert H, Haensch, Wilfried and Jamsek, Damir (2008) An 8T-SRAM for Variability Tolerance and Low-Voltage Operation in High-Performance Caches. IEEE J Solid-State Circ 43(4):956–963(2008)
13. Islam A, Hasan M (2012) A technique to mitigate impact of process, voltage and temperature variations on design metrics of SRAM cell. Microelectron Rel 52(2):405–411
14. Lin S, Kim Y-B, Lombardi F (2010) Design and analysis of a 32 nm PVT tolerant CMOS SRAM cell for low leakage and high stability. Integration 43:176–187
15. Gupta S, Gupta K, Calhoun BH, Pandey N (2018) Low-power near threshold 10T SRAM bit cells with enhanced data-independent read port leakage for array augmentation in 32-nm CMOS 66(3):978–988
16. Singh P, Vishvakarma SK (2017) Ultra-Low Power High Stability 8T SRAM for Application in Object Tracking System. IEEE Access 6(11):2279–2290

17. Do AT et al (2011) An 8 T differential SRAM with improved noise margin for bit interleaving in 65 nm CMOS. IEEE Trans Circ Syst I, Reg Paper 58(6):1252–1263
18. Seevinck E, List F, Lohstroh J (1987) Static-noise margin analysis of MOS SRAM cells. IEEE J Solid-State Circ 22(10):748–754
19. Gierczynski N, Borot B, Planes N, Brut H (2007) A new combined methodology for write margin extraction of advanced SRAM. In: Proceedings IEEE international conference on microelectron. Test Struct. (ICMTS), 97–100 (2007)

Conformal Omni Directional Antenna for GPS Applications

Shravya Krishna Komati and Damera Vakula

Abstract In the proposed paper, design of an omnidirectional antenna using directional antennas is proposed at L2 band(1.2268 GHz.) for GPS applications. The Omnidirectional pattern is obtained using directional antennas by arranging them in such a way that their far field patterns interfere constructively and create an Omnidirectional pattern. To validate our design proposal Omnidirectional antenna is designed using directional antennas in simulation tool. The simulated results are observed, which illustrates good performance of proposed antenna. Bandwidth obtained for designed antenna is 45.4 MHz

1 Introduction

Applications of patch antennas are increasing rapidly, because of their easy fabrication process, light in weight, low cost and low volume [1]. Therefore, patch antennas are suitable for wireless communications and satellite communications [2, 3]. Omnidirectional antennas are used in radio broadcasting applications, in mobile devices such as FM radio, cell phone, walkie-talkie. Omnidirectional antennas are also used in base stations which generate signal in all directions in order to achieve the communication with the user. Flush disc antenna can be used to achieve nearly omnidirectional pattern [4].

Omni-directional antennas have many useful applications; One of the limitations of omnidirectional is less gain to quality factor ratio when compared with directional antennas [5]. Other limitation is less bandwidth which can be increased by connecting two dipole arms using parallel strip line [6].

S. K. Komati (✉) · D. Vakula
National Institute of Technology, Warangal, India
e-mail: shravyakrishnakomati@gmail.com

D. Vakula
e-mail: vakula@nitw.ac.in

T. Laxminidhi et al. (eds.), *Advances in Communications, Signal Processing, and VLSI*, Lecture Notes in Electrical Engineering 722,
https://doi.org/10.1007/978-981-33-4058-9_28

In the proposed work omnidirectional radiation pattern is achieved by using antennas with directional radiation pattern. As directional radiators have high gain, the proposed antenna in this work has omnidirectional pattern in addition to high gain.

2 Design Procedure of Patch Antenna

A microstrip patch antenna consists of ground plane and a patch; ground plane and patch are conducting in nature and these are separated by a dielectric medium called substrate. Size of the ground plane is large when compared to the size of the patch, usually ground plane is considered extended to infinity.

Patch dimensions depend on the resonant frequency and dielectric medium. Design equations for the length and width of the patch are given below. Geometry of the rectangular patch is shown in Fig. 1.

Width

$$W = \frac{c\sqrt{2/(+1)}}{2fr} \tag{1}$$

W—Width of the patch.
C—Velocity of an electromagnetic wave.
Fr—Resonant frequency.
Effective refractive index

Fig. 1 Patch of the antenna

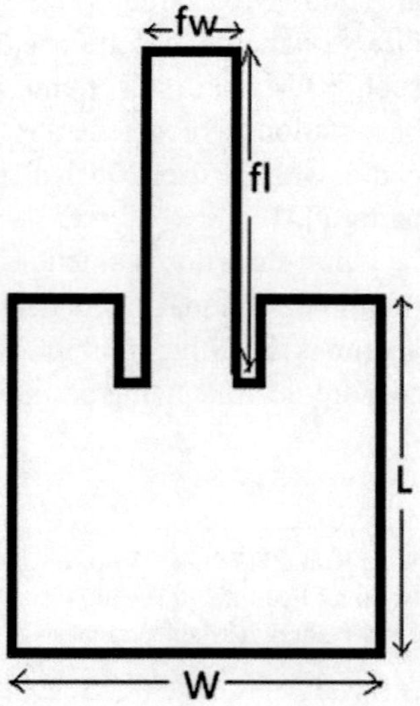

$$\varepsilon_{reff} = \frac{\varepsilon_r + 1}{2} + \frac{\varepsilon_r - 1}{2}\left[1 + 12\frac{h}{W}\right]^{1/2}, W/h > 1 \tag{2}$$

Length

$$\frac{\Delta L}{h} = 0.412\frac{\left(\varepsilon_{reff} + 0.3\right)\left(\frac{W}{h} + 0.264\right)}{\left(\varepsilon_{reff} - 0.258\right)\left(\frac{W}{h} + 0.8\right)} \tag{3}$$

$$L\frac{c}{2fr * \left(\varepsilon_{reff}\right)^{\left(\frac{1}{2}\right)} - 2\Delta L} \tag{4}$$

3 Design of Omnidirectional Antenna

Omnidirectional antenna can be designed using multiple directional antennas. In the proposed work the omnidirectionality is attempted with 2, 4 and 8 elements. The best results are obtained by considering 4 elements arranged in a square geometry. Because of this square shape arrangement, each antenna radiates in a particular direction which results in constructive interference of radiation pattern to achieve omnidirectional pattern. The substrate used for this is FR-4 which contains relative permittivity of 4.4 and a thickness of 4.5 mm. Dimensions of rectangular patch antenna designed for L2 band is shown in Table 1. The arrangement of the antennas to get Omnidirectional pattern is shown in Figs. 2 and 3.

Table 1 Geometrical values of proposed antenna

Parameter	Value(mm)
Length of the patch (L)	56
Width of the patch (W)	38
Length of the ground plane (Lg)	112
Width of the ground plane (Wg)	76
Width of the feed (fw)	12.5
Thickness of substrate(h)	4.5
Thickness of patch (t)	0.1
Length of the feed (fl)	40.5

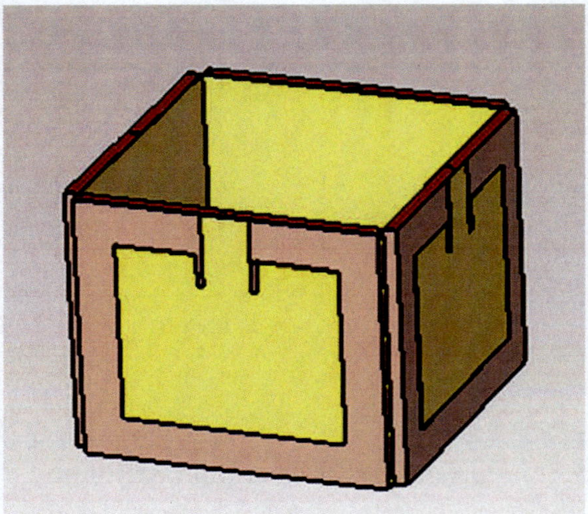

Fig. 2 Omnidirectional antenna using four directional antennas

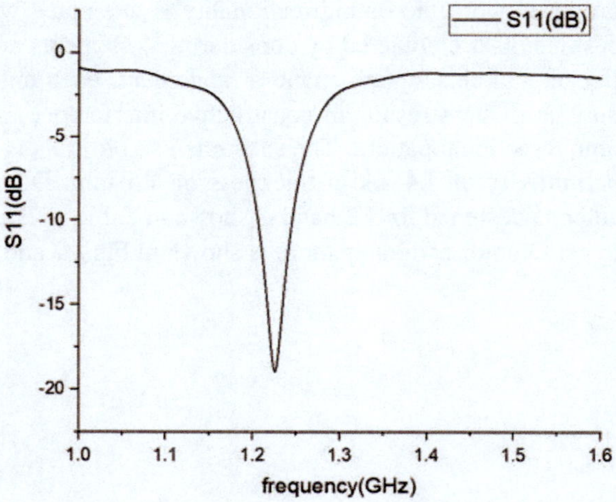

Fig. 3 S11 plot for patch antenna

4 Results and Discussion

The proposed rectangular patch antenna designed at 1.2268 GHz is simulated using CST. The reflection coefficient (dB) is illustrated in Fig. 3 which shows a −10 dB impedance bandwidth of 45.4 MHz. The directional radiation patterns are shown in Figs. 4 and 5. The 3 dB beam width is obtained as 90° and 203.2° along E&H planes respectively. Gain for the rectangular patch antenna is 1.8dBi.

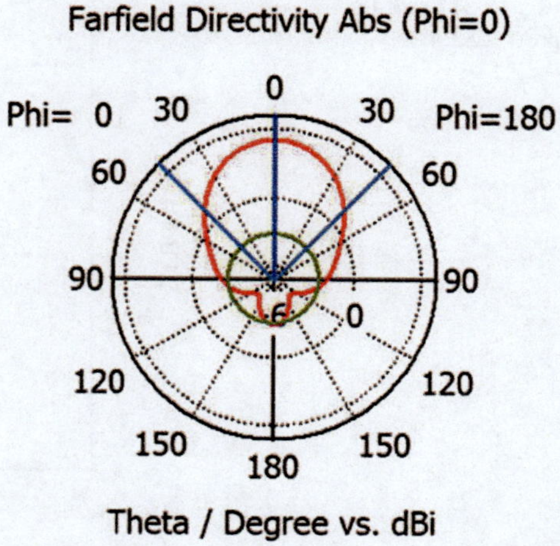

Fig. 4 H-plane pattern for patch antenna

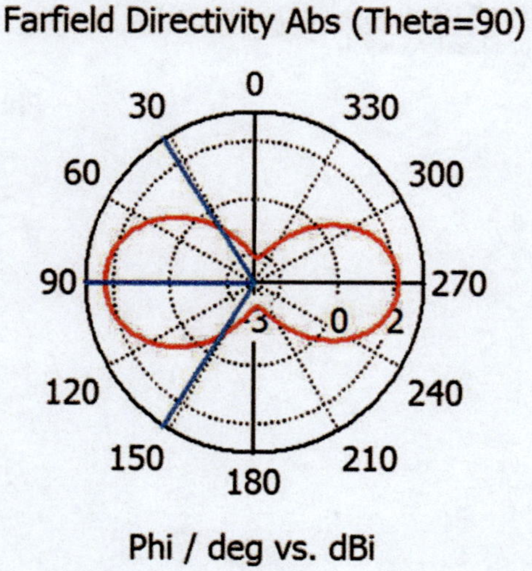

Fig. 5 E-plane pattern for patch antenna

The four-directional antennas arranged in the square geometry is simulated by exciting all the antennas. The reflection coefficient in dB at all the 4 ports is given in Fig. 6. The −10 dB bandwidth is 45.4 MHz as represented in Fig. 6. The simulated E-plane and H-plane patterns for the square geometry are shown in Figs. 7 and 8. The E-plane radiation pattern is the figure of eight. And H- plane radiation pattern is nearly circle. This establishes that the radiation pattern of the square geometry is omnidirectional. Gain for the omnidirectional is 1.65 dBi.

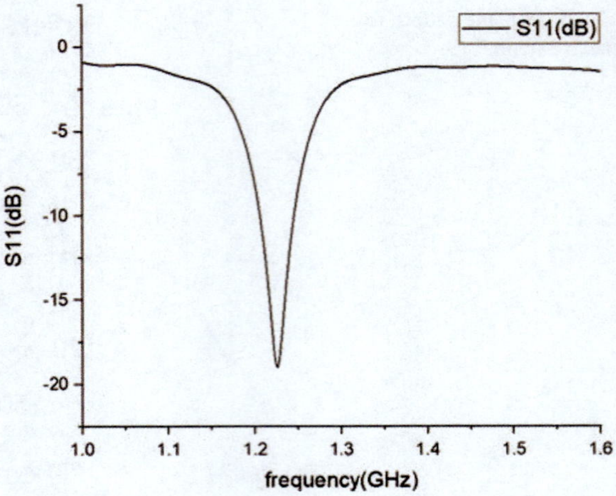

Fig. 6 S11 plot for port1,2,3&4

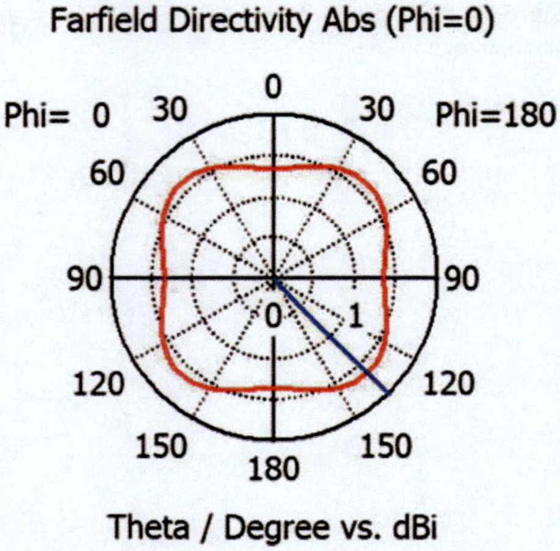

Fig. 7 H-plane pattern for 4 patch antennas

5 Conclusion

In this paper four directional patch antennas are used to generate an Omnidirectional pattern with a resonant frequency of 1.2268 GHz and a bandwidth of 45.4 MHz. The antenna has gain of 1.65dBi. The proposed antenna can be used as omnidirectional antenna at L2 band for GPS applications.

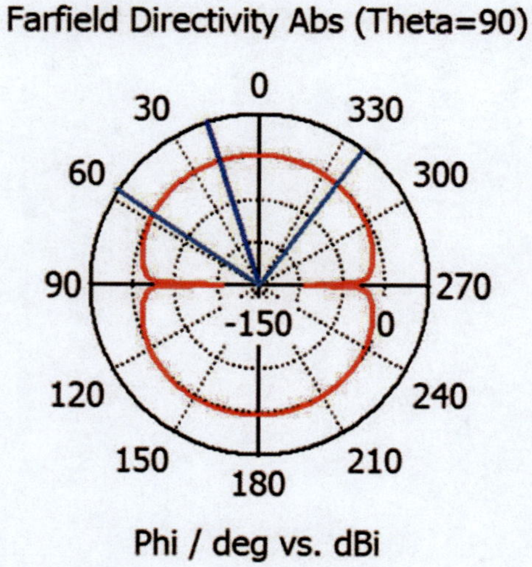

Fig. 8 E-plane pattern for 4 patch antennas

References

1. Torres DL, Sumba LP, Bermeo JP, Cuji DA (2017) Dual band rectangular patch antenna with less return loss for WiMAX and WBAN applications in 2017 IEEE second ecuador technical chapters meeting (ETCM) ,16–20 Oct 2017, Ecuador
2. MT Islam MN Shakib N Misran TS Sun 2009 Broadband microstrip patch antenna Eur J Sci Res 27 2 174 180
3. Y Bhomia A Kajla D Yadav 2010 Slotted right angle triangular microstrip patch antenna Int J Electr Eng Res 2 3 393 398
4. Wang X, Qin F, Wei G, Xu J (2012) Design of low profile wideband omnidirectional antenna, 8th IEEE, IET international symposium on communication systems, networks and digital signal processing, July 18–20 2012, Poznan University of Technology Poznan, Poland
5. W Geyi 2003 Physical Limitations of Antenna IEEE Trans Anten Propag 51 8 2116 2121
6. Wu D, Yin Y, Guo M , Shen R, Wideband dipole antenna for 3G base stations, 2005 IEEE international symposium on microwave, antenna, propagation and EMC technologies for wireless communications Proceedings, Beijing, China

Total Internal Reflection Quasi-Phase-Matching-Based Broadband Second Harmonic Generation in Isotropic Semiconductor: A Comparative Analysis

Minakshi Deb Barma, Sumita Deb, and Satyajit Paul

Abstract This paper presents a comparison in various non-parallel configurations of isotropic semiconductor materials for broadband second harmonic generation using total internal reflection quasi-phase-matching technique. The effect of conversion yield-limiting factors, namely, the surface roughness, Goos-Hanchen (GH) shift, absorption loss and the destructive interference effect due to nonlinear law of reflection have been considered in the analysis. Finally a comparative analysis as regard to the performance parameters, viz., conversion efficiency and 3-dB bandwidth as well as operating wavelength zone has been prepared. The important applications have also been highlighted with special reference to quantum cascade laser.

Keywords Broadband second harmonic generation · Total internal reflection quasi-phase-matching · Isotropic semiconductor · Random quasi-phase-matching

1 Brief Background

The concept of quasi-sphase-matching (QPM) with zigzag total internal reflection (TIR) paths in a plane parallel slab was first suggested by Armstrong et al. [1] in the year 1962, which has subsequently been investigated in GaAs, ZnSe, and ZnS slabs for resonant- QPM second harmonic generation (SHG) by Boyd and Patel [2] as well as by Komine et al. [3]. In the year 2004, Haïder et al. [4] has extended the same technique as nonresonant QPM towards difference frequency generation (DFG) in isotropic semiconductor like ZnSe and GaAs. Here "Resonant" refers to the situation in which the distance between two consecutive reflection bounces is exactly an odd multiple of the coherence length whereas "nonresonant" is the situation in which the distance is not equal to an odd multiple (neither equal to an even multiple) of the coherence length. Then in the year 2006, Haïder [5] proposed the concept of fractional

M. D. Barma (✉) · S. Deb · S. Paul
Department of Electrical Engineering, National Institute of Technology, Agartala, Barjala, Jirania 799046, Tripura (West), India
e-mail: minakshi_nita@rediffmail.com

T. Laxminidhi et al. (eds.), *Advances in Communications, Signal Processing, and VLSI*, Lecture Notes in Electrical Engineering 722,
https://doi.org/10.1007/978-981-33-4058-9_29

QPM technique where the interaction length between consecutive bounces is less than even one coherence length which aimed at much higher conversion efficiency.

Later on a group of researchers has shown that the random motion of the relative phases of the interacting waves in highly transparent polycrystalline materials can be an effective strategy for achieving efficient phase matching in isotropic materials [6]. This method named as random QPM technique has certain unique advantages like i) linear dependence of the conversion yield with sample thickness, ii) eliminate the need to select material orientation or specific polarization configuration, and iii) wavelength-dependent resonant size for the polycrystalline grains [6]. Random quasi-phase-matching thus steered itself as a low-cost technique with extremely loose frequency selectivity which in turn makes it of particular interest for generating optical radiations with ultra-wide spectral tunability.

Next, Myriam Raybaut et al. [7] have reported an experimental study of the SHG conversion efficiency for Fresnel phase matching as a function of a gallium arsenide sample length which shows a strong deviation from the classical quadratic law, explaining that the low conversion yield demonstrated by the experimental data as compared to the analytical values is due to the noncollinearity induced by nonlinear law of reflection as reported by Bloembergen and Pershan [8].

After the theoretical and experimental approach on TIR QPM-based single frequency conversion technique, a new concept of broadband frequency conversion using the same technique has been analytically demonstrated by Saha et al. [9–16] in various isotropic slab configurations which results in flatter spectra of the generated second harmonic bandwidth (BW) in comparison to a parallel slab, although the conversion efficiency will be lower in case of their proposed configurations which are highlighted in subsequent sections.

2 Various Slab Configurations for Broadband Generation

2.1 *Tapered Isotropic Slab*

In their introductory paper, Saha et al. [9] have analytically illustrated the concept of broadband SHG using TIR QPM technique in the mid-infrared region in a tapered slab configuration made of either GaAs or ZnSe as shown in Fig. 1.

In this configuration, when fundamental optical radiation having a center frequency ω is incident at an angle θ_1 with respect to the normal on the incident slab end face and when θ_1 is greater than the critical angle for the range of input frequencies, then the collimated optical radiation will undergo TIR inside the slab. Since the input fundamental optical radiation incident on the tapered slab is a broadband source rather than a single frequency, and the interaction length between successive bounces goes on increasing as the input collimated fundamental laser radiation propagates through the slab, it may so happen that one interaction length in between successive bounces may coincide with an odd multiple of the coherence length for a

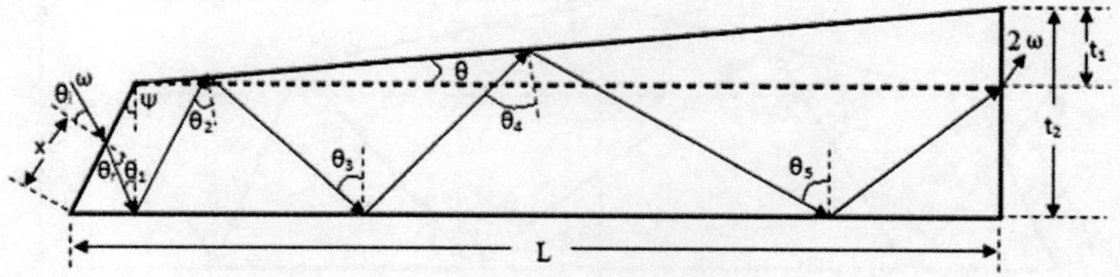

Fig. 1 Geometry of tapered semiconductor slab showing SHG

particular frequency available in the input broadband source, whereas another interaction length may coincide with an odd multiple of the coherence length of another frequency of the input broadband source and so on, thereby resulting in a flatter SH broadband spectra.

Then in the year 2011, same group has again shown the generations of broadband SH in the near-infrared region in a tapered ZnSe slab using the technique of TIR-based random QPM [10]. As reported by the authors, the phase shifts of the interacting waves vary randomly during their propagation inside the slab, thereby giving rise to a situation identical to random QPM.

Later they have proposed fractional TIR QPM technique in the same slab configuration made of ZnTe material whose refractive index shows minimum deviation over a wide wavelength range resulting in increased coherence length which leads to fractional QPM scenario, thereby ensuring highly efficient SH output [11].

Using the same concept of fractional TIR QPM technique in the same slab of ZnTe material, they have incorporated for the first time, the destructive interference effect arising due to the nonlinear law of reflection which shows a heavy drop in the peak conversion efficiency with a slight increase in the 3 dB BW [12].

2.2 *Double Tapered Slab*

After the analytical description of broadband SHG using TIR QPM technique in a tapered isotropic slab, the same concept has been extended in a double tapered configuration made of GaAs in order to achieve an extremely wide 3 dB BW as compared to single tapered configuration [13].

Due to the double tapered structure as shown in Fig. 2, the angle of incidence and the length between successive bounces will go on increasing with the propagation of the input broadband radiations up to forward section length, L_1, after which both the parameters will go on decreasing in the reverse section length, L_2, till the beam emerges out of the slab. This provides the flexibility to widen the 3 dB BW by optimizing L_1 and L_2 [13].

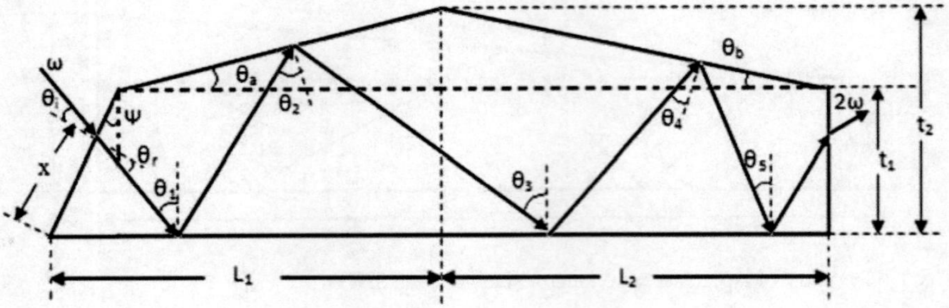

Fig. 2 Geometry of double tapered semiconductor slab showing SHG

2.3 *Trapezoidal Slab with Elliptical Upper Surface*

In Ref. [14], the authors have proposed an extremely wide broadband frequency converter with appreciable conversion efficiency in a trapezoidal isotropic slab with an elliptical upper surface as shown in Fig. 3a, made of GaAs using the same technique of SHG. Here the angle of incidence and the length between consecutive bounces will go on increasing with the propagation of the fundamental optical radiation throughout the length of the crystal until the beam crosses the center of the ellipse which will then subsequently decrease as the beam further propagates through the slab.

2.4 *Parabolic Profiled Trapezoidal Slab*

In their next paper [15], the authors have demonstrated further the broadband SHG in a trapezoidal slab with a parabolic profiled upper surface made of same material, i.e., GaAs in order to achieve good conversion efficiency and flat wide BW (Fig. 3b).

2.5 *Reverse Tapered Slab*

After analytical description of broadband SHG using TIR QPM technique in various isotropic slab configurations, viz., tapered, double tapered, trapezoidal slab with elliptical upper surface and parabolic profiled upper surface, we have proposed a highly efficient broadband SH generator in the mid-infrared region by using a reverse tapered semiconductor slab configuration (as shown in Fig. 4) made of ZnTe, using fractional TIR QPM technique [16]. The analysis also includes the destructive interference effect due to nonlinear law of reflection. The Rayleigh range which determines the depth of focus of the beam has also been taken care of in the analysis. Apart from being fractional, the highly efficient SH output can be an attribute of the reverse tapered view as well, where the incident angle and the interaction length

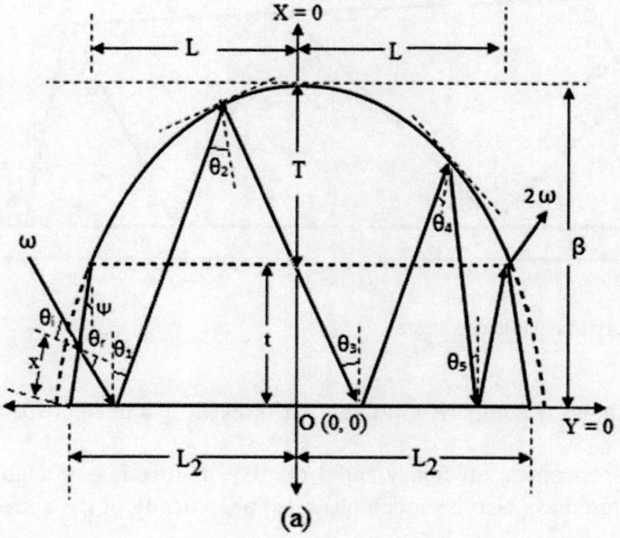

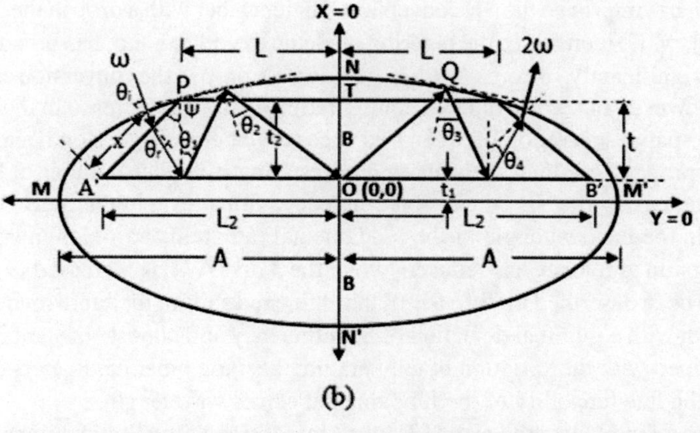

Fig. 3 Geometry of trapezoidal slab with an (**a**) elliptical upper surface [14] and (**b**) parabolic profiled upper surface

between two consecutive bounces will go on decreasing with the propagation of the input broadband radiations inside the slab resulting in greater number of bounces as compared to the earlier tapered configuration [9] for the same slab dimensions. This results in SH conversion efficiency as high as twice the earlier case [16].

The polarization configuration used in all the above analysis is *ppp* as because in *ppp* the net phase shift quickly approaches π over a wide range of incidence angle [17]. The effect of conversion yield-limiting factors, namely, the surface roughness, Goos-Hanchen (GH) shift and absorption loss have been taken care of in all the analysis.

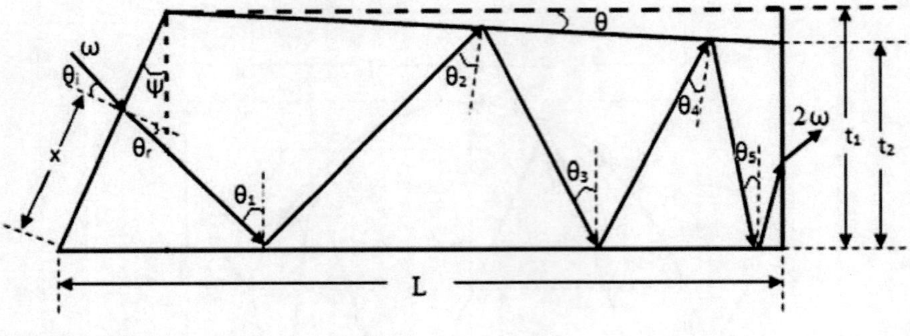

Fig. 4 Geometry of reverse tapered slab

Table 1 shows a comparison of the performance parameters for various slab configurations.

The peak conversion efficiency and 3 dB BW, as functions of slab parameters, and temperature have also been demonstrated analytically in the above-mentioned works [9–16].

In single tapered configurations, it has been observed that increasing the crystal slab length has improved the SH conversion efficiency but with a drop in the 3 dB BW [9–11, 16], while increase in the tapering angle on the other hand, has broadened the 3 dB BW significantly, of course with a considerable drop in the conversion efficiency [9–11]. However, in reverse tapered configuration [16], with increase in the tapering angle, the spatial separation between two successive bounces goes on decreasing as the beam propagates along the slab length, resulting in higher number of bounces, thereby, increasing the SH conversion efficiency with a drop in the 3 dB BW. With decrease in the incident angle of the fundamental radiation at air-slab interface, the SH conversion efficiency has reduced, while the 3 dB BW has increased [9, 11, 16]. It has also been described in Ref. [9, 10] that a fine red shift of the fundamental center wavelength with a minimal deviation of SH efficiency and almost constant BW have been obtained with the variation of temperature, keeping other parameters constant, ensuring the fine tunability of the fundamental center wavelength.

Since in a double tapered slab [13], trapezoidal slab with elliptical upper surface [14] or parabolic profiled upper surface [15], the consecutive bounce lengths do not follow any symmetrical increase or decrease, rather it is a combination of increasing and subsequent decreasing bounce length, the effect of variation in slab length is not uniform. In a double tapered slab [13], a sharp dip in the central region of the 3 dB BW spectra with two peaks has been observed with increasing slab length which then becomes a single peak point after further increase of the same. However, increase in the length of the trapezoidal slab with elliptical upper surface [14] or parabolic profiled upper surface [15] first results in an increase of the SH efficiency which is subsequently followed by a drop in the same parameter.

The variation of tapering angle in a double tapered slab [13] has shown a non-uniform change in the performance parameters. Initially there is a splitting of the 3 dB BW spectra into two separate peaks with a dip in the central region which in turn is

Table 1 Comparison of the performance parameters for various slab configurations when an input beam intensity of 10 MW/cm^2 has been considered

Sl. No	Configuration	Material	Scheme of SHG	Slab dimensions	Whether nonlinear law of reflection has been included?	Peak conversion efficiency (%)	Fundamental center wavelength (μm)	3-dB BW (nm)
1	Tapered [9]	GaAs	TIR QPM	L = 30 mm, t_1 = 400 μm, t_2 = 402 μm, ψ = 0.03934 rad (GaAs), = 0.2934 rad (ZnSe) x = 300 μm (GaAs) = 100 μm (ZnSe)	No	1.052	7.911	187
		ZnSe				1.043	5.807	196
2	Tapered [10]	ZnSe	Random TIR QPM	L = 10 mm, t_1 = 300 μm, t_2 = 305 μm ψ = 0.5934 rad x = 50 μm	No	0.02	4.05	557
3	Tapered [11]	ZnTe	Fractional TIR QPM	L = 30 mm, t_1 = 400 μm, t_2 = 405 μm, ψ = 0.4 rad, x = 100 μm	No	19.6	8.583	193
4	Tapered [12]	ZnTe	Fractional TIR QPM	L = 4.5 mm, t_1 = 90 μm, t_2 = 95 μm	Yes	3.73	–	490
5	Double tapered [13]	GaAs	TIR QPM	L = 30 mm, t_1 = 400 μm, t_2 = 405 μm, ψ = 0.6 rad, x = 270 μm	No	1.929	9.146	574

(continued)

Table 1 (continued)

Sl. No	Configuration	Material	Scheme of SHG	Slab dimensions	Whether nonlinear law of reflection has been included?	Peak conversion efficiency (%)	Fundamental center wavelength (μm)	3-dB BW (nm)
6	Trapezoidal slab with an elliptical upper surface [14]	GaAs	TIR QPM	L = 40 mm, t = 400 μm, T = 4 μm, ψ = 0.393 rad, x = 300 μm	No	1.08	9.215	770
7	Trapezoidal slab with parabolic profiled upper surface [15]	GaAs	TIR QPM	L = 36 mm, t = 400 μm, ψ = 0.393 rad, x = 300 μm, T = 4 μm	No	1.105	– —	642
8	Reverse tapered [16]	ZnTe	Fractional TIR QPM	L = 10 mm, t_1 = 400 μm, t_2 = 405 μm, ψ = 0.4 rad, x = 100 μm	Yes	1.51	8.32	205

followed by a drop in the conversion efficiency. However, with further increase in the tapering angle, the separate peaks combine together to give a single peak efficiency point. The variation of thickness (T) between the upper edge of the slab and the vertex or directrix of the ellipse or parabola, respectively, has also been studied in Ref. [14, 15] which shows that increasing T has decreased the peak conversion efficiency with a significant rise in the 3 dB BW.

Since there is a combination of increasing and then decreasing angle of incidence at each TIR bounce point as described in Ref. [14, 15], increase in the angle of incidence at the air-material interface of trapezoidal slab with an elliptical upper surface results in a minimal drop of the peak conversion efficiency, whereas the 3 dB BW first shows a rise in its value after which it is accompanied by a heavy drop with further increase of same [14]. However, there is an increase of both peak conversion efficiency and 3 dB BW with increase in angle of incidence for the same slab material with parabolic profiled upper surface [15].

All the slab configurations as used in Ref. [13–15] show a red shift of the center wavelength of the fundamental beam with an increase in temperature which is followed by almost negligible drop in the SH conversion efficiency. However, there is a considerable drop in the 3 dB BW.

3 Conclusion

This paper shows an overview of the broadband SHG using TIR QPM technique in various slab configurations considering the effect of conversion yield-limiting factors. The table presented in this paper gives a complete information about the performance parameters for different configurations, dimensions and materials which will enable one to select a broadband frequency converter of desired BW in the mid-infrared region finding useful applications in spectroscopy, materials processing, chemical and biomolecular sensing. Since the broadband frequency converters described in Ref. [9, 16], fall in the two atmospheric transmission windows of 3–5 μm and 8–13 μm, it can be important for atmospheric, security and industrial applications such as remote explosive detection, countermeasures against heat-seeking missiles and convert communication systems [17]. In mid-infrared region, water molecules exhibit a set of strong absorption lines which causes strong mid-IR absorption in human skin, thereby, facilitating numerous biomedical applications in laser surgery, tissue ablation and photo dermatology [18]. The level of absorption can be varied significantly to fine-tune the machining process by tuning the wavelength of the mid-IR source. Additional mid-IR laser applications include particle acceleration and uv/x-ray pulse generation.

The proposed broadband converters [9–16] may come forward in the same line with Quantum cascade (QC) lasers, which are semiconductor lasers that emit in the mid- to far-infrared portion of the electromagnetic spectrum. Since QC lasers utilize intersubband transitions within a multiple quantum-well structure, they offer excellent design flexibility because the staircase of intersubband transitions can be designed to obtain particular emission wavelengths [17]. The high optical power output, tuning range and room temperature operation make QC lasers useful for various applications which include remote and point sensing of environmental gases and pollutants in the atmosphere [19], free space optical communication [20, 21], infrared (IR) countermeasures, metal detection, astronomical science and medical science [22]. The broadband frequency converters as discussed in this paper seem to have certain features quite comparable to the QC lasers which are highlighted below:

1. A single QC laser offers tremendous advantage of being able to be tuned from one wavelength to the other in a precise manner [23] by changing the direction of the electrical current flowing into the laser or by controlling the external temperature for which the laser is operating. The proposed broadband converters of Ref. [9, 13–15] also allow fine tuning of the fundamental center wavelength of the output 3 dB BW by changing the operating temperature which shows a minimal change in the SH conversion efficiency and a small deviation in the 3 dB BW, in the range of few nanometer. However, the converter of Ref. [10] allows fine tuning of the fundamental

center wavelength of the output 3 dB BW by varying the angle of incidence which shows a fine blue shift of the center wavelength.

2. QC Lasers have higher optical power than other types of mid and far-infrared lasers [23]. It has been observed that SHG using fractional QPM technique gives highly efficient SH output which can be achieved using ZnTe as the slab material [11, 13, 16]. The converters mentioned in Ref. [13, 16] show high value of conversion efficiency even in the presence of all efficiency limiting factors, viz., surface roughness, absorption loss, GH shift and destructive interference effect arising due to nonlinear law of reflection. Moreover, the damage threshold of ZnTe is very high (about 100 GW/cm^2 [24]) ensuring higher optical power of the generated 3 dB BW. This might extend the use of the converter in the sensing of gases like methane which requires high optical power as compared to other gases [23].

3. The ability of QC lasers to operate in continuous wave mode at room temperature is another factor that aids in its better gas sensing [23]. The broadband frequency converters [9–16] have also been analyzed in continuous wave mode at room temperature.

Thus, due to the above features, the broadband SHG-based frequency converters as described in this paper can come out as strong candidates in the field of nonlinear optics particularly in the mid-IR region of the optical spectrum.

References

1. Armstrong JA, Bloembergen N, Ducuing J, Pershan PS (1962) Interactions between light waves in a nonlinear dielectric. Phys Rev 127:1918–1939
2. Boyd GD, Patel CKN (1966) Enhancement of optical second harmonic generation (SHG) by reflection phase matching in ZnS and GaAs. Appl Phys Lett 8:313–315
3. Komine H, Long WH Jr, Tully JW, Stappaerts EA (1998) Quasi-phase-matched second-harmonic generation by use of a total-internal-reflection phase shift in gallium arsenide and zinc selenide plates. Opt Lett 23:661–663
4. R. Haïder, Kupecek P, Rosencher E, Triboulet R, Lemasson P (2003) New mid-infrared optical sources based on isotropic semiconductors (Zinc selenide and gallium arsenide) using total internal reflection quasi phase matching. Opto-Electron Rev 11(2):155–160
5. Haïder R (2006) Fractional quasi- phase matching by Fresnel Birefringence. Appl Phys Lett 88:211102- 1- 211102- 3
6. Baudrier-Raybaut M, Haïder R, Kupecek P, Lemasson P, Rosencher E (2004) Random quasi phase matching in bulk poly crystalline isotropic nonlinear materials. Nature 432:374–376
7. Raybaut M, Godard A, Toulouse A, Lubin C, Rosencher E (2008) Nonlinear reflection effects on Fresnel phase matching. Appl Phys Lett 92:121112/1-121112/3
8. Bloembergen N, Pershan PS (1962) Light waves at the boundary of nonlinear media. Phys Rev 128:606–622
9. Saha A, Deb S (2011) Broadband second harmonic generation in a tapered isotropic semiconductor slab using total internal reflection quasi phase matching. Opt Comm 284:4714–4722
10. Saha A, Deb S (2011) Broadband second-harmonic generation in the nearinfrared region in a tapered zinc selenide slab using total internal reflection quasi-phase matching. Jan J Appl Phys 50:102201
11. Deb S, Saha A (2013) Highly efficient broadband second harmonic generation in a tapered zinc telluride slab using total internal reflection fractional quasi phase matching. Optik 124:2428–2431

12. Deb S, Saha A (2015) Fractional quasi phase matched broadband second harmonic generation in a tapered zinc telluride slab using total internal reflection considering the effect of nonlinear law of reflection. Optik 126:3371–3375
13. Saha A, Deb S (2014) Broadband second-harmonic generation in a doubletapered gallium arsenide slab using total internal reflection quasiphase matching. Optik 125:6861–6866
14. Banik S, Deb S, Saha A (2013) Analysis of broadband second harmonic generation in a trapezoidal isotropic semiconductor slab with an elliptical upper surface using total internal reflection quasi phase matching. Opt Comm 287:196–202
15. Banik S, Das U, Deb S, Saha A (2013) Numerical analysis of broadband second harmonic generation using TIR-QPM in a parabolic profiled isotropic semiconductor slab. Opt Comm 295:180–187
16. Deb Barma M, Deb S, Saha A (2015) Broadband fractional total internal reflection quasi phase matching based second harmonic generation in reverse tapered isotropic semiconductor considering the effect of nonlinear law of reflection. J Nonlin Optic Phys Mater 24:1550041 (2015)
17. Editorial, Innovative mid-infrared laser technologies are anticipated to broaden the applications of existing mid-infrared laser sources and bring unexpected scientific discoveries. Nat Photonic 6:407 (2012)
18. Rudy CW (2014) Mid-IR lasers: power and pulse capability ramp up for mid-IR lasers. Laser Focus World 50(5), May 2nd 2014
19. Normand E, Howieson I, McCulloch M, Black P (2007) QUANTUM-CASCADE LASERS: quantum-cascade lasers enable gas sensing technology. Laser Focus World 43(4), April 1st 2007
20. Chuanwei L et al (2015) Free-space communication based on quantum laser. J semiconductors 36(9):094009-1-094009-4
21. Capasso F, Paiella R, Martini R, Colombelli R, Gmachl C, Mysers TL, Taubm M, Williams R, Bethea C, Unterrainer K, Hwang H, Sivco DL, Cho AY, Sergent AM, Liu HC, Whittaker ED (2002) Quantum cascade lasers: ultrahigh-speed operation, optical wireless communication, Narrow Linewidth and Far-Infrared Emission. IEEE J Quantum Electr 38(6):511–529
22. Serebryakov VA, Boiko EV, Petrishchev NN, Yan AV (2010) Medical applications of mid-IR lasers. Problems and prospects. J Optic Technol 77(1):6–17
23. http://www.york.cuny.edu/yorkscholar/v3/rumala (Accessed June, 2016)
24. Yuan T, Xu JZ, Zhang XC (2004) Development of terahertz wave microscopes. Infrared Phys Technol 45(5–6):417–425

Automated Bayesian Drowsiness Detection System Using Recurrent Convolutional Neural Networks

Mahima Jeslani and Naidu Kalpana

Abstract Drowsiness and fatigue in drivers account for the considerable proportion of motor vehicular accidents. Measuring the level of drowsiness in drivers in real time can help to alert them ahead of time in order to prevent road accidents. In this paper, we propose an Automated Bayesian Drowsiness Detection (ABDD) system, a hybrid deep recurrent convolutional learning framework for identifying the drivers' drowsiness in real time. ABDD system uses the recurrent sequence model jointly trained with a convolutional model to learn crucial high-level perceptual representations and temporal dynamics from the input videos in order to make Bayesian drowsiness predictions. Estimates of uncertainty are crucial as they allow us to decide whether or not to trust the model's decision. The proposal for the use of a hybrid deep neural network learning scheme capable of modeling important features exclusive to drowsiness in the visual data is one salient contribution of this paper. The ABDD system is trained and evaluated on the NTHU drowsy Driver Dataset (NDD), the state-of-the-art driver fatigue detection video dataset and achieves a classification accuracy (ACC) of 94.2 on two-class classification. Comprehensive assessment of the proposed method on the video dataset is provided along with an extensive qualitative comparison against the latest techniques.

1 Introduction

In recent years, distraction in drivers due to drowsiness has been a particularly significant cause for motor accidents. The German Council for Road Safety (DVR) claims that twenty-five percent of road traffic accidents result from transitory driver drowsiness [1]. Further, the Traffic Safety Council of the US (NHTSA) reported an aggregate of 795 fatalities in 2017 due to drowsy driving-related crashes [2]. These

M. Jeslani · N. Kalpana (✉)
Department of ECE, NIT, Warangal, India
e-mail: kalpana@nitw.ac.in

M. Jeslani
e-mail: jmahima@student.nitw.ac.in

T. Laxminidhi et al. (eds.), *Advances in Communications, Signal Processing, and VLSI*, Lecture Notes in Electrical Engineering 722,
https://doi.org/10.1007/978-981-33-4058-9_30

alarming statistics surely indicate a pressing need to develop a real-time driver assistance method that reduces the amount of drowsiness in drivers to prevent accidents. Different methods employed to measure the drowsiness and fatigue in drivers have been classified as follows:

(a) Vehicle-based measures—The driving pattern of the vehicle is monitored using several metrics such as reduced control on the wheel and acceleration pedal, steering wheel motion, and the unbalanced lane drifting of vehicle [3, 4]. Steering wheel movement patterns have proved useful in monitoring fatigue in drivers [4–6]. For instance, Berglund et al. in [7] have exploited variables such as steering wheel torque and lateral acceleration as indications for sleepiness in drivers.
(b) Physiological measures—Bio-electrical time signals like Electrocardiogram (ECG) [8], Electroencephalogram (EEG) [9] and Electrooculography (EOG) [10] are studied to register the physiological activity of the driver as an aid to detect drowsiness.
(c) Behavioral measures—Behavioral assessment of the driver is used to interpret levels of drowsiness. Features including eye-blinking pattern, head movement, and yawning are monitored using computer vision to identify the beginning of fatigue in drivers [11–15].

Indicators like rapid and constant blinking, head swinging, or nodding are considered to imply the onset of drowsiness [16, 17]. Besides, PERCLOS [18], which is an estimate of eyelid closure is thought-out as a decisive measure of drowsiness and has been deployed in real-time driver assistance products at the consumer level. Also known as signs of drowsiness are subtle facial features such as brow-raising, jaw-dropping, and eye-blinking. Until recently, the majority of vision-based drowsiness recognition systems analyze features such as head movements and facial features which are hand-crafted in nature. Such features have a restricted viability in practical scenarios incorporating situations like drivers with sunglasses and different variations in luminosity. On the contrary, deep learning techniques have been more efficient in learning effective features related to drowsiness in practical situations.

Choi et al. in [19] devised an algorithm for gaze detection utilizing Convolutional Neural Network (CNN) features. Dwivedi et al. in [20] adopted a shallow three-layered CNN network for drowsiness detection using representation learning on extracted facial region of the driver while driving. [21] makes use of three dissimilar deep neural networks for different hand-crafted features to detect fatigue. Weng et al. [22] recommended the use of hierarchical belief networks deep enough to incorporate high-level head and facial feature representations as essential drowsiness-related symptoms. Model compression technique has been proposed in [23] to deploy a deep learning inspired drowsiness detection system on an embedded board for daily use vehicles.

In order to avoid accidents, drowsiness of the driver is to be detected at the earliest. For this reason, proposed algorithm reduces the required computational time to recognize the driver's drowsiness in the following ways: (a) Facial region of

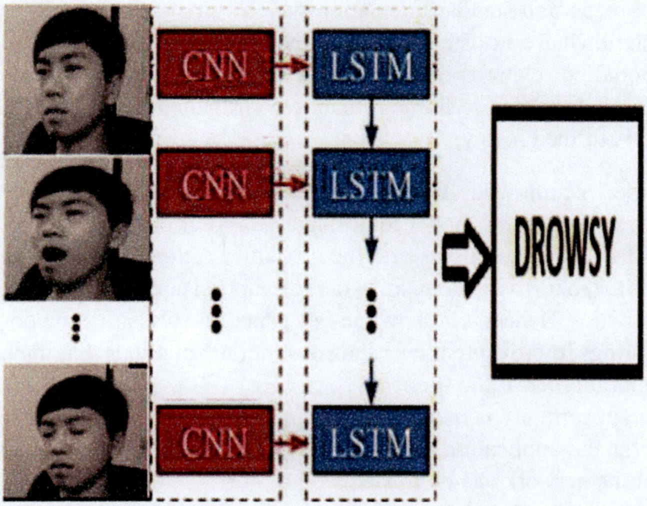

Fig. 1 The figure depicts the outline of the deep recurrent convolutional network [21] used in the proposed automated system for drowsiness detection. The input frames go through a deep CNN architecture followed by a Long-term recurrent network where temporal dependencies in the previously learnt convolutional features are learnt. The final classification is made to decide the intensity of driver's drowsiness

the driver alone is extracted in each frame (of video) using Haar Cascade Classifiers [24]. (b) Moreover, video processing time is lessened further by applying image prepossessing measures beforehand.

This paper uses the facial region of the driver as the primary inputs to the proposed 'Automated Bayesian Drowsiness Detection' (ABDD) system since eye-blinking and yawning have proved to be the best descriptors for drowsiness detection in videos so far [11, 12]. The proposed ABDD system indicated in Fig. 1 comprises of deep hierarchical visual feature extractor Inception V3 network [25] to extract spatial features. Subsequently, ABDD employs long-term recurrent network (or Long Short-Term Memory networks—LSTM) for long-range temporal learning in videos. Thus, ABDD system introduces an end-to-end deep learning technique that depends on computer vision to detect the driver's drowsiness.

As the network is trained end-to-end in the proposed method, it becomes advantageous for the vision-based task of drowsiness detection. Unlike other contemporary models [23, 26] that accept a fixed visual input representation and perform straightforward averaging in temporal domain for sequential analysis, ABDD system uses recurrent convolutional models which are deep enough to efficiently learn spatial and temporal feature representations. In Drowsiness detection where static and shallow temporal methods have been used before, LSTM-style RNN (Recurrent Neural Network) used in this paper can lead to functional detection improvements when there is ample learning data. The proposed network uses dropout at test time to make Bayesian predictions on the video data. These predictions are then used to compute the mean and standard deviation which can aid the user in deciding if a certain prediction can be trusted. Several previous works [21, 22] focused on learning dimensional correlations of static images that are disadvantageous due to the lack of class labels related to frame semantics. The salient features of the proposed technique are outlined below:

(a) This paper recommends using a hybrid learning framework deep enough to detect drowsiness and model long-range temporal clues in differential-length input video sequences along with the miniature patterns of spatial motion.
(b) LSTM is taken up to illustrate long-range temporal cues in addition to the spatial characteristics. Hence, it is shown in this paper that classification done based on LSTM brings improvement over the conventional methods that intend to ignore the temporal order of the frames.
(c) Bayesian uncertainty is included in the prediction of the model which is highly critical for the application of drowsiness detection. This suggests whether the result of the network can be trusted well or not.
(d) Through a comprehensive comparison of experimental results, it is revealed that the proposed method outperforms several previously proposed techniques. Moreover, a decent competitive performance is achieved on the specific NDD (i.e., NTHU drowsy Driver Dataset [22]) benchmark.

In this paper, network is trained on the NDD dataset consisting of videos of drivers while driving at different levels of consciousness [22].

The paper is divided as follows: Sect. 1 explains the details of NDD dataset used for training. Section 2 gives details about the proposed ABDD system. Section 3 brings about the experimental results and Sect. 4 draws the conclusions.

2 Dataset

The NTHU drowsy Driver Dataset (NDD) has videos compiled by the NTHU Computer Vision Lab [22]. NDD consists of videos of 36 subjects having different ages and distinct ethnicities recorded while they are driving in a simulated environment with a driving wheel and pedals. For each subject, a separate video compilation is provided for Alert and Drowsy driving in five varying situations like (a) Bare faces, (b) Faces wearing normal glasses, (c) Bare faces at night, and (d) Faces wearing glasses at night, and (e) Faces with sun glasses. The videos are made available in a 640×480 resolution AVI format. Active Infrared (IR) illumination module is used to collect video data at night. The videos are 1.5 min long consisting of sequences with high PERCLOS drowsiness-related symptoms like yawning, eye-blinking, and head-nodding and 1 min long non-drowsiness-related combinations with low PERCLOS actions like talking, laughing, and active gazing. The original compilation of 360 videos of [22] is quite raw and needs comprehensive video pre-processing for effective learning of drowsiness features. Hence, these videos are clipped into sequences

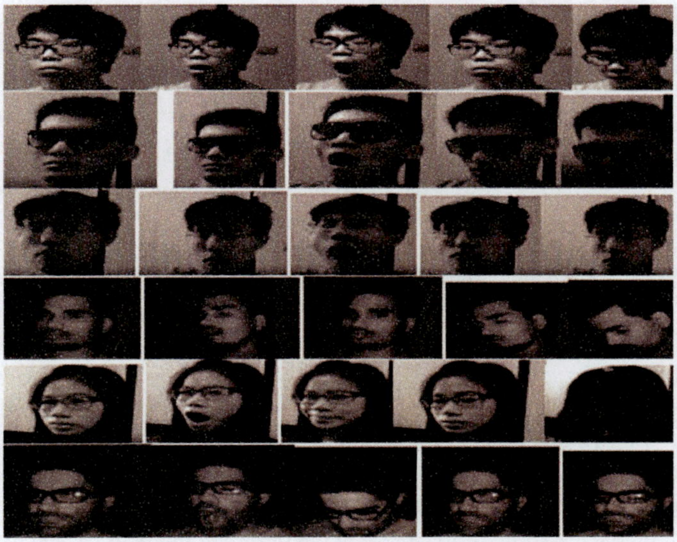

Fig. 2 The figure depicts the sequences of frame from the videos of NDD Dataset proposed in [22]. Five different categories have been considered in the dataset precisely: **a** Bare faces, **b** Faces wearing normal glasses, **c** Bare faces at night, **d** Faces wearing glasses at night, and **e** Faces with sun glasses. The proposed ABDD system learns crucial spatial-temporal features in frame sequences as shown in the figure to detect drowsiness

corresponding to the important scenarios showing drowsiness and non-drowsiness-related symptoms as described above. The resulting dataset summarizes around 1400 video clips, each of 15–20 sec duration. Some example video sequences of different settings are shown in Fig. 2. These videos are taken in different environmental conditions which gives them global robustness to background and environmental variations.

3 Automated Bayesian Drowsiness Detection System

This section elucidates the details of the proposed 'Automated Bayesian Drowsiness Detection' (ABDD) System for drivers. This system uses a recurrent convolutional architecture for the classification process. It consists of a deep Inception-based network as a spatial feature extractor followed by a LSTM network which provides an improved recognition of visual representations over long-range recursion. The proposed approach provides a novel optimized strategy towards end-to-end mapping of input pixels to predict static output. Figure 1 illustrates the essence of the proposed approach.

3.1 Convolutional Feature Extractor

The proposed ABDD system's CNN base is the Inception V3 [25] model, an inception-based convolutional architecture. The original network is trained on a subset of the ImageNet [27] dataset consisting of 1.2M images from the ILSVRC-2012 [28] classification challenge. Compared to the current deep networks [29, 30]; Inception V3 is simpler and computationally efficient in large-scale visual perception tasks. The architecture and parameters of the network are summarized in Fig. 3. This facilitates an improved and accelerated training on the relatively small driver drowsiness detection (NDD) dataset used and avoids overfitting as well. The spatial CNN features are learnt over the sampled individual frames from the input videos from the CNN model using transfer learning approach. The input videos are decoupled into static individual frames. Each visual input frame x_t passes through $\phi\ V(.)$ feature transformation with Inception V3 network parameters V to produce a $\phi\ V(x_t)$ fixed-length vector representation. The ϕ visual transformation feature matches the activations in the deep network's last layer. $\phi\ V(.)$ exist invariantly and independently at every step of the time, thereby helping to make vital convolutional deductions that are parallel to all input steps.

3.2 Temporal Modeling with Long-Short-Term Memory

The spatial stream CNN's training process considers a static visual frame at a time while rejecting the frames' temporal order. To deal with this limitation; LSTM network that has been successfully applied to several recurrent applications of speech

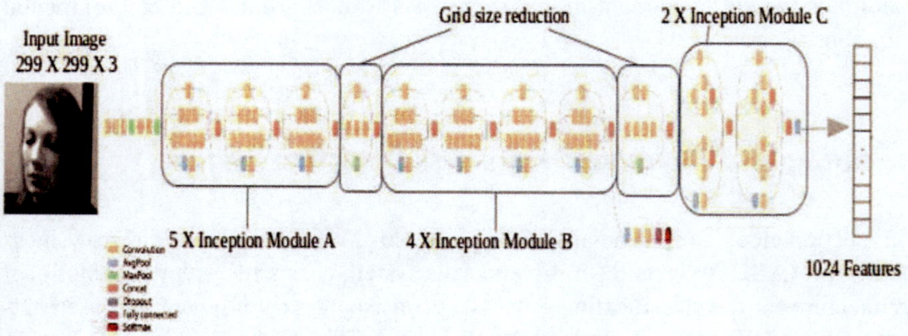

Fig. 3 The figure shows the architectural representation of the 42-layered deep Inception V3 [25] network used by ABDD system. It is complex neural network design heavily architectured in order to improve classification. The deep layers incorporate three Inception design-based complex modules named A, B, and C. The input image (299 × 299 × 3) goes through multiple convolutional and pooling layers. The output from just before the final classification layer in the original architecture is extracted as a 1024 feature tensor later used by subsequent recurrent network

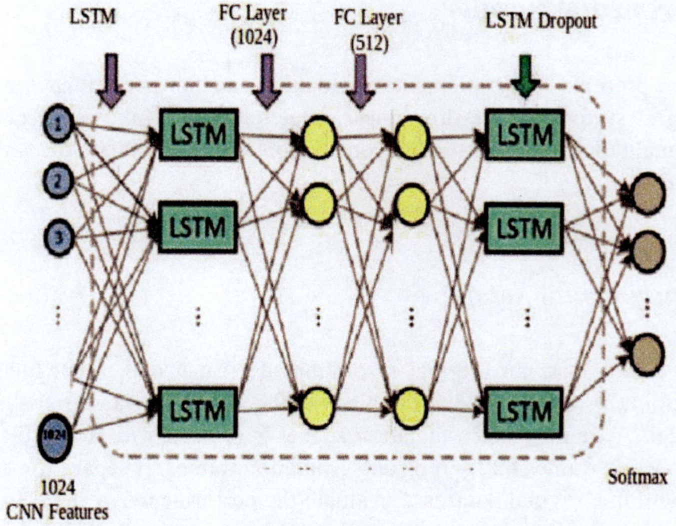

Fig. 4 The recurrent model used by the ABDD system is a deep LSTM network inspired from [31]. The first layer has 1024 features input from each frame in a video from the dataset. It is a 4096-wide LSTM layer along with a dense layer, with a dropout of 0.5 in between. The final Fully Connected (FC) layer performs classification based on temporal dependencies learnt from the convolutional features over a video

recognition [32] and image captioning [33, 34] is used. Using LSTM network ensures to configure boundless adaptive information related to drowsiness in video sequences. Traditionally, Recurrent Neural Networks suffer with the major problem of 'vanishing gradient'. LSTMs are a variation to the classic recurrent architecture with addition of memory units, updation of which can be controlled thus making it effective in numerous extensive collection of sequential frameworks. The input descriptions in the ongoing time increment x_t are mapped to the output term z_t through a train of hidden state $h\{.\}$, and therefore the LSTM learning transitions are to be consecutive. The video frames are passed through the CNN stream sequentially and the outputs of ϕ V from the forward pass are stacked as a feature map for each video. These characteristics are transferred to the LSTM model that follows. Using the hidden state from the last layer, prediction score is attained, respectively, for every time segment with softmax transformation. The network is as described in Fig. 4. In our case, end-to-end LSTM network fine tuning is gaining and yielding comparable results as the contemporary CNNs. By training a deep LSTM model, the use of a recurrent network gets rid of the complicated multi-step pipelines to detect drowsiness as used in [21].

4 Experimental Results

The section presents the implementation details and the performance evaluation of the ABDD system for the drowsiness detection task on the NDD dataset. An elaborate qualitative and quantitative comparison with the state-of-the-art methods is presented.

4.1 Experimental Setup

The input videos have an original resolution of 640×480. Each image frame extracted from the dataset is downsampled to 299×299 for training the inception-based model. A training and evaluation subset is created with a 80:20 split ratio resulting in video frames for training and evaluation frames. A separate test subset is provided with the original dataset to evaluate the performance of the network. The deep CNN network is trained on GTX Tesla GPU system with the following training parameters: 4000 training steps, 0.01 learning rate, batch size of 10 and dropout of 0.5, and Adam regularization. The experimental results from training are outlined in the next section. In order to learn the temporal features by the subsequent recurrent LSTM network, a compilation of features learnt by the convolutional network is obtained for each video in the dataset. For this purpose, videos are forward passed through the trained Inception network. Subsequently, feature output is obtained from the last convolutional layer which is then accumulated in cache memory. With 1024 features learned from each input frame, the dimension of an individual feature set corresponds to $(* \times * \times 1024)$. The features are forwarded to the deep LSTM network as described in sub-sections 3.1 and 3.2. Being a static network, the input features are prepared by zero-padding to get input dimension of $1000 \times 1000 \times 1024$. The deep LSTM model learns useful temporal dynamics over the convolutional features, the results of which are compiled in the next section.

4.2 Performance

Efficiency of the proposed ABDD system is assessed on the NDD dataset. The results of convolutional network are shown by means of the classification accuracy (ACC) over the test videos and the area covered by the precision-recall curve (AUC). Figure 5 elucidates the performance of the Inception network over five different categories considering these two classification performance measures.

It is observed that (a) The network performs well in the daylight conditions like Bareface and Glasses. (b) The reduction in the accuracy of the network in the case of Sunglasses is because of its inability to learn essential drowsiness features such as eye-blinking pattern and gaze activity due to covered eye region. (c) The drop

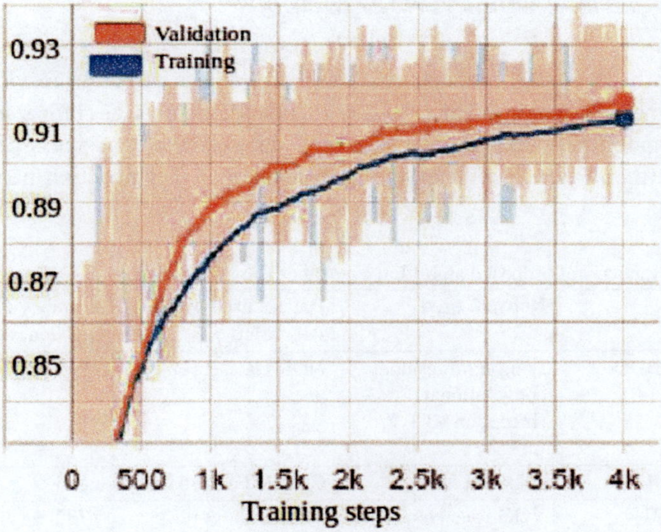

Fig. 5 Classification accuracy of the CNN Inception network [25] over 4000 training steps. The validation and training accuracies are obtained as 0.918 and 0.91, respectively

Table 1 This Table shows the accuracy of Drowsiness detection on the test dataset for different subjects

ID	004	022	026	030	Average
ACC	0.6621	0.9533	0.9725	0.9876	0.942

in the network's performance on the night-vision videos is expected because of lack of luminosity and unnecessary shadows in the input frames due to camera settings. Figure 5 illustrates the classification accuracy of the deep CNN network on training and validation videos over 4000 training steps. The network achieves a classification accuracy of 0.91 over the training dataset and 0.918 validation accuracy. The subsequent LSTM network is fed with the 1024 size feature space from the prior network to learn essential temporal dependencies in the input videos. The network is evaluated over the test dataset from the NDD dataset, the results of which are compiled in Table 1.

The overall classification accuracy obtained by using ABDD system is 0.942 ACC. As expected our system is shown to obtain more significant improvements when long-term recurrent learning is combined with the conventional convolutional learning for drowsiness detection in videos.

4.3 Comparison with Previous Methods

There have been several attempts made for drowsiness detection in videos using feature learning by deep networks [20–23, 26]. Table 2 depicts the comparative results of the compiled networks used over several datasets for drowsiness detection in drivers. Further, Fig. 6 elucidates the performance of the Inception network over

Table 2 Advantages of proposed algorithm over other known algorithms

Approach	Network used	Dataset used by the Algorithm	Accuracy of Drowsiness Detection
Proposed ABDD system	Long-term residual convolutional (Inception v3 + LSTM)	NDD Dataset [22]	0.942
Reddy et al. [23]	Alex Net	Custom Facial dataset	0.895
Park et al. [21]	VGG Face Net, Alex Net, Flow Image Net	NDD Dataset [22]	0.732
Dwivedi et al. [20]	Shallow CNN	Custom Facial dataset	0.9233
Jabbar et al. [26]	MLP	NDD dataset	0.8712
Weng et al. [22]	Hybrid Deep Belief Network	NDD dataset [22]	0.7520

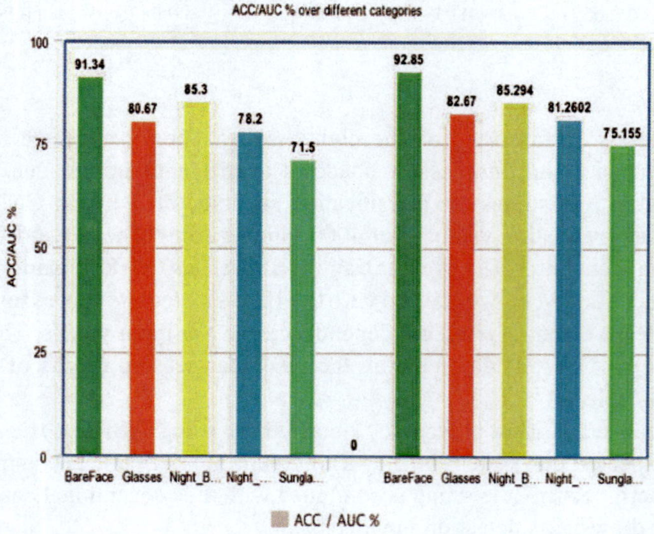

Fig. 6 The figure shows the performance accuracy of the Inception V3 on the NDD dataset over fiver different categories (BareFace, Glasses, Bareface-Night, Glasses-Night, and Sunglasses). The accuracy is determined in terms of classification accuracy over the test dataset (ACC) and the area covered by the precision-recall curve (AUC)

five different categories considering these two classification performance measures. Clearly the proposed algorithm outperforms the preceding networks trained on the same NDD dataset [21, 22, 26] owing to its novel approach of temporal feature learning for drowsiness detection. The inception model used by the proposed system is far more accurate and efficient than the VGGFacenet, AlexNet used in [21], hence the improvement in accuracy. The facial landmark detection by [20] constraints its network from learning essential facial features apart from predefined landmark features. Moreover, the proposed double-deep network is capable of learning more number of useful features over both spatial and temporal domain from the videos. This is why, accuracy is increased in the proposed system over the preceding network's performance by a wide margin.

5 Conclusion

The proposed ABDD System uses deep architecture based on recurrent convolutional network for detecting the drivers' drowsiness. This network is based on deep learning of facial features exclusive to drowsiness from input video stream. The framework first uses deep Inception network to identify and learn crucial facial features given an input frame of driver using spatial learning. It is then followed by a recurrent network build upon it that accepts the feature maps and further decomposes them to temporal sub-spaces to extract the long-range dependencies within the frames in videos. This approach gets rid of the static learning of spatial features for drowsiness deployed by several previous papers and attempts to infuse temporal learning with classic convolutional training to learn effective features of drowsiness in humans via input videos. The proposed model also produces uncertainty in addition to the mean estimates which have been shown to perform superior to the contemporary methods. Thus, ABDD system accomplishes a competitive performance on the current dataset for driver drowsiness. Predicting drowsiness with more accuracy in drivers helps in preventing the vehicular accidents to a larger extent. Identifying drowsiness in real time using non-facial features like head movements and body gestures would be a promising future direction for driver's drowsiness detection because the system would be able to combine its capabilities to learn facial expressions along with other hand-crafted features over time.

References

1. Husar P (2010) Eyetracker Warns against Momentary Driver Drowsiness, 2010. http://www.fraunhofer.de/en/press/research-news/2010/10/eye-tracker-driver-drowsiness.html
2. Drowsy Driving NHTSA reports (2017). https://www.nhtsa.gov/risky-driving/drowsy-driving
3. Liu CC et al (2009) Predicting driver drowsiness using vehicle measures: Recent insights and future challenges J Saf Res 40(4):239–245

4. Forsman PM et al (2013) Efficient driver drowsiness detection at moderate levels of drowsiness. Accident Anal Prevent 50:341–350
5. Deram P (2004) Vehicle based detection of inattentive driving for integration in an Adaptive Lane Departure Warning System-Distraction detection. Master thesis, Royal Institute of Technology, 2004
6. Jarek K et al (2009) Steering wheel behavior based estimation of fatigue. In: Proceedings of Fifth International Driving Symposium on Human Factors in Driver Assessment, Training and Vehicle Design
7. Berglund J (2007) In-vehicle prediction of truck driver sleepiness-steering related variables. Master thesis, Linkoping University, 2007
8. Khushaba RN, Kodagoda S, Lal S, Dissanayake G (2011) Driver Drowsiness Classification Using Fuzzy Wavelet-Packet-Based Feature-Extraction Algorithm. IEEE Trans Biomed Eng 58(1):121–131
9. Mardi Z et al (2011) EEG-based drowsiness detection for safe driving using chaotic features and statistical tests. J Med Sig Sens 1(2)
10. Chia Chieh T et al (2005) Development of vehicle driver drowsiness detection system using electrooculogram (EOG). In: Proceedings of the CCSP, Kuala Lumpur, Malaysia, 2005, pp 165–168
11. Fan X et al (2009) Yawning Detection Based on Gabor Wavelets and LDA. In: Journal of Beijing University of Technology, 2009
12. Zhang Z et al (2010) A new real-time eye tracking based on nonlinear unscented Kalman filter for monitoring driver fatigue. In: J Control Theory Appl 8(2)
13. Omidyeganeh M et al (2016) Yawning detection using embedded smart cameras. IEEE Trans Instrum Meas 65(3)
14. Oyini Mbouna R et al (2013) Visual analysis of eye state and head pose for driver alertness monitoring. In: IEEE Trans Intell Transp Syst 14(3)
15. Mittal A et al (2016) Head movement-based driver drowsiness detection: A review of state-of-art techniques. Proc ICETECH, Coimbatore
16. Lee SJ et al (2011) Real-time gaze estimator based on driver's head orientation for forward collision warning system. In: IEEE Trans Intell Transp Syst 12(1)
17. Horng W-B et al (2004) Driver fatigue detection based on eye tracking and dynamic template matching. Proc ICNSC, Taiwan, May
18. Sommer D et al (2010) Evaluation of PERCLOS based current fatigue monitoring technologies. Proc IEEE Eng Med Biol Soc, Buenos Aires
19. Choi I-H et al (2016) Real-time categorization of driver's gaze zone using the deep learning techniques. Proc IEEE BigComp, Hong Kong, pp 143–148
20. Dwivedi K et al (2014) Drowsy driver detection using representation learning. Proc Gurgaon, IEEE IACC, pp 995–999
21. Park P et al (2016) Driver Drowsiness Detection System Based on Feature Representation Learning Using Various Deep Networks. Proceedings ACCV 2016 workshop, Taiwan, Nov 2016, pp 154–164
22. Weng, C-H et al (2016) Driver drowsiness detection via a hierarchical temporal deep belief network. In: Proceedings ACCV 2016 workshop, Taiwan, Nov 2016, pp 117–133
23. Reddy B et al (2017) Real-time driver drowsiness detection for embedded system using model compression of deep neural networks. Proc Honolulu, HI, IEEE CVPRW, pp 438–445
24. Viola P, Jones M (2001) Rapid object detection using a boosted cascade of simple features. Proc IEEE CVPR, Kauai, HI, USA
25. Szegedy C et al (2016) Rethinking the inception architecture for computer vision. Proc Las Vegas, NV, IEEE CVPR, pp 2818–2826
26. Jabbar R et al (2018) Real-time driver drowsiness detection for android application using deep neural networks techniques. Procedia Comput Sci 130:400–407
27. Deng J et al (2009) ImageNet: A large-scale hierarchical image database. Proc Miami, FL, IEEE CVPR, pp 248–255

28. Russakovsky O et al (2015) ImageNet large scale visual recognition challenge. Int J Comput Vis 115(3)
29. Krizhevsky A et al (2017) ImageNet classification with deep convolutional neural networks Commun ACM 60(6)
30. Liu S, Deng W (2015) Very deep convolutional neural network based image classification using small training sample size. Proc Kuala Lumpur, IAPR ACPR, pp 730–734
31. Zhu W et al (2016) Co-occurrence feature learning for skeleton based action recognition using regularized deep LSTM networks. In: Proceedings AAAI conference on artificial intelligence, Phoenix, Arizona
32. Graves A et al (2013) Speech recognition with deep recurrent neural networks. Proc IEEE ICASSP, Vancouver, Canada, May
33. Donahue J et al (2017) Long-term recurrent convolutional networks for visual recognition and description. IEEE Trans Pattern Anal Mach Intell 39(4):677–691
34. Venugopalan S (2015) Translating videos to natural language using deep recurrent neural networks. In: Proceedings conference of the north american chapter of the association for computational linguistics: human language technologies. Denver, Colorado, Jun, p 2015

Enhancement of Bandwidth and VSWR of Double Notch E-Shaped Inset-Fed Patch Antenna

Pavada Santosh and Prudhvi Mallikarjuna Rao

Abstract This paper presents the design and performance evaluation of Double Notch E-Shaped Inset-Fed patch antenna for high frequency applications. In recent times, microstrip patch antennas are gaining a lot of importance for enhanced communication. In this work, a single notch E-shaped patch antenna and a double notch E-shaped patch antenna are simulated and the performance characteristics are compared, in which double notch E-shaped patch antenna exhibits better performance characteristics like Bandwidth, Return loss, and VSWR. An Inset feed is used as a feeding technique for an efficient power transfer. RT Duriod is used as a substrate which has a dielectric constant of 2.2. CST 2015 electromagnetic tool has been used to obtain the simulated results. The results are presented at the end.

Keywords CST microwave studio · Bandwidth · VSWR · Return loss · Inset feed · Microstrip patch antenna

1 Introduction

Nowadays Microstrip patch antennas have evolved so much in the field of wireless communications. Their main applications are unlimited for the reason that they are light weighted, small sized, and are very less complicated when manufacturing the prototype. One disadvantage with these antennas is that they provide a very narrow bandwidth [1]. The latest investigations have established many ways to overcome these drawbacks, and vast approaches have been made to modify the shape of the patch, experimenting with different substrate materials, and also by using different techniques for feeding the antenna [2]. To design an antenna, conducting and nonconducting materials are used [3]. The feeding technique plays a vital role in efficient

P. Santosh (✉) · P. Mallikarjuna Rao
ECE Department, AU College of Engineering (A), Visakhapatnam 530003, AP, India
e-mail: santosh.pavada@gmail.com

P. Mallikarjuna Rao
e-mail: pmraoauece@yahoo.com

T. Laxminidhi et al. (eds.), *Advances in Communications, Signal Processing, and VLSI*, Lecture Notes in Electrical Engineering 722,
https://doi.org/10.1007/978-981-33-4058-9_31

power transfer. For a proper matching of impedance between the feedline and the patch, the element is triggered for radiation modes using different feeding techniques. Due to maximum current at the middle of the patch, the impedance will be null at the center and because of zero current at the open circuit edge, the edge closer to the source has infinite impedance [4]. Hence, by moving the feed point closer to the center of the patch, an input impedance of 50 ohms is selected for maximum power transfer.

2 Inset Feed

The convenience with this feeding is that it exhibits a consistent design because the width of the operating strip is small when compared to the patch. The main operation of this inset cut in the antenna is matching the impedance to the feedline without any additional matching element [5]. By a proper selection of dimensions and inset cut position, an efficient matching can be accomplished.

3 Design Equations

From this standard antenna design equations, the following parameters of rectangular patch antenna have been calculated [6].

$$W = \frac{c}{2f_r}\sqrt{\frac{2}{\in_r + 1}} \tag{1}$$

$$\in_{reff} = \frac{\in_r + 1}{2} + \frac{\in_r - 1}{2}\left(1 + 12\frac{h}{w}\right)^{\frac{-1}{2}} \tag{2}$$

$$L_{eff} = \frac{c}{2f_r\sqrt{\in_{reff}}} \tag{3}$$

$$L = L_{eff} - 2\Delta L \tag{4}$$

$$\Delta L = 0.421h\frac{(\in_r + 0.3)\left(\frac{w}{h} + 0.264\right)}{(\in_r - 0.258)\left(\frac{w}{h} + 0.8\right)} \tag{5}$$

Table 1 Design Parameters of Double Notch E-Shaped Inset-Fed Patch Antenna

S.No.	Antenna Parameters	Double Notch E-shaped Antenna
1	Resonant frequency (f_r)	14 GHz
2	Length of the patch (L)	6 mm
3	Width of the patch (W)	8 mm
4	Dielectric constant ($\in_r$)	2.2
5	Feed line Length & Width	6.89 mm & 1.78 mm
6	Substrate (length x breadth)	100 × 100 mm
7	Substrate height	1.5 mm

4 Design Parameters

Design parameters of Double Notch Inset-fed E-shaped patch antenna at 14 GHz are calculated based on the following standard antenna design equations, which are presented in the Sect. 3.

5 Single Notch E-Shaped Patch Antenna at 14 GHz

In many applications such as satellite communications, spacecraft, and aircraft, these microstrip patch antennas are being used extensively. For the fabrication of an antenna, a patch, feedline, ground, and a high conducting metal is used. While designing the patch antenna, the return loss is kept as low as possible for a good transmission of power. Depending upon the application, we choose the material for a substrate. A substrate material is chosen based on the loss tangent and dielectric constant which varies with frequency and temperature. Permittivity and substrate thickness shows the electrical properties of the antenna. In order to minimize the harmonics at the higher order modes, a notch on the radiating patch is proposed. A notch on the patch controls the flow of current on the top of patch [7].

The Fig. 1 represents the single notch E-shaped patch antenna which is resonating at 14 GHz. This antenna is operated in the Ku band which has a wide range of applications mainly satellite applications. From Figs. 2, 3, and 4, we can witness that the single notch inset-fed patch antenna gives a bandwidth of 1.3 GHz, VSWR value of 1.87, return loss of −19.30 dB, and gain of 2.66 dB, respectively.

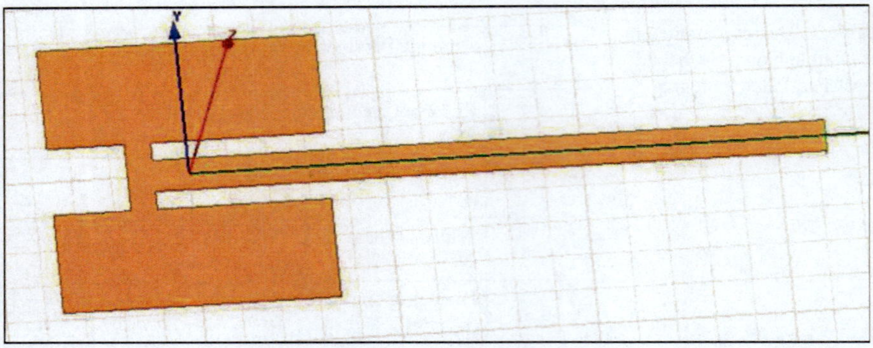

Fig. 1 Single notch E-shaped patch Antenna

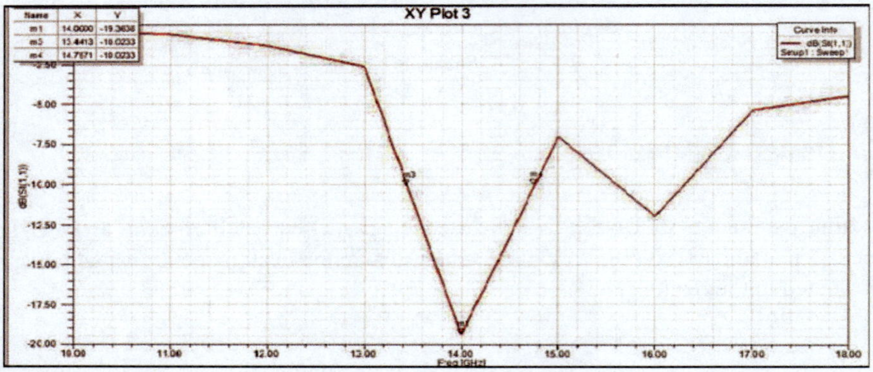

Fig. 2 Return Loss of Single notch Antenna

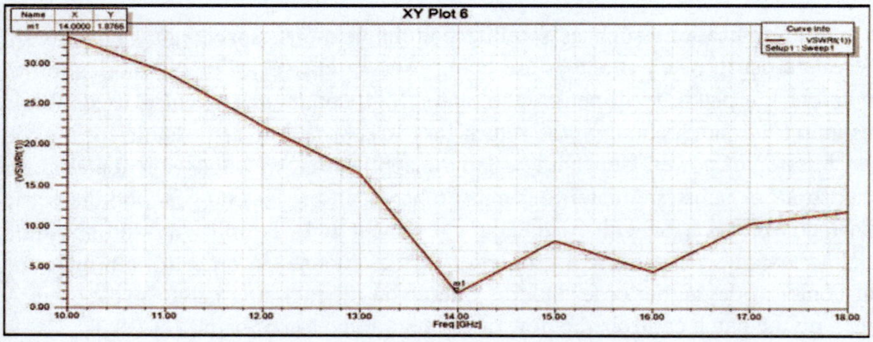

Fig. 3 VSWR of Single notch Antenna

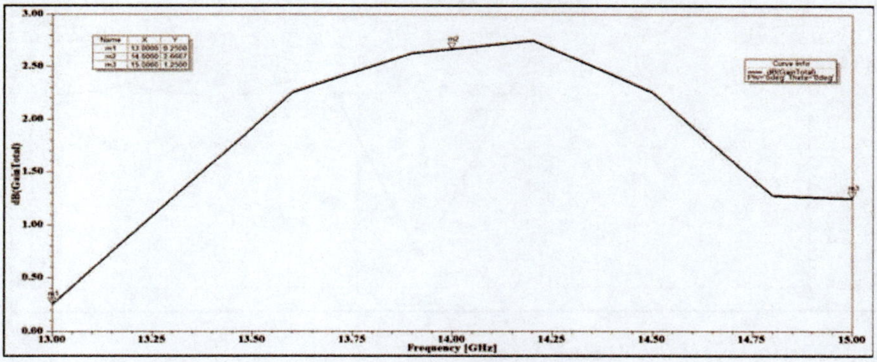

Fig. 4 Gain of Single notch Antenna

6 Double Notch E-Shaped Patch Antenna at 14 GHz

Figure 5 represents the Double Notch E-shaped patch antenna which is resonating at 14 GHz in Ku band. From Figs. 6, 7, and 8, we can observe that a bandwidth of 1.7 GHz, VSWR value of 1.09, return loss of −26.70 dB and a gain of 2.35 dB has been achieved. It shows that double notch inset-fed patch antenna gives better performance characteristics when compared to single notch inset-fed patch antenna.

From the Tables 2, 3, and 4, it is observed that the bandwidth, VSWR, and Return Loss of Double Notch E-shaped Inset-fed antenna exhibit better results than the Single Notch E-shaped Inset-fed antenna.

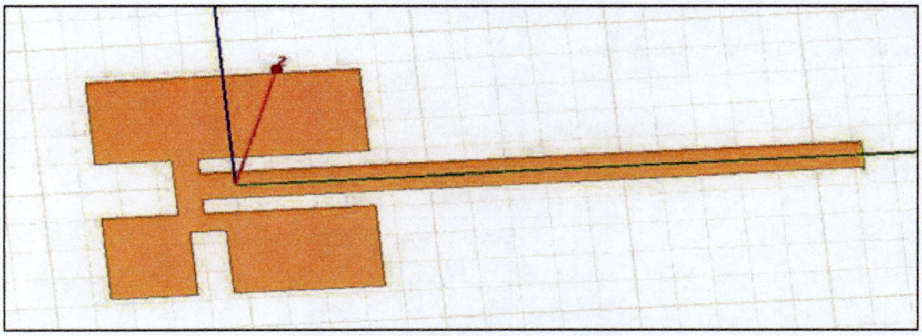

Fig. 5 Double notch E-shaped patch Antenna

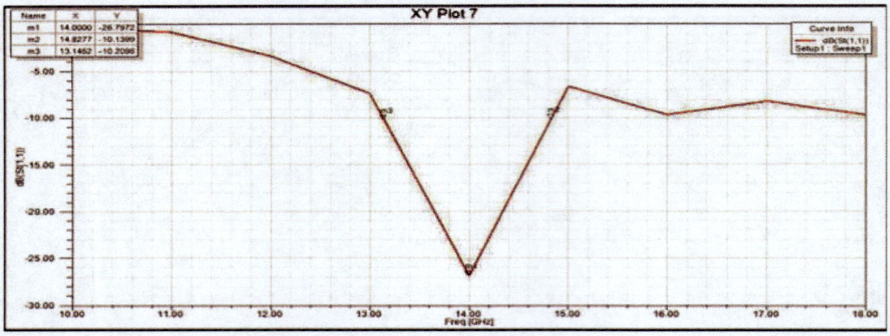

Fig. 6 Return Loss of Double notch Antenna

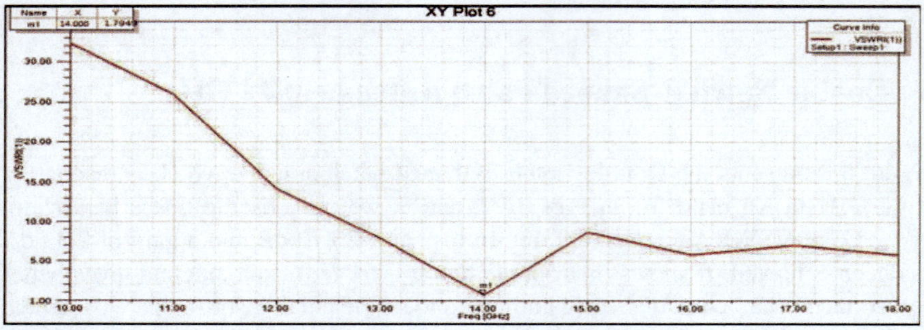

Fig. 7 VSWR of Double notch Antenna

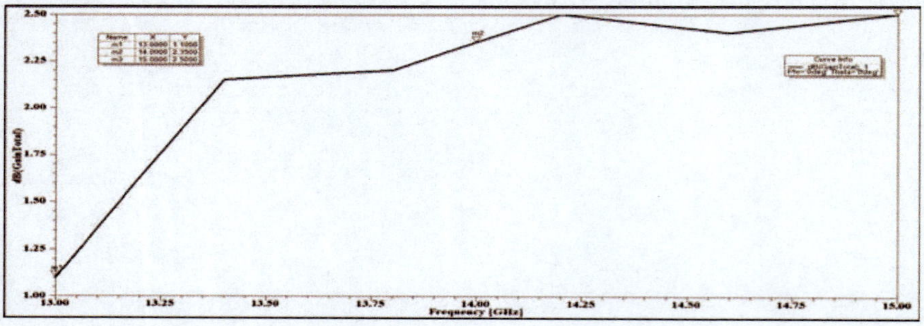

Fig. 8 Gain of Double notch Antenna

Table 2 Bandwidth

S. No.	Antenna	Value (GHz)
1	Single Notch E-shaped Inset-fed antenna	1.3
2	Double Notch E-shaped Inset-fed antenna	1.7

Table 3 VSWR

S. No.	Antenna	Value
1	Single Notch E-shaped Inset-fed antenna	1.87
2	Double Notch E-shaped Inset-fed antenna	1.79

Table 4 Return Loss

S. No.	Antenna	Value (dB)
1	Single Notch E-shaped Inset-fed antenna	−19.36
2	Double Notch E-shaped Inset-fed antenna	−26.79

Table 5 Gain

S. No	Antenna	Gain(dB)
1	Single Notch E-shaped Inset-fed antenna	2.66
2	Double Notch E-shaped Inset-fed antenna	2.35

7 Conclusion

In this present work, a Single and Double notch E-shaped patch antennas are designed and simulated at 14 GHz for high frequency applications. With the proposed Double notch E-shaped patch antenna, better results have been obtained for the parameters like Bandwidth, Return loss, VSWR and gain. In comparison with Single notch E-shaped patch antenna, the Double notch E-shaped patch antenna gives 30.76% improvement in bandwidth, return loss is decreased from −19.36 to −26.79 dB and VSWR value is reduced from 1.87 to 1.79. These simulated results show that Double notch E-shaped patch antenna exhibits better performance characteristics. The designed antenna is used for fixed satellite communication services. This proposed patch antenna will be fabricated and the experimental results shall be compared with respect to simulated results for the validation of the antenna in future due course.

References

1. Ammann M (1997) Design of rectangular microstrip patch antenna for the 2.4 GHz Band. Appl Microw Wireless: 23–34
2. Soh PJ, Rahim MKA, Asrokin A, Abdul Aziz MZA (2007) Design, modeling, and performance comparison of feeding techniques for a microstrip patch antenna. J Technol 47:103–120 (Universiti technology Malaysia)
3. Carver KR, Mink JW (1981) Microstrip antenna technology. IEEE Trans Antennas Propagat AP-29:2–24
4. Deshmukh AA, Kumar G (2001) Compact broadband gap coupled shorted L-shaped micro strip antennas. In: International Symposium on IEEE antennas and propagation, vol. 1. IEEE, Baltimore, Maryland, pp. 106–109

5. Basilio LI, Khayat MA, Williams JT, Long SA (2001)The dependence of the input impedance on feed position of probe and microstrip line-fed patch antennas. IEEE Trans Antennas Propag AP-49:45–47
6. Balanis CA (2016) Antenna theory analysis and design, 2nd edn. John Wiley & Sons, Microstrip Antenna Design Handbook. Artec House Inc. Norwood, MA 1–68, 253-316
7. Rahim RA, Hassan SIS, Malek F, Junita MN (2012) A 2.45 GHz Harmonic suppression rectangular patch antenna. In: 2012 International symposium on computer applications and industrial electronics (ISCAIE 2012) 3–4 Dec 2012, Kota Kinabalu Malaysia

Multiple Adjacent Bit Error Detection and Correction Codes for Reliable Memories: A Review

K. Neelima and C. Subhas

Abstract The memories and registers are the critical components in a processor which are prone to errors like single bit upset or multiple bit upsets due to radiation effects. The error detection and correction codes are used to recover the memory from storing erroneous data or address which ensures reliability in operation. This review paper projects a brief of the codes that handle the errors in memories. The error detecting and correcting codes are capable of correcting errors from one bit to three adjacent bits. They operate based on the parity bits generated from data bits which are used to encode the data for transmission which are based on XOR gates. The decoding process varies for various methods which have syndrome calculation and error masking capabilities. The syndrome specifies the location of error if its value is non-zero. The error masking capability is achieved at a trade off with additional hardware. As the number of parity or redundant bits increases, multiple errors can be detected or corrected. The EDAC codes are assessed based on the metrics like the delay, number of redundant bits, number of errors detected, number of errors corrected, etc. Among the EDAC Codes, the matrix codes with QAEC decoding prove to be a better choice for memories.

Keywords Adjacent errors · Error detection and correction codes · Random errors · Redundancy

1 Introduction

The evolution in technology ensures fast and smart devices but due to downscaling in device dimensions, there exists a chance to produce radiation particles that induce

K. Neelima (✉)
Department of ECE, JNTUA, Anantapuramu 515002, India
e-mail: neelumtech17@gmail.com

C. Subhas
Professor and Head of ECE Department, Vice-Principal, JNTUA College of Engineering, Kalikiri 517234, India
e-mail: schennapalli@gmail.com

T. Laxminidhi et al. (eds.), *Advances in Communications, Signal Processing, and VLSI*, Lecture Notes in Electrical Engineering 722,
https://doi.org/10.1007/978-981-33-4058-9_32

bit upsets in critical elements like memories and registers. As memories and registers are the key components in any processor, then for reliable applications the device components need to be ensured for error free operation. This requirement initiated the development of error detecting and correcting codes, especially for reliable applications like smart non-invasive wearable medical devices, memories and registers used while launching a satellite, etc.

The error detection and correction codes are usually evaluated based on metrics like redundancy or number of parity bits used, logic depth of the encoder and decoder circuits, size or area of the circuit, worst-case delay of the encoder and decoder circuits, number of errors detected, number of errors corrected, power dissipated, etc. Always there exists a trade off among the parameters like area, power dissipation and delay, i.e., the challenge is to reduce the power dissipation while maintaining less delay and area. Also, the requirement is that the redundancy must be kept low and the code has to detect or correct multiple errors.

As the memory is most influenced by the ionizing radiation effects which could create faults like stuck-at faults, address faults, transient faults, neighbourhood pattern sensitive faults and coupling faults which manifest as errors in data read from memory. The error mitigation can be performed in two ways, i.e., one by treating the memory as a matrix directly and perform the error detection and correction for both random and adjacent errors . The other by reading the data from the memory, rearranging it as a matrix and perform error detection and correction for evaluating its reliability.

In Literature, there exist various codes that have varying redundancy and errors correction capability. The Sect. 2 describes the various codes described in literature. The metrics used for evaluating the error detecting and correcting codes are mentioned in Sect. 3. The Sect. 4 consolidates the performance of the various codes in terms of various metrics. The scope of evolution for future EDAC codes is projected in Sect. 5 and finally the paper is concluded.

2 Error Detection and Correction Codes

2.1 Hamming Code

Hamming codes are proposed by R.W. Hamming that represent linear block error correcting codes. They can perform only single error correction and double error detection. The conditions imposed are parity bits have to be located at the respective places only. The code length is $n-2r-1$ and error correcting capability is $t = 1$(dmin $= 3$). The lexicographic matrix (12, 8) Hamming code is shown in Fig. 1.

The condition imposed on generator and parity check matrices of SEC-DED-DAEC codes are (i) all columns in H must be different and non-zero, (ii) all columns in H must have odd-weight with data columns whose weight is greater than one

Fig. 1 The lexicographic matrix for the shortened Hamming code (12, 8)

$$H=\begin{pmatrix} 0&0&0&0&0&0&1&1&1&1&1&1\\ 0&0&0&1&1&1&1&0&0&0&0&1\\ 0&1&1&0&0&1&1&0&0&1&1&0\\ 1&0&1&0&1&0&1&0&1&0&1&0 \end{pmatrix}$$

Fig. 2 Extended hamming Code for (21, 16)

$$H=\begin{pmatrix} 0&1&0&1&0&1&0&1&0&1&0&1&0&1&0&1&0&0&0&0&1\\ 1&1&1&1&1&1&0&0&1&0&0&0&0&0&0&1&1&0&1&0&1\\ 1&1&1&0&0&0&1&1&0&1&0&0&0&0&1&0&0&1&0&1&0\\ 1&0&0&1&1&0&1&1&0&0&1&0&0&1&0&1&1&1&0&0&0\\ 0&0&1&0&1&1&1&0&0&1&0&0&1&1&0&1&0&0&1&1&0 \end{pmatrix}$$

and (iii) the sum of two adjacent columns must be non-zero in H and (iv) all bit sequence chosen must be different in both rows and columns [1].

2.2 *Extended Hamming Code*

The Extended Hamming Codes use an additional parity bit to have Triple Adjacent Error detection capability. The condition is to get an even weight syndrome, for which, the Columns are arranged in weight order as odd odd even odd odd even… and so on. This code places the smallest weighted column value with the expected weight, if the expected weight is even, then it ignores the result obtained from the combination of any previous consecutive three columns.

Consider the data bits with 16-bits, 32-bits and 64-bits, the m−1 rows of the first (k/2 + m−1) columns of the check matrices correspond to the matrix of (k/2) data bits. The additional parity bit verifies the parity of the code word excluding the bit itself by either independently or by modifying the matrix as shown in Fig. 2. This can be used to correct up to 1/4th of adjacent errors in the encoded data bits.

The extended hamming code [1] can be further modified for high order error bits by using selective placement.

2.3 *Hsiao Code*

Hsiao codes [1] are optimized Hamming codes that require each column has to be different and non-zero for parity check matrix, every column contains an odd number of 1 s, the total number of 1 s should be minimum and the number of 1 s in each row should be nearly equal to the average number as shown in Fig. 3.

b0	b1	b2	b3	b4 u0	b5 u1	b6 u2	b7 u3	Encoding formulas	r0	r1	r2	r3	r4 u0	r5 u1	r6 u2	r7 u3	Syndrome bits
1	0	0	0	0	1	1	1	b0=u1+u2+u3	1	0	0	0	0	1	1	1	s0=r0+r5+r6+r7
0	1	0	0	1	0	1	1	b1=u0+u2+u3	0	1	0	0	1	0	1	1	s1=r1+r4+r6+r7
0	0	1	0	1	1	0	1	b2=u0+u1+u3	0	0	1	0	1	1	0	1	s2=r2+r4+r5+r7
0	0	0	1	1	1	1	0	b3=u0+u1+u2	0	0	0	1	1	1	1	0	s3=r3+r4+r5+r6

Fig. 3 The parity check matrix for an (8,4) Hsiao Code

The Hsiao codes are capable of detecting anyone random error but can correct up to 1/8th of adjacent errors of corrupted data word. Further, the Hsiao Codes are modified for double adjacent error correcting codes [2].

2.4 *Matrix Codes*

These codes are capable of correcting all single errors and multiple errors with a distance $\leq$L, where L$-$1 represents the maximum horizontal distance between the errors in the data word. As the errors in the data word can occur in a group of L consecutive columns, then by rearranging the word into a matrix [3], it can correct up to a distance of four -1 adjacent errors, where Ci and Dj are parity bits as shown in Fig. 4 that are evaluated as

$$C1 = X_1 \oplus X_3 \oplus X_5 \oplus X_7$$
$$C2 = X_2 \oplus X_4 \oplus X_6 \oplus X_8$$
$$D1 = X_1 \oplus X_{17}$$
$$D9 = X_9 \oplus X_{25}$$

Several modifications are made in the representation of matrix mapping of the code retrieved from memory [4, 5] which could correct from 1/8th to 1/4th of the errors in the retrieved data.

Fig. 4 Logical organization of 32 bit Matrix Codes

X1	X2	X3	X4	X5	X6	X7	X8	C1	C2
X9	X10	X11	X12	X13	X14	X15	X16	C3	C4
X17	X18	X19	X20	X21	X22	X23	X24	C5	C6
X25	X26	X27	X28	X29	X30	X31	X32	C7	C8
D1	D2	D3	D4	D5	D6	D7	D8		
D9	D10	D11	D12	D13	D14	D15	D16		

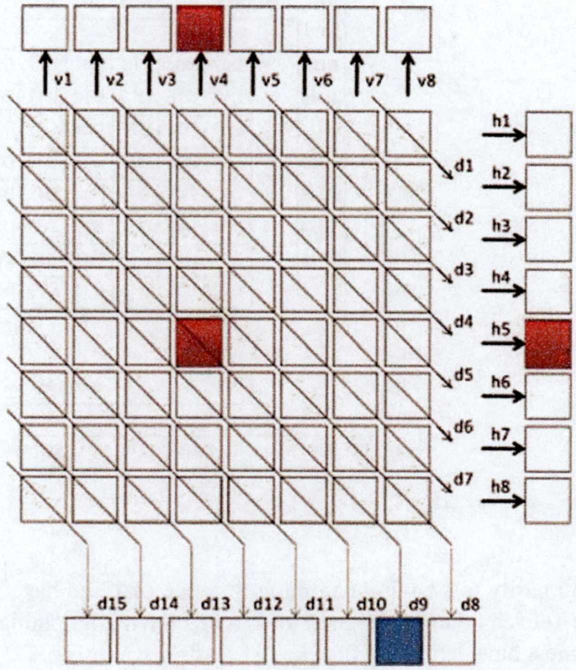

Fig. 5 HVD Code along with error detection and correction

2.5 HVD Codes –Horizontal, Vertical and Diagonal Codes

The Horizontal, Vertical and Diagonal codes use more number of parity than other codes but they ensure the adjacent error corrections which are well suited for Memories. The procedure is to map the read memory bits using an address into a matrix as shown in Fig. 5, then the horizontal, vertical and diagonal bits are calculated using xor function for encoding. These encoded parity bits are again given to the decoder along with the message bits retrieved from the memory. Again the HVD Parity bits are calculated and compared with for every memory location using all three basic directions, called as 3D HVD Codes [6]. Hence, by using these parity bits, errors can be corrected up to 1/4th of the adjacent errors in the entire memory.

The 4D Codes [7] are modified versions of the 3D HVD codes by using horizontal, vertical, forward diagonal (slash diagonal) and backward diagonal (back-slash diagonal) parity bits as shown in Fig. 6, but they could not correct more than the error correction capability of 3D HVD Code.

2.6 Ultrafast Codes

The ultrafast codes [8, 9] are the modified versions of Hsiao Codes where the constraints in the development of H-matrix are like (i) each column must be assigned

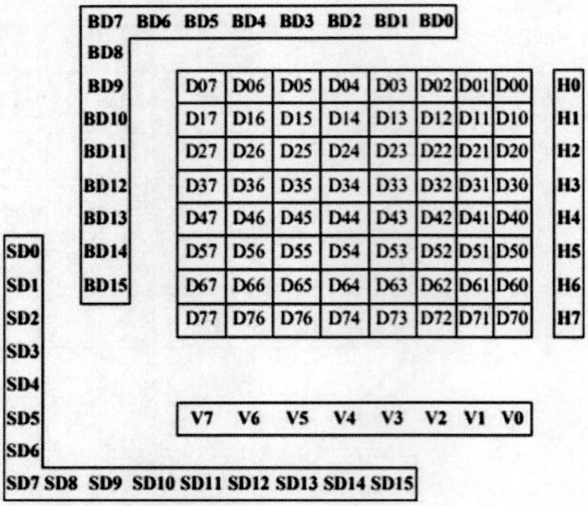

Fig. 6 4D codes

to parity bits having hamming distance of 1 and have to be different and non-zero, (ii) each column assigned to data bits having hamming distance of 2, (iii) each row must have hamming distance of 3 as shown in Fig. 7.

The encoding formulae for ultrafast codes are shown in Fig. 8.

The formulae for syndrome calculation are shown in Fig. 9, where the XOR function is used to evaluate the syndrome bits of received bits.

The formulae used for calculating the decoded data are

$$
\begin{aligned}
u0 &= (s0.s1) \oplus r8 \\
u1 &= (s0.s2) \oplus r9 \\
u2 &= (s1.s3) \oplus r10 \\
u3 &= (s2.s4) \oplus r11 \\
u4 &= (s3.s5) \oplus r12
\end{aligned}
$$

$$
H = \begin{pmatrix}
1 & 0 & 0 & 0 & 0 & 0 & 0 & 0 & 1 & 1 & 0 & 0 & 0 & 0 & 0 & 0 \\
0 & 1 & 0 & 0 & 0 & 0 & 0 & 0 & 1 & 0 & 1 & 0 & 0 & 0 & 0 & 0 \\
0 & 0 & 1 & 0 & 0 & 0 & 0 & 0 & 0 & 1 & 0 & 1 & 0 & 0 & 0 & 0 \\
0 & 0 & 0 & 1 & 0 & 0 & 0 & 0 & 0 & 0 & 1 & 0 & 1 & 0 & 0 & 0 \\
0 & 0 & 0 & 0 & 1 & 0 & 0 & 0 & 0 & 0 & 0 & 1 & 0 & 1 & 0 & 0 \\
0 & 0 & 0 & 0 & 0 & 1 & 0 & 0 & 0 & 0 & 0 & 0 & 1 & 0 & 1 & 0 \\
0 & 0 & 0 & 0 & 0 & 0 & 1 & 0 & 0 & 0 & 0 & 0 & 0 & 1 & 0 & 1 \\
0 & 0 & 0 & 0 & 0 & 0 & 0 & 1 & 0 & 0 & 0 & 0 & 0 & 0 & 1 & 1
\end{pmatrix}
$$

Fig. 7 A (16, 8) Ultrafast Code H-Matrix for parity generation

b0	b1	b2	b3	b4	b5	b6	b7	b8	b9	b10	b11	b12	b13	b14	b15	Encoding Formulae
								u0	u1	u2	u3	u4	u5	u6	u7	
1	0	0	0	0	0	0	0	1	1	0	0	0	0	0	0	b0=u0⊕u1
0	1	0	0	0	0	0	0	1	0	1	0	0	0	0	0	b0=u0⊕u2
0	0	1	0	0	0	0	0	0	1	0	1	0	0	0	0	b0=u1⊕u3
0	0	0	1	0	0	0	0	0	0	1	0	1	0	0	0	b0=u2⊕u4
0	0	0	0	1	0	0	0	0	0	0	1	0	1	0	0	b0=u3⊕u5
0	0	0	0	0	1	0	0	0	0	0	0	1	0	1	0	b0=u4⊕u6
0	0	0	0	0	0	1	0	0	0	0	0	0	1	0	1	b0=u5⊕u7
0	0	0	0	0	0	0	1	0	0	0	0	0	0	1	1	b0=u6⊕u7

Fig. 8 Encoding Formulae for Ultrafast Code

b0	b1	b2	b3	b4	b5	b6	b7	b8	b9	b10	b11	b12	b13	b14	b15	Decoding Formulae
								u0	u1	u2	u3	u4	u5	u6	u7	
1	0	0	0	0	0	0	0	1	1	0	0	0	0	0	0	b0=r0⊕r8⊕r9
0	1	0	0	0	0	0	0	1	0	1	0	0	0	0	0	b0=r1⊕r8⊕r10
0	0	1	0	0	0	0	0	0	1	0	1	0	0	0	0	b0=r2⊕r9⊕r11
0	0	0	1	0	0	0	0	0	0	1	0	1	0	0	0	b0=r3⊕r10⊕r12
0	0	0	0	1	0	0	0	0	0	0	1	0	1	0	0	b0=r4⊕r11⊕r13
0	0	0	0	0	1	0	0	0	0	0	0	1	0	1	0	b0=r5⊕r12⊕r14
0	0	0	0	0	0	1	0	0	0	0	0	0	1	0	1	b0=r6⊕r13⊕r15
0	0	0	0	0	0	0	1	0	0	0	0	0	0	1	1	b0=r7⊕r14⊕r15

Fig. 9 Syndrome Calculations for Ultrafast Codes

$$u5 = (s4.s6) \oplus r13$$
$$u6 = (s5.s7) \oplus r14$$
$$u7 = (s6.s7) \oplus r15$$

which are capable of correcting 1/4th of the erroneous data bits and only two erroneous parity bits.

2.7 *QAEC (Quarternary Adjacent Error Correction) Codes*

The QAEC Codes [10] are capable of correcting a maximum of four adjacent errors as shown in Fig. 10. The decoder complexity and delay are optimized by minimizing the total number of ones and by minimizing the number of ones in the heaviest row in the parity check matrix.

Also to reduce the cost of run time and improve the performance, the recursive backtracking algorithm is used which is a function of weight restriction and recording.

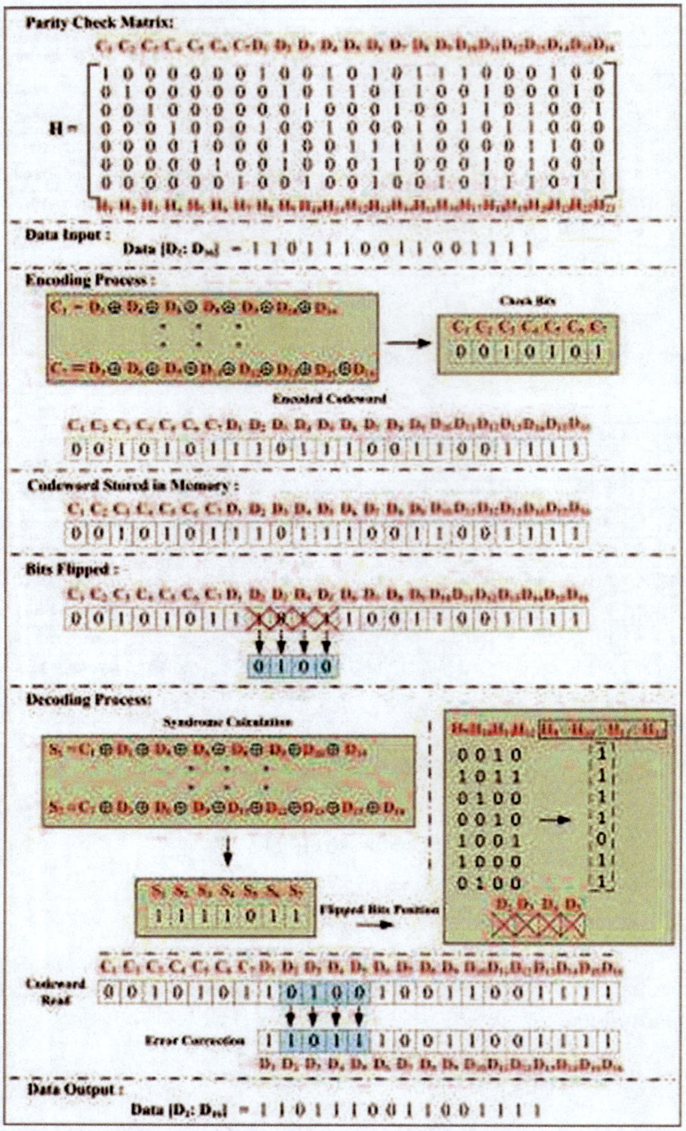

Fig. 10 QAEC Codes with Encoding and Decoding Process

The other codes include the systematic codes [11], BCH Codes (Bose Chaudhuri Hocquenghem Codes), Low Density parity Check (LDPC) Codes, Turbo Codes [12], Orthogonal Latin Square Codes [13], Reed Solomon Codes [14], Golay Codes [15], etc., along with some modifications which can correct two adjacent and two random error bits.

3 Evaluation Metrics

The Metrics that can be used to compare the designs of EDAC Codes for memories are given as below.

3.1 Power Dissipation

Power Dissipation represents the amount of power dissipated by the designed logic circuit which is evaluated by using power simulation performed by the tool. The power dissipation must be as minimum as possible for a good design.

3.2 Size of the Circuit

The size of the circuit is usually represented by the number of two inputs XOR gates as required by the H-matrix which can be reduced by minimizing the multiple output logic.

3.3 Delay

The delay of the logic circuit indicates the measure of the worst-case delay that can happen in the circuit. It must be as minimum as possible.

3.4 Cost Function

The overall cost function gives a measure of the tradeoff between the power dissipation, delay and size of the circuit.

$$\begin{aligned} Overall\ Cost\ Fnction =& (Weight_{Power} * Power\ Dissipation) \\ &+ (Weight_{Circuit} * Circuit\ Size) + (Weight_{Delay} * Circuit\ Delay) \end{aligned}$$

$$Weight_{Power} + Weight_{Circuit} + Weight_{Delay} = 1$$

3.5 Number of Redundant Bits Used

The number of redundant bits is the sum of the number of parity bits used to encode and decode the data bits.

3.6 Correction Coverage

The correction coverage represents the measure of the percentage of total number of faulty bits corrected against the total number of faults injected in the data word.

$$\%C_{Corrected} = \frac{Errors_{Corrected}}{Errors_{Injected}} \times 100$$

3.7 Detection Coverage

The detection coverage represents the measure of the percentage of total number of faulty bits detected against the total number of faults injected in the data word.

$$\%C_{Detected} = \frac{Errors_{Corrected} + Errors_{Detected}}{Errors_{Injected}} \times 100$$

3.8 M Coverage

The M Coverage give the ratio of the product of corrected and detection coverage to the product of area, power, delay and number of redundancy bits.

$$M = \frac{C_{Corrected} \times C_{Detected}}{Area \times Power \times Delay \times Redundancy}$$

4 Performance Evaluation of EDAC Codes

The designs are functionally verified by using the Xilinx ISIM Simulation Tool. The designs are modelled in Verilog HDL and synthesized for Spartan3E FPGA for XC32500E Device with package FG320 and speed grade of -5. Before storing the

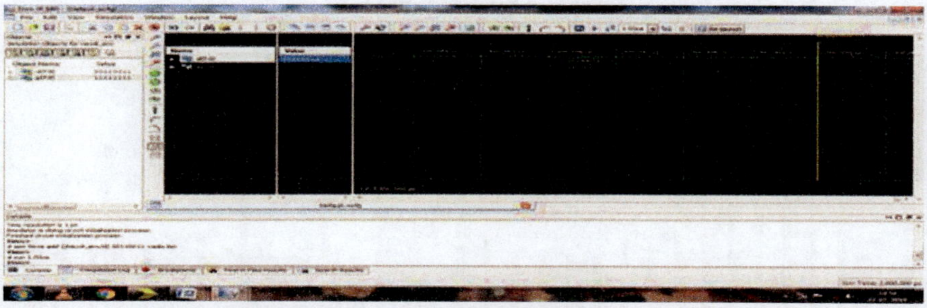

Fig. 11 Simulation Result of 8-bit New Matrix Code encoder

data, the parity bits are calculated. The decoding process is performed when the data bits are read from the memory.

The simulation result of 8-bit New Matrix Code Encoder is shown in Fig. 11, where for the data input of 00110011, the corresponding parity bits were obtained as 11111111.

Figure 12 shows the 8-bit New Matrix Code Decoder where for the same data input two adjacent errors were detected and corrected but only one random error was detected and corrected. In the similar manner, all the codes were developed and verified. The tabulated results are shown as the codes that were capable of detecting at least two adjacent errors.

Figure 13 shows the comparison of number of redundant bits used with percentage, from figure, as the number of data bits increase, the number of redundant bits also increase.

Figure 14 shows the comparison of % Fault Coverage for N/4 adjacent errors in data bits stored. The 100% Fault Coverage (includes both detection and correction coverages) is obtained only for HVD Codes and Matrix Codes.

From Figs. 13 and 14, it is clear that the existing EDAC codes can correct to a maximum of 1/4th of the adjacent error data bits, i.e., 16/64 = 1/4th of adjacent

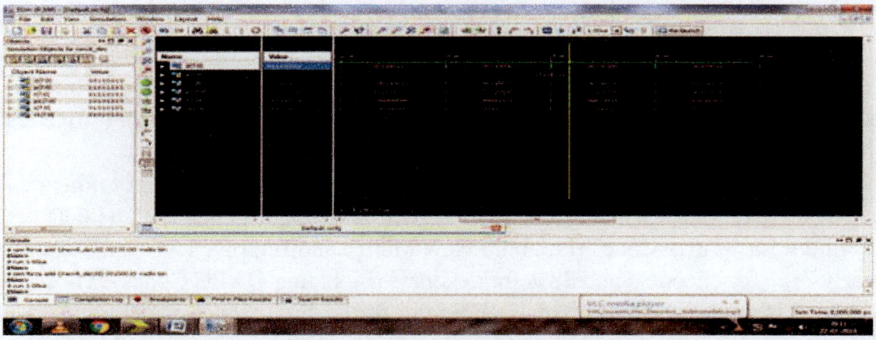

Fig. 12 Simulation Result of 8-bit New Matrix Code decoder

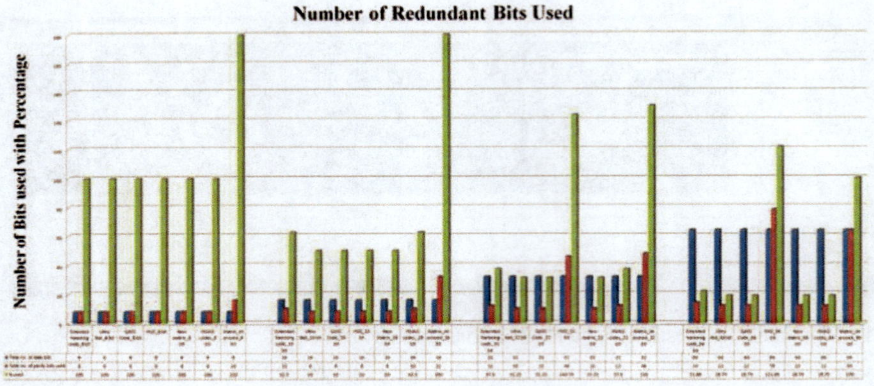

Fig. 13 Comparison Results of Number of Redundant Bits used for the data bits with percentage

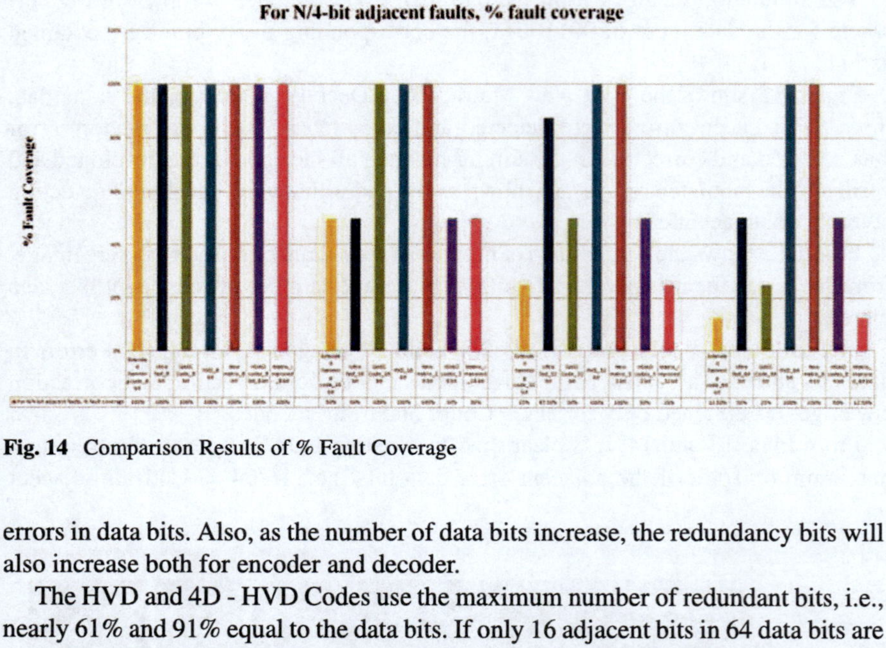

Fig. 14 Comparison Results of % Fault Coverage

errors in data bits. Also, as the number of data bits increase, the redundancy bits will also increase both for encoder and decoder.

The HVD and 4D - HVD Codes use the maximum number of redundant bits, i.e., nearly 61% and 91% equal to the data bits. If only 16 adjacent bits in 64 data bits are erroneous, then the fault coverage is 100% but if the number of adjacent erroneous bits increase, then the fault coverage degrades.

Table 1 shows the comparison of various EDAC Codes with respect to the area and delay parameters when synthesized to Spartan 3E FPGA device. The delay is minimum for Matrix Codes (i.e., both New Matrix and Improved Matrix Code) to nearly 31% for encoders and 38% for decoders (including QAEC Codes). The area occupied in terms of LUTs used is minimum for 64-bit encoder of QAEC Codes (29 out of 9312), i.e., 67% LUTs when compared to the highest occupied for 64-bit encoder of 4D – HVD Code (90 out of 9312). The area occupied in terms of LUTs

Table 1 Comparison of various EDAC Codes with Respect to number of LUTs and Slices used and Combinational Path Delays

Name of the code with No. of data bits		No. of LUTs (Out of 312)		No. of slices (out of 4656)		Combinational path delay	
		Enc	Dec	Enc	Dec	Enc	Dec
Extended hammind code	8	4	22	2	13	7.858ns	11.472ns
	16	8	33	4	19	9.045ns	12.833ns
Utra fast	8	2	4	4	7	5.847ns	6.991ns
	16	8	15	4	8	7.850ns	9.378ns
	32	16	26	9	15	7.850ns	9.378ns
	64	32	62	18	31	7.850ns	9.378ns
QAEC code	16	15	32	9	18	9.317ns	13.949ns
	32	22	35	12	21	10.342ns	14.200ns
	64	29	38	15	24	11.365ns	15.126ns
HVD	8	9	29	5	17	7.850ns	9.440ns
	16	13	56	7	16	7.850ns	10.795ns
	32	29	108	17	61	9.305ns	10.984ns
	64	69	207	39	119	9.313ns	12.033ns
4D-HVD	64	90	366	51	210	9.317ns	13.467ns
New Matrix	8	8	14	4	7	7.824ns	9.378ns
	16	16	28	9	15	7.824ns	9.378ns
	32	32	54	18	29	7.824ns	9.378ns
	64	64	112	37	60	7.824ns	9.378ns
HSIAO codes	22/16/6	19	64	11	36	10.564ns	15.159ns
	40/32/7	34	121	17	69	9.344ns	17.642ns
Matrix improved	8	8	32	4	18	7.824ns	10.801ns
	16	16	64	9	37	7.824ns	10.801ns
	32	24	80	14	46	7.824ns	10.879ns
	64	32	192	18	110	7.824ns	10.871ns

used is minimum for 64-bit decoder of QAEC Codes (38 out of 9312,) i.e., 89.6% LUTs when compared to the highest occupied for 64-bit decoder of 4D – HVD Code (366 out of 9312). Similarly, the area occupied in terms of Slices used is minimum for 64-bit encoder of QAEC Codes (15 out of 4656), i.e., 70.5% LUTs when compared to the highest occupied for 64-bit encoder of 4D – HVD Code (51 out of 4656). Also, the area occupied in terms of Slices used is minimum for 64-bit decoder of QAEC Codes (24 out of 9312), i.e., 88.5% LUTs when compared to the highest occupied for 64-bit decoder of 4D–HVD Code (210 out of 9312). Hence, the matrix codes along with ultrafast decoding prove to be a better choice for high speed memory access with built-in error detection and correction capability.

5 Conclusion

The erroneous data retrieval happens in memories and registers of a processor due to memory faults manifested as errors caused by radiation effects. The EDAC codes can be used to recover the memory from storing erroneous data or address which ensures the reliability in operation, especially for satellite or jet propulsion systems. This paper consolidates the EDAC codes that are capable of correcting errors from one bit to three adjacent bits. For error detection and correction, the data from memory has to encoded prior to writing into the memory and decoded after reading from the memory. The decoding process varies for various EDAC Codes which have syndrome calculation and error masking capabilities. The syndrome specifies the location of error. The error masking capability is achieved at the cost of additional hardware. As the number of parity or redundant bits increases, more number of errors can be detected or corrected which is limited by hamming distance. The error detection and correction codes are assessed based on the metrics like delay, number of redundant bits, number of errors detected, number of errors corrected, etc. From the review of EDAC Codes, the matrix codes with QAEC decoding prove to be a better choice for memories.

References

1. Sanchez-Macian A, Reviriego P, Maestro JA (2014) Hamming SEC-DAED and extended hamming SEC-DED-TAED codes through selective shortening and bit placement. IEEE Trans Device Mater Reliab 14(1):574–576. Doi: https://doi.org/10.1109/tdmr.2012.2204753
2. Dutta A, Touba NA (2007) Multiple bit upset tolerant memory using a selective cycle avoidance based SEC-DED-DAEC Code. In: 25th IEEE VLSI Test Symmposium (VTS'07), 0-7695-2812-0/07, IEEE. Doi: https://doi.org/10.1109/vts.2007.40
3. Sunita MS, Bhaaskaran K (2013) Matrix code based multiple error correction technique for N-bit memory data. Int J VLSI Des Commun Syst (VLSICS) 4(1):30–37
4. Argryides et al (2011) Matrix codes for reliable and cost efficient memory chips. IEEE Trans VLSI Syst 19:420–428. https://doi.org/10.1109/tvlsi.2009.2036362
5. Maheswari T, Sukumar P (2015) Error detection and correction in SRAM cell using decimal matrix code. IOSR J VLSI Signal Process (IOSR-JVSP) 5(1):9–14. Doi: https://doi.org/10.9790/4200-05120914
6. Tambatkar S, Menon SN, Sudarshan V, Vinodhini M, Murty NS (2017) Error detection and correction in semiconductor memories using 3D parity check code with hamming code. In: International conference on communication and signal processing, 6–8 April 2017, India, pp. 0974-0978. Doi: https://doi.org/10.1109/iccsp.2017.8286516
7. Tawar V, Gupta R (2015) A 4-Dimensional parity based data decoding scheme for EDAC in communication systems. Int J Res Appl Sci Eng Technol (IJRASET) 3(IV):183–191
8. Saiz-Adalid LJ, Gil P, Ruiz JC, Gracia-Moran J, Gil-Tomas D, Baraza-Calvo JC (2016) Ultrafast error correction codes for double error detection/correction. In: 12th European dependable computing conference, pp 108–116. Doi: https://doi.org/10.1109/edcc.2016.28
9. Reviriego P, Martinez J, Pontarelli S, Maestro JA (2014) A method to design SEC-DED-DAEC codes with optimized decoding. IEEE Trans Device Mater Reliab 14(3):884–889. Doi: https://doi.org/10.1109/tdmr.2014.2332364

10. Li J, Reviriego P, Xiao L, Argyrides C, Li J (2018) Extending 3-Bit burst error-correction codes with quadruple adjacent error correction. IEEE Trans Very Large Scale Integr (VLSI) Syst 26(2):221–229. Doi: https://doi.org/10.1109/tvlsi.2017.2766361
11. Gulliver TA, Bhargava VK (1993) A systematic (16, 8) code for correcting double errors and detecting triple adjacent erros. IEEE Trans Comput 42(1):109–112. Doi: https://doi.org/10.1109/12.192220
12. Alwan MH, Singh M, Mahdi HF (2015) Performance comparison of turbo codes with LDPC codes and with BCH codes for forward error correcting codes. In: IEEE Student conference on research and development (SCOReD). Doi: https://doi.org/10.1109/scored.2015.7449398
13. Reviriego P, Pontarelli S, Evans A, Maestro JA (2015) A Class of SEC-DED-DAEC Codes derived from orthogonal Latin Square Codes. IEEE Trans Large Scale Integr (VLSI) Syst 23(5):968–972. Doi: https://doi.org/10.1109/tvlsi.2014.2319291
14. Namba K, Lombardi F (2015) A single and adjacent symbol error correcting parallel decoder for reed–solomon codes. IEEE Trans Device Mater Reliab 15(1):75–81. https://doi.org/10.1109/tdmr.2014.2379513
15. Revriego P, Liu S, Xiao L, Maestro JA (2016) An efficient single and double-adjacent error correcting parallel decoder for the (24, 12) extended golay code. IEEE Trans Very Large Scale Integr (VLSI) Syst 24(4):1603–1606. Doi: https://doi.org/10.1109/tvlsi.2015.2465846

Optimisation of Cloud Seeding Criteria Using a Suite of Ground-Based Instruments

P. Sudhakar, K. Anitha Sheela, and M. Satyanarayana

Abstract All fresh-water, whether on the surface or underground, comes from the atmosphere in the form of precipitation. Nevertheless, a large volume of water present in the clouds is never transformed into precipitation on the ground. This has prompted researchers to explore the possibility of augmenting water supplies by the use of "cloud seeding" technique to initiate and accelerate the precipitation process. The seeding technique is expected to provide an increase in precipitation from the cloud and provide rain almost immediately at the targeted region/ location. This is done by dispersing suitable substances into the cloud that serve as cloud condensation or ice nuclei. Although many projects around the world have successfully demonstrated a considerable increase in precipitation due to seeding, majority of the projects, however, yielded inconclusive results on precipitation enhancement [1]. The reason for this inconsistency is that the physical mechanisms of aerosol effects on cloud and precipitation development are much more complex than anticipated earlier. There are many ongoing operational cloud seeding programs and the number has been steadily increasing with time. Despite this, there is still a great need for more intensive FIELD experiments to standardize the cloud seeding technology for increased reliability and enhancement of precipitation from clouds. The technology of rain enhancement is based on the science of cloud physics with major linkages reaching

P. Sudhakar (✉)
Department of ECE, Geethanjali College of Engineering & Technology, Cheeryal (V), Keesara (M), Hyderabad 501301, India
e-mail: sudhakar.lidar@gmail.com

K. A. Sheela
Department of ECE, JNTUH College of Engineering Hyderabad (Autonomous), Hyderabad 500085, India
e-mail: kanithasheela@gmail.com

M. Satyanarayana
Ananth Technologies Ltd (ATL), Hyderabad 500081, India

Department of ECE, VNR Vignan Jyothi Institute of Engineering & Technology, Hyderabad 500090, India

M. Satyanarayana
e-mail: drsatyanarayana.malladi@gmail.com

T. Laxminidhi et al. (eds.), *Advances in Communications, Signal Processing, and VLSI*,
Lecture Notes in Electrical Engineering 722,
https://doi.org/10.1007/978-981-33-4058-9_33

into mesoscale and boundary layer meteorology, weather forecasting, diffusion and turbulence, physical chemistry, aerosol physics, statistics, and instrumentation [2–4]. In this paper, we present the details of the multi-wavelength dual polarization lidar being used and the methodology to monitor the various cloud parameters involved in the precipitation process.

Keywords Cloud seeding · Precipitation · Rain enhancement · Dual polarization lidar

1 Statement of the Proposed Research and Method of Implementation

The success of the cloud seeding process can be predicted with confidence if the processes and the chain of events involved, leading to precipitation are fully understood and monitored in real time. It is proposed to monitor the cloud characteristics and measure these events in real time to decide on the go/ no go criteria for seeding to increase the efficacy in inducing and increasing precipitation. One of the important criteria for the timing of the seeding is the knowledge of microphysical properties of the clouds in real time when the seeding of the material is done in the clouds. The direct monitoring of the ice water balance of the clouds will yield valuable information on the right conditions for seeding to obtain efficient precipitation. Background aerosol, water vapor, and other meteorological parameters also play a crucial role in the precipitation process involved in cloud seeding operations [5]. The main goal of the proposed work is to precisely measure the physical conditions and chain of events in precipitation development in real time using a suite of ground-based instruments consisting of cloud radar, polarization diversity and aerosol lidar, automatic weather station, and other supporting equipment. A series of field experiments are to be conducted in the real-time conditions of the clouds. From these experiments, the right conditions of the clouds for seeding will be optimized. From the data obtained on the optimized seeding conditions, it is proposed to make the cloud seeding technology as a regular operational and reliable technique for rain enhancement.

2 Objective of the Research

One of the main goals of the proposed research on rain enhancement is to firmly establish the physical chain of events in precipitation development so that the perturbations both intentional and inadvertent can be understood and quantified through field experiments. The research addresses the common fundamental understanding of aerosol, cloud, and precipitation processes that helps progress in other important research areas including quantitative precipitation forecasting. The research also envisages fundamental research on seeding materials through experimental investigations [6].

3 Detailed Methodology

Field experimentations are required for monitoring the nearby aerosols, meterological parameters, and cloud chattels in the real-time atmosphere for optimizing the standards for seeding to enhance the precipitation at the stressed region. Towards this, it is proposed to conduct a series of cloud seeding experiments in arid and semiarid regions to establish the optimized conditions of the cloud for seeding. It is proposed to use also a Multifunctional Dual Polarization Llidar in measurements of the microphysical characteristics of clouds, aerosols, and water vapor in the laboratory [7, 8]. The Lidar provides simultaneous measurements on cloud structure, particles, humidity, temperature, and also, more importantly, the background aerosol system which plays an important role in cloud formation and precipitation. The formation of the cloud droplets from the aerosols after seeding depends on many factors of the seeding material including the chemical composition, number concentration, size distribution, etc. The meteorology and topography of the location are also important for seeding operations. As such the following commercially available ground-based instruments are proposed to be used for real-time monitoring and optimization of seeding methods.

4 Multi-Wavelength Dual Polarization LIDAR and Cloud RADAR

- It is proposed the above for integrated measurements of cloud characteristics and precipitation process and to measure the Temporal and Spatial variation of size distribution functions of aerosols, microphysical properties of Clouds and Water Vapor, the Nd: YAG dual polarization lidar will be used [9].
- The system is transportable and can be setup within 24 h at the required location.
- It can make the measurements both day and night by having sufficient signal-to-noise ratio, even in broad daylight conditions.
- The required software for characterizing the aerosols and clouds and measurement of water vapor with good vertical resolutions in real time is provided as required for the cloud seeding program.
- The whole equipment will be fitted on a Mobile Platform in a Truck with a facility to enclose the equipment in an air-conditioned environment along with a Power Generator facility to carry out the research experiments at different locations.
- It is proposed to conduct an investigation on the clouds using Multi-wavelength Dual Polarization Lidar and Cloud Radar (35 GHz/ 95 GHz) during the period of cloud seeding operations over the targeted areas.
- Measurement of the size and distribution of cloud condensation nuclei (CCN) that is released artificially after burning the flares, containing Silver iodide/ Dry ice/ Calcium Chloride are carried out.

- The size distribution can be optimized after different trials of success.
- The occurrence frequency of various clouds will be derived using the Cloud RADAR data collected during the period of cloud seeding program [10].

5 Integration of Data Obtained by LIDAR and Cloud RADAR Simultaneously in the Field for Optimization of Seeding Methods

- The results obtained from the simultaneous observations of both RADAR and LIDAR can be used to understand the influence of size distribution of aerosols on the size of the cloud droplets.
- As the Multi-wavelength Dual polarization lidar able to distinguish the type of aerosol present at different altitudes the data of different types of aerosols present at different altitudes can be used to understand the influence of the various types of CCN on the rainfall enhancement process. The results obtained in the entire observations will be analyzed and the conditions for optimization of cloud seeding techniques will be arrived.

Automatic Weather Station (AWS)

Standard commercially available AWSs are to be located at different stations in the target area to obtain real-time information on various meteorological parameters when the cloud seeding field experiments are conducted

Seeding methodology

It is proposed to use the aircraft and drone platforms to seed the clouds with selected materials at the appropriate altitude and time. Aircraft seeding is a well established procedure and is being applied for the purpose in many countries successfully. Recently, drones are being used with additional advantages successfully for seeding operations in USA.

It is thought-provoking to recognize that wide-ranging information is accessible in ancient *VEDIC* literature on the usage of '*HOMAM*' for inducing rain at the vital sites. The method comprises the spreading of various materials into the cloud region by burning the material in fire ponds, especially setup on the ground. It is proposed to conduct simultaneously the seeding operations from the ground using this technique also. This technique is expected to be very expedient and cost-effective for seeding clouds.

6 Conclusion

The design details of a dual polarization lidar used for characterizing the microphysical properties of clouds are described. The methodology used in the characterization of clouds to understand the precipitation processes in cloud seeding techniques is described in this research article.

References

1. Zoljoodi M, Didevarasl Ali (2013) Evaluation of cloud seeding project in Yazd Province of Iran using historical regression method. Nat Sci 5(6):1006–1011. doi:http://dx.doi.org/10.4236/ns.2013.59124
2. Krishnakumar et al (2014) Lidar investigations on the optical and dynamical properties of the cirrus clouds in the UTLS region at a tropical station Gadanki, India (13.5^{O}N, 79.2^{O}E)". J Appl Remote Sens 8(1):083659. doi: https://doi.org/10.1117/1.jrs.8.083659
3. Krishnakumar V, Satyanarayana MV, Radhakrishnan SR, Mahadevan Pillai VP, Raghunath K, Venkat Ratnam M et al (2011) Investigations on the physical and optical properties and their role in the nucleation of cirrus clouds using lidar at Gadanki (13.50 N, 79.20E). J Appl Remote Sens 5:053567. doi: https://doi.org/10.1117/1.3662877
4. Meenu S, Rajeev K, Parameshwaran K (2011) Regional and vertical distribution of semitransparent cirrus clouds and cloud top altitudes over tropical Indian region derived from CALIPSO data. J Atmos Solar Terr Phys 73(13):1967–1979. doi: https://doi.org/10.1016/j.jastp.2011.06.007
5. Radhakrishnan SR, Satyanarayana M, Presennakumar B, Mahadevan Pillai VP, Murthy VS (2008) Measurement of water vapor in the lower atmosphere at a coastal station, Trivandrum (80 N, 770E) using a high resolution Raman lidar. Indian J Radio Space Phys 37:353–359
6. Drofa AS, Ivanov VN, Rosenfeld D, Shilin AG (2010) Studying an effect of salt powder seeding used for precipitation enhancement from convective clouds. Atmos Chem Phys 10(1): 8011–8023. doi: https://doi.org/10.5194/acp-10-8011-2010
7. Parameswaran K et al (2003) Lidar observations of cirrus cloud near the tropical tropopause: temporal variations and association with tropospheric turbulence. Atmos Res 69(1–2): 29–49. doi: http://dx.doi.org/10.1016/j.atmosres.2003.08.002
8. Sunilkumar SV et al (2003) Lidar observations of cirrus clouds near the tropical tropopause. Atmos Res 66(3):203–207. http://dx.doi.org/10.1016/S0169-8095(02)00159-X
9. Sassen K (2000) Lidar backscatter depolarization technique for cloud and aerosol research. In: Mishchenko ML, Hovenier JW, Travis LD (eds.) Light scattering by nonspherical particles: theory, measurements, and geophysical applications. Academic Press, San Diego, CA, p 393
10. Göke S, Ochs III HT, Raube RM (2007) Radar analysis of precipitation initiation in maritime versus continental clouds near the Florida coast: Inferences concerning the role of CCN and giant nuclei J Atmos Sci 64:3695-3707

Periodic Octagon Split Ring Slot Defected Ground Structure for MIMO Microstrip Antenna

F. B. Shiddanagouda, R. M. Vani, and P. V. Hunagund

Abstract In this work, a periodic octagon split ring slot defected ground structure for MIMO (Multiple Input Multiple Output) microstrip antenna is proposed. The prototype of MIMO microstrip antenna consists of four similar rectangular microstrip antenna elements with a partition of $\lambda/4$ distance. The antennas are printed on a 1.6mm thick FR-4 substrate with an overall dimension of 62.8 X 60 mm^2. To improve the antenna parameters, the proposed MIMO microstrip antenna elements are etched with narrow rectangular edge slit and ground plane defected with periodic octagon split ring slot defected ground structure (POSRSDGS). The proposed MIMO microstrip antenna resonates at dual frequency points, i.e., 4.1GHz, 5.9GHz with a bandwidth of 88MHz and 454MHz along with minimum return loss of −22.7dB and −19.02dB, respectively. The envelope correlation coefficient (ECC) is lower than the acceptable limit across the dual operating bands. Mutual coupling coefficient (MCC) at dual resonating frequency points are −36.21dB and −42.93dB, respectively. The simulated and fabricated results are found in good agreement and ideal for wireless communication applications.

Keywords Microstrip Antenna · MIMO · POSRSDGS · Bandwidth · MCC · ECC

1 Introduction

Industry and academic sectors are predicted, future wireless communication scenarios that, 7 trillion wireless devices are serving 7.5 billion peoples. High data rates and low error rates are the essential requirments for the future generation of

F. B. Shiddanagouda (✉)
Department of ECE, Vignan Institute of Technology and Science, 508284 Hyderabad, India
e-mail: siddu.kgp09@gmail.com

R. M. Vani
Department of USIC, Gulbarga University, 585106 Kalaburagi, India

P. V. Hunagund
Department of Applied Electronics, Gulbarga University, 585106 Kalaburagi, India

T. Laxminidhi et al. (eds.), *Advances in Communications, Signal Processing, and VLSI*,
Lecture Notes in Electrical Engineering 722,
https://doi.org/10.1007/978-981-33-4058-9_34

wireless communication systems. To achieve these requirements, many researchers recommended enabling key technology such as adding many antennas in multiple input multiple output (MIMO) array configuration [1–7].

The following sections of the paper is explained in a systematic way of proposed MIMO antenna design, results and discussions, and finally followed by conclusion.

2 Antenna Design

The prototype of conventional MIMO microstrip antenna (CMMA) consists of a four similar rectangular microstrip patch antenna elements with a partition of λ/4 distance. The antennas are embedded on FR-4 substrate with loss tangent 0.02 and thickness is 1.6mm, respectively. Separate 50 Ω microstrip feed line were excited to four patch antenna elements.The simulated and fabricated CMMA is shown in Figs.1 and 2 and dimensions are represented in Table 1, respectively.

The study carried by conventional MIMO microstrip antenna (CMMA) loaded with periodic octagon split ring slot defected ground structure (POSRSDGS) and radiating patch elements are etched with narrow rectangular slit to improve the antenna parameters and **it is named as proposed MIMO microstrip antenna (PMMA)**. Periodic means repetition of defected ground structure unit cells with finite space. The distributions of periodic defected ground structures in the ground plane have drawn much attention for their extensive applications in antenna design [8, 9]. Figure 3 shows the optimized unit cell of octagon split ring slot defected ground structure (OSRSDGS). The geometry was obtained through simulation and final dimensions are given in Table 2. The corresponding OSRSDGS unit cell is distributed periodically in the CMMA ground plane, which are aligned with an equal spacing of '*S'*= 10 mm, vertically aligned to both radiating edges and four OSRSDGS unit cells are placed horizontal direction with a space of 'T'=7 mm between each adjacent vertical OSRSDGS and center gap between the horizontal OSRSDGS maintained with a space of 'R'=12 mm. The complete ground plane look like "H" shape periodic structure. Then radiating patch elements are etched with narrow rectangular

Fig. 1 Geometry of CMMA

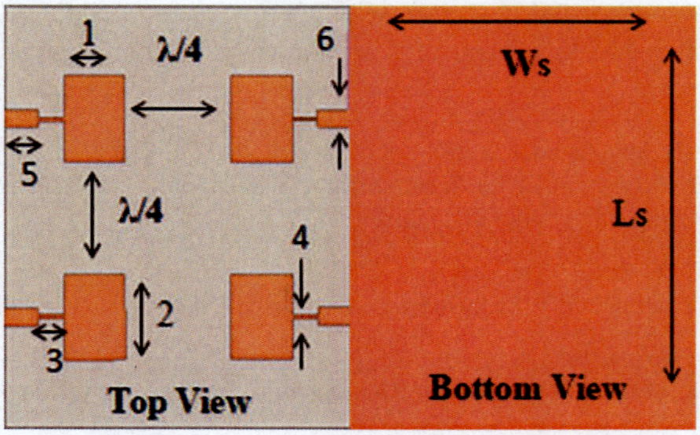

Fig. 2 Photograph of CMMA

Table 1 Dimensions of CMMA

Parameters	Ls	Ws	1	2
Dimensions (mm)	60	62.8	11.35	15.25
Parameters	3	4	5	6
Dimensions (mm)	4.9	0.5	6.15	3.06

Fig. 3 Geometry of OSRSDGS

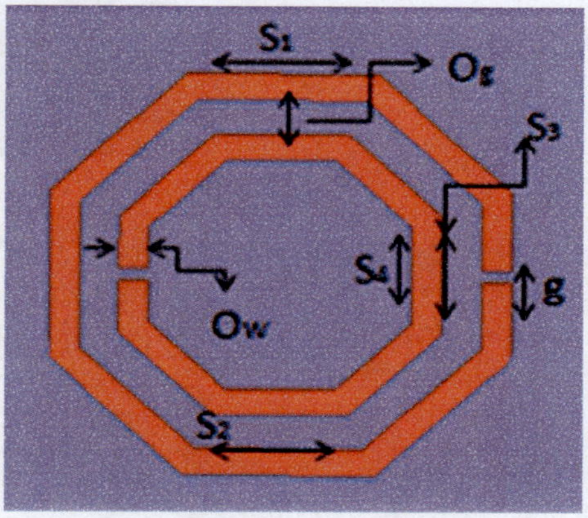

Table 2 Dimensions of OSRSDGS

Parameters	Dimensions (mm)
S1, S2, S3, S4	4, 3.5, 2.8, 2.3
Og	0.845
g	0.3
Ow	0.6

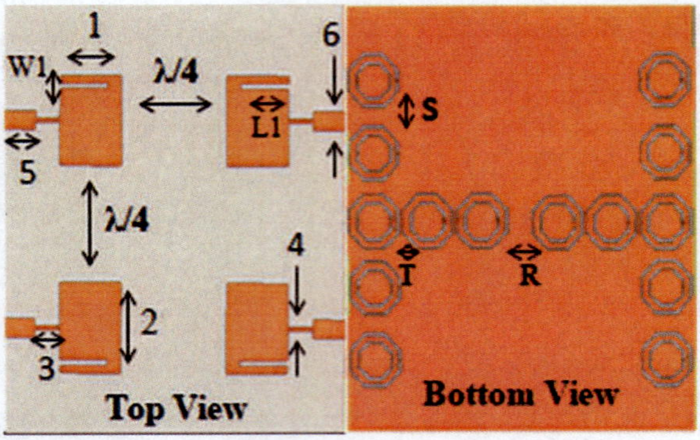

Fig. 4 Geometry of PMMA

Fig. 5 Photograph of PMMA

edge slit. The dimensions of the narrow rectangular slits are (L1, W1) = (10 mm, 0.5 mm), are cut on the left side of the radiating inset edge patch at a distance of 1 mm from the non radiating edges are taken. Figures 4 and 5 shows the geometry and photograph of PMMA.

3 Results and Discussions

In this work, using ANSYS HFSS 15.0 Electromagnetic simulation software, antennas were designed. The whole experimental works of designed antennas are carried out by using German make Rohde and Schwarz (R&S) Vector Network Analyzer (VNA) of ZVK model (10MHz to 40MHz). The return loss characteristics of simulated and fabricated CMMA are shown in Fig. 6. The antenna resonates at

Fig. 6 Return Loss characteristics of CMMA

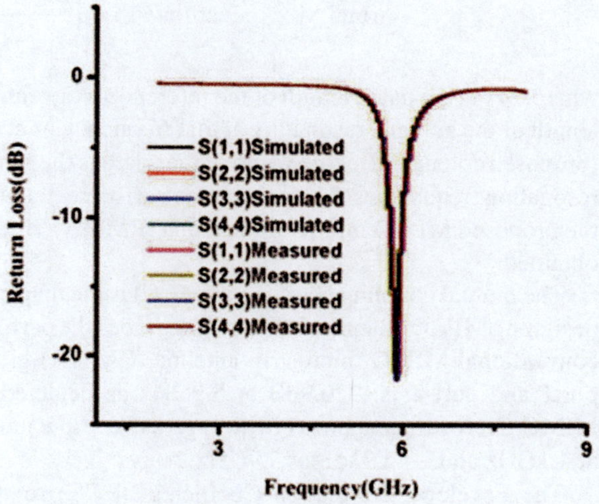

5.9GHz. The return loss obtained at the resonating frequency is equal to −21.3 dB and bandwidth equal to 204 MHz is obtained.

The return loss characteristics of simulated and fabricated PMMA are shown in Fig. 7. The antenna resonates at dual band frequency points, i.e., 4.1GHz and 5.9GHz with bandwidths of 88MHz and 454MHz along with minimum return loss of −22.7dB, and −19.02dB, respectively. Hence, by implementing the narrow rectangular slit, as well as POSRSDGS suppress, the unwanted surface wave and to control harmonics in PMMA to enhance the parameters interns of dual band resonance, bandwidth enhancement as compare to CMMA, as well as virtual size reductions are obtained. So from the Eq. (1), virtual size reduction of antenna is calculated.

Fig. 7 Return Loss characteristics of PMMA

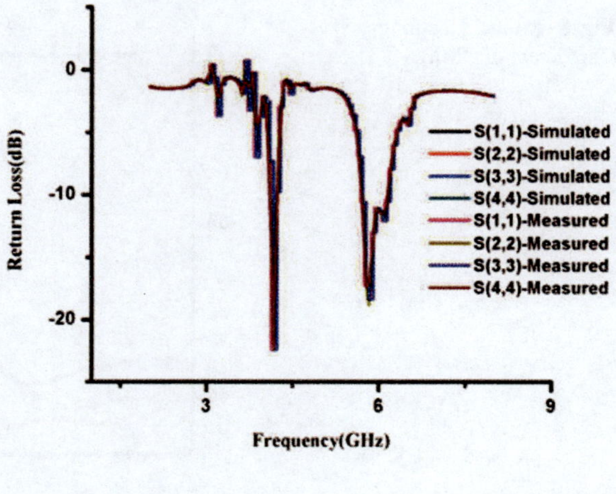

$$\text{Virtual Size reduction}(\%) = \left(\frac{L_C - L_{RA}}{L_C}\right) x100 \quad (1)$$

where L_{RA} is the patch length of the reference (conventional) antenna, L_C is the patch length of the antenna resonating at that frequency or at reduced resonant frequencies (proposed antenna). But the width of the patch is the same at both designed and actual resonating frequencies. So that by using narrow rectangular edge slit and POSRSDGS the proposed MIMO microstrip antenna (PMMA) virtual size reduction of 30.5% is obtained.

The mutual coupling coefficient (MCC) is the major factor to be considered while designing MIMO antennas because it degrades the performance of the system. Hence, conventional MIMO microstrip antenna (CMMA) gives mutual coupling between port1 and port 2 is −20.9dB at 5.9GHz as depicted in Fig. 8 and the proposed MIMO microstrip antenna (PMMA) gives very low mutual coupling, i.e., −36.21dB at 4.1GHz and −42.93dB at 5.9GHz, respectively, is depicted in Fig. 9.

The Envelope Correlation Coefficient (ECC) for the CMMA and PMMA are shown in Figs. 10 and 11, respectively. It decides how much the communication channels are isolated. ECC can be estimated of individual elements from the S-parameters [7]. So from the Eq. (2), CMMA achieved 0.001 ECC at 5.9GHz and PMMA dual resonating frequency points, i.e., 0.35 at 4.1GHz and 0.01 at 5.9GHz.

$$\rho = \frac{\left|S_{11}^{*}\, S_{12} + S_{21}^{*}\, S_{22}\right|^{2}}{\left(1 - |S_{11}|^{2} - |S_{21}|^{2}\right)\left(1 - |S_{22}|^{2} - |S_{12}|^{2}\right)} \quad (2)$$

Fig. 8 Mutual Coupling Coefficient of CMMA

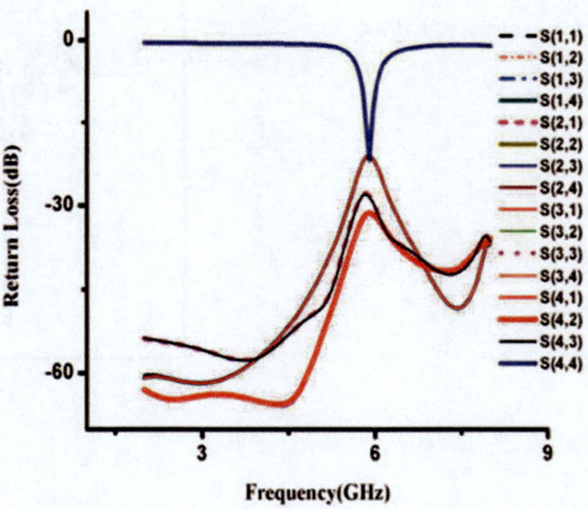

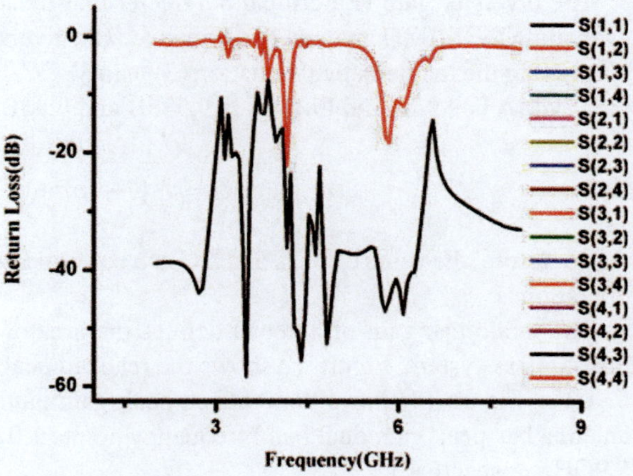

Fig. 9 Mutual Coupling Coefficient of PMMA

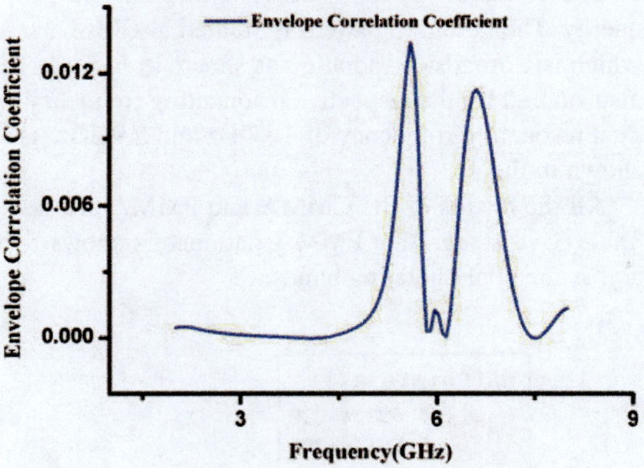

Fig. 10 Envelope Correlation Coefficient of CMMA

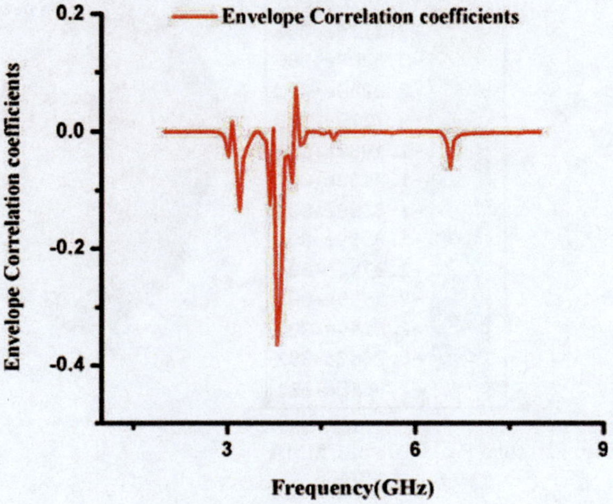

Fig. 11 Envelope Correlation Coefficient of PMMA

The diversity gain is a critical parameter that must be taken into account while evaluating the MIMO antenna performance. The diversity gain (DG) has been calculated using the mathematical equation (3) using ECC. The obtained diversity gain of the CMMA is 9.9dB and PMMA is 9.53dB and 9.9dB, respectively.

$$DG = \sqrt[10]{\left(1 - |\rho|^2\right)} \quad (3)$$

Therefore, the value of ECC and DG can confirm PMMA is acceptable for MIMO operation.

The total peak gain of antenna defines the area of coverage and link budget of the wireless system. Figure 12 shows the realized peak gain of CMMA is 5.69dB at 5.9GHz. Figure 13 shows the realized peak gain pattern of PMMA. The designed antenna has peak gain dual band frequency points 1.07dB at 4.1GHz and 4.14dB at 5.9GHz, respectively.

The radiation pattern decides how antenna propagates the electromangnetic energy. The radiation pattern is studied at CMMA resonating frequency of 5.9GHz which is a broadside radiation as shown in Fig. 14. The PMMA radiation pattern is also studied for the respective resonanting frequency points. The radiation pattern at dual resonating frequency of 4.1GHz and 5.9GHz which is a broad side radiation as shown in Fig.15.

All the results of the CMMA and PMMA are summarized in Table 3. From the Table 3, it is seen that PMMA parameters shows that there is an acceptable limit across the dual operating bands.

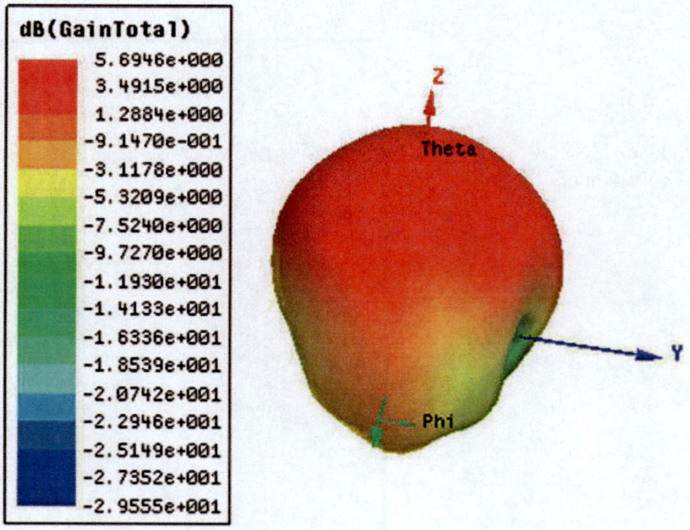

Fig. 12 Total Peak Gain of CMMA

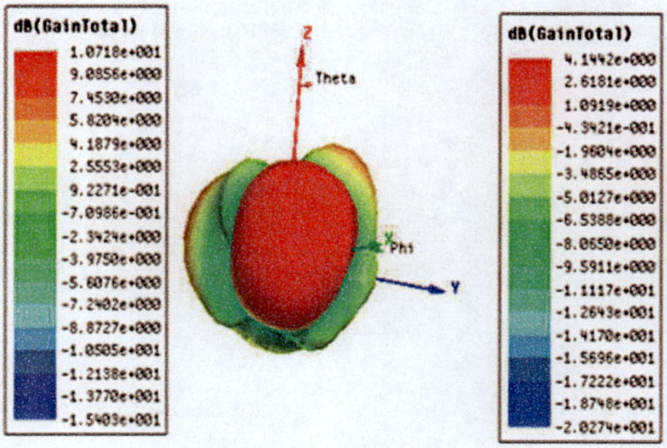

(a) 1.07dB at 4.1GHz

(b) 4.14 dB at 5.9GHz

Fig. 13 Total Peak Gain of PMMA

Fig. 14 Radiation Pattern of CMMA

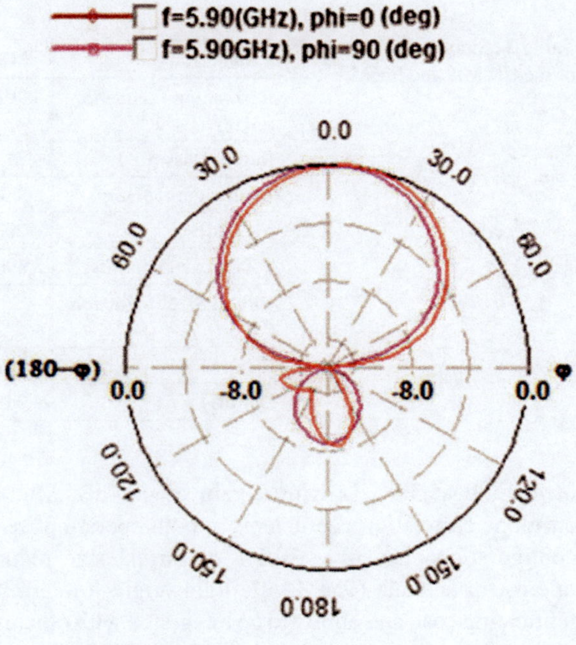

4 Conclusion

A periodic spit ring slot defected ground structure of MIMO microstrip antenna is presented. The antenna resonates at 4.1GHz and 5.9GHz frequency. The antenna offers 88MHz bandwidth at 4.1GHz with a total peak gain of 1.07dB and 454MHz

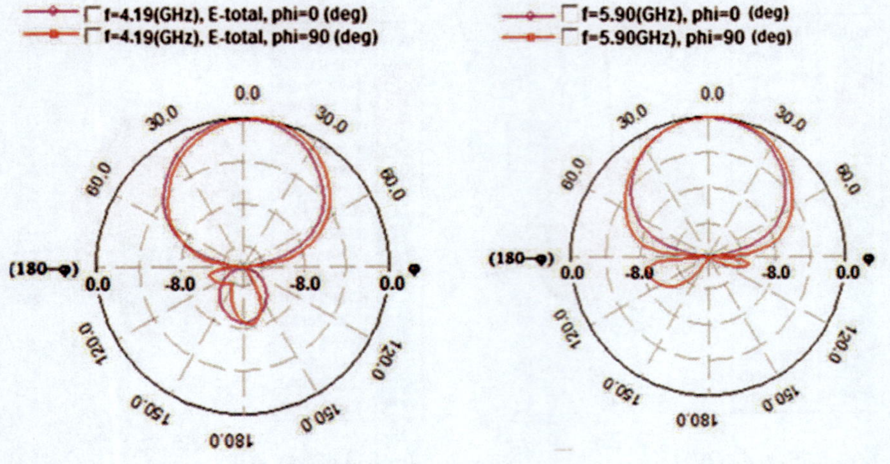

(a) 4.1GHz (b) 5.9GHz

Fig. 15 Radiation Pattern of PMMA

Table 3 Summarized results of the CMMA and PMMA

Parameters	CMMA	PMMA	
Resonating Frequency (GHz)	5.9GHz	4.1GHz	5.9GHz
Return Loss (dB)	−21.3dB	−22.7dB	−19.02dB
Bandwidth (MHz)	204MHz	88MHz	454MHz
MCC(dB)	−20.9dB	−36.21dB	−42.93dB
Total Peak Gain (dB)	5.69dB	1.07dB	4.14dB
Virtual Size Reduction (%)	–	30.5%	
ECC	0.001	0.35	0.01
DG(dB)	9.9dB	9.53dB	9.9dB

bandwidth at 5.9GHz with a gain of 4.14dB. Mutual coupling coefficients and envelope correlation coefficients at both operating frequencies are significant. This configuration also gives 30.5% of virtual size reduction. The proposed MIMO microstrip antenna (PMMA) is lightweight low profile, planar configuration, low fabrication cost, and ability to be integrated with other microwave circuits. Hence the designed antenna parameters shows that, PMMA is ideal for wireless communication applications.

Acknowledgement We would like to thank Indian Institute of Technology Kharagpur and Gulbarga University Kalaburagi for providing licensed versions of HFSS simulation software as well as antenna measurements research facility. We are also thankful to Atlantic Engineering Enterprise ECIL, Hyderabad for providing the PCB fabrication facility.

References

1. Robert J (1981) Microstrip array technology. IEEE Trans Antenna Propag AP-29(1)
2. Pozar DM (1992) Microstrip antennas and arrays on chiral substrates. IEEE Trans Antenna Propag AP-40(10)
3. Balanis A (1993) Theory of Antennas. IEEE Trans Antenna Propag AP-41(9)
4. Zahn R, Shutie P (1995) Advanced antenna technologies for X band SAR. IEEE Trans Antenna Propag AP-95(2)
5. Han MS, Choi J (2010) Compact multiband MIMO antenna for next generation USB dongle applications. IEEE Trans Antenna Propag AP-2(10)
6. Chi YJ, Chen FC (2012) 4-port quadric-polarization diversity antenna with novel feeding network. In: Proceedings of the Antenna and propagation Conference
7. Wu D, Cheung S (2013) Design of a printed multiband MIMO antenna. In: Proceedings of the EuCAP Conference
8. Kumar M, Nath V (2016) Analysis of low mutual coupling compact multi band compact antenna and its array using defected ground structure. Eng Sci Technol Int J. ISSN: 2215-0986
9. Alhegazi A, Azawan N (2018) Compact UWB filtering antenna with controllable WLAN band rejection using defected microstrip structure. Int J Electromag Radio Eng 27(1)

Multipath Delay Cell-Based Coupled Oscillators for $\Sigma\Delta$ Time-to-Digital Converters

R. S. S. M. R. Krishna, Amrita Dikshit, Ashis Kumar Mal, and Rajat Mahapatra

Abstract The multipath delay cell-based coupled ring oscillator for $\Sigma\Delta$ time-to-digital converters (TDC) is presented in this work. Unlike the traditional analog integrators, the ring oscillator integrator (ROI) can achieve theoretically infinite dc gain at low frequency. Moreover, the ROI inherently possesses the dynamic element matching (DEM) for shaping the mismatch between the delay cells. Two new multipath designs for coupled oscillators are proposed, which are simulated in UMC65, UMC40 and UMC28 CMOS technologies using Cadence Virtuoso Tools. Because of multipath techniques, the effective propagation delay is reduced well below the technology limit. This enables to achieve fine time resolution with stable phase noise performance for $\Sigma\Delta$ TDC.

Keywords Sigma-Delta time-to-digital converter ($\Sigma\Delta$ TDC) · Time-domain · Coupled oscillator · Ring Oscillator Integrator (ROI) · Multipath · Jitter · Fan-out of 4 (FO4) delay · Coupling factor · Dynamic Element Matching (DEM)

This research was supported in part by Visvesvaraya PhD Scheme for Electronics & IT, MeitY, Govt. of India, Grant No: PhD-MLA/4(29)/2015-16/01 and Special Manpower Development Programme for Chips-to-System Design, MeitY, Govt. of India, Grant No: 9(1)/2014-MDD (Vol III).

R. S. S. M. R. Krishna (✉) · A. Dikshit · A. K. Mal · R. Mahapatra
Department of Electronics and Communication Engineering, National Institute of Technology Durgapur, Durgapur 713209, India
e-mail: ssmr.krishna@ieee.org

A. Dikshit
e-mail: ad.17p10333@mtech.nitdgp.ac.in

A. K. Mal
e-mail: akmal@ece.nitdgp.ac.in

R. Mahapatra
e-mail: rajat.mahapatra@ece.nitdgp.ac.in

T. Laxminidhi et al. (eds.), *Advances in Communications, Signal Processing, and VLSI*, Lecture Notes in Electrical Engineering 722,
https://doi.org/10.1007/978-981-33-4058-9_35

1 Introduction

In this modern era of electronics, the scaling of process technology has a high impact on the integrated circuits in terms of system performance, fabrication cost, and power consumption. The scaling is usually done to increase the integration density of the chip with enhanced functionality. The digital circuits whose performance is not affected by the feature size of the device take full advantage of the scaling. With the reduced dimensions of the transistors, these circuits possess high switching speed, less power dissipation and occupy less area [1, 2].

On the contrary, the efficiency of the analog circuits is profoundly degraded with the technology scaling because of the decrease in supply voltage and the intrinsic gain but almost constant threshold voltage [3]. The reduction in the feature size of the transistor leads to various second-order effects such as channel length modulation, velocity saturation, mobility degradation, threshold voltage variations and hot-carrier effects, etc. So with the advancement in the process technology, the time-domain circuits are emerging as the best alternative to combat the scaling-induced performance deterioration of the analog and mixed-signal circuits [1, 4]. The time interval (T_{in}) between the rising edges of two digital signals, namely the *Start* and the *Stop* represents the time variable, as shown in Fig. 1. In a more mature sense, it is the pulse width modulation of the analog signal that involves [2]. The time-to-digital converter (TDC) is a functional block of utmost importance in most of the analog and mixed-signal processing circuits because of its digital-friendly behaviour. It measures the time interval and gives the digital representation of the time input. Initially, it was used in atomic and high-energy physics experiments as a precise time interval measurement unit that involves laser ranging such as in LIDAR (light detection and ranging) for time-of-flight estimation. With the technological advancements, the applications have been extended to all-digital phase-locked loops (ADPLLs) where the TDC acts as a phase-frequency detector (PFD), time-domain analog-to-digital converters (ADCs), temperature sensors and Serializer/Deserializer (SerDes) [2].

The Fan-out of 4 (FO4) delay is a technology-dependent parameter [5] used in the CMOS process, for quantifying the digital cell. As the technology scales down, the FO4 delay also becomes finer, leading to fast switching of the device. From Table 1, it is clear that, as the process scales from UMC65 to UMC28, the FO4 delay decreases from 31.35 ps to 18.12 ps.

Various researchers have investigated the design of ring oscillators using multipath techniques [6, 7]. In [8], the precise delay generation with the improved resolution

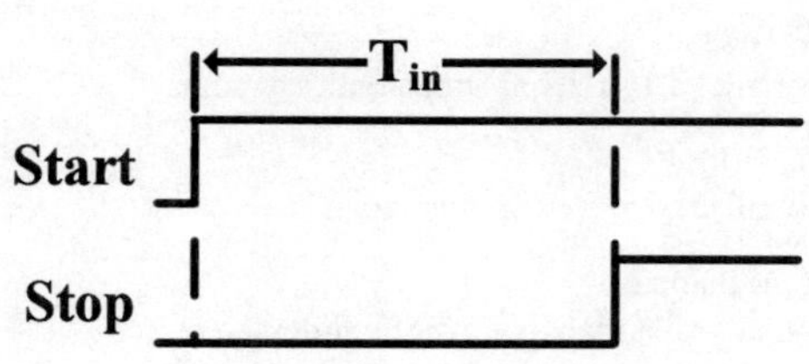

Fig. 1 The typical representation of time signal as the difference between the *Start* and *Stop* events

Table 1 FO4 delay of the inverter in UMC technologies

Technology[a]	FO4 Delay (ps)
UMC65	31.35
UMC40	23.35
UMC28	18.12

[a]United Microelectronics Corporation

has been shown, which had used an array oscillator with M number of coupled ring oscillators. It has achieved the delay resolution of 101ps at 141MHz operating frequency. The oscillation frequency will be lessened if more output phases are involved. A high-speed ring oscillator was demonstrated in [9] for multiphase clock generation using the skewed delays, which had generated the oscillation frequency 50% higher than the conventional ring oscillators. In [10], low power and high-resolution multiphase generation system for coupled oscillators has been presented, which has achieved the resolution of 32ps at 490MHz with the phase accuracy of -1.0 to 0.8 LSB. The first order noise-shaped time-to-digital converter (TDC) using multipath gated ring oscillator was demonstrated in [7], which has attained a resolution of 6ps with the oversampling rate of 50 MS/s. In [11], the second-order noise-shaped time-to-digital converter (TDC) using switched ring oscillator (SRO) and gated switched ring oscillator (GSRO) was presented, which has shown high OSR of 400 MS/s.

The subject of the multipath techniques to reduce the propagation delay has not been adequately addressed in the literature. In our present work, we have focused on the study of the multipath delay cell using the coupled ring oscillators in a more qualitative manner. The coupling factor is analysed and compared with the theoretical value. The paper is organised as follows. In Sect. 2, the details are given about the ring oscillator as an integrator with theoretically infinite dc gain at the sufficiently low operating frequency. Section 3 comprises of the multipath delay cell architecture for two topologies. The simulation results for UMC65, UMC40 and UMC28 CMOS technologies are compared and analysed in Sect. 4, and lastly, the conclusions are drawn in Sect. 5.

2 Ring Oscillator as an Integrator

The ring oscillators are widely used in phase-locked loops (PLLs), TDCs, time-domain ADCs, temperature sensors and SerDes to generate precise delays at high operating frequency because of their high delay linearity [1, 2]. If the delay of each inverter of the ring oscillator is identical, then the total oscillation period is uniformly divided into precise sub-delays. The inverter delay is equal to the period of oscillation divided by twice the number of inverters in the ring oscillator [5]. Although ring

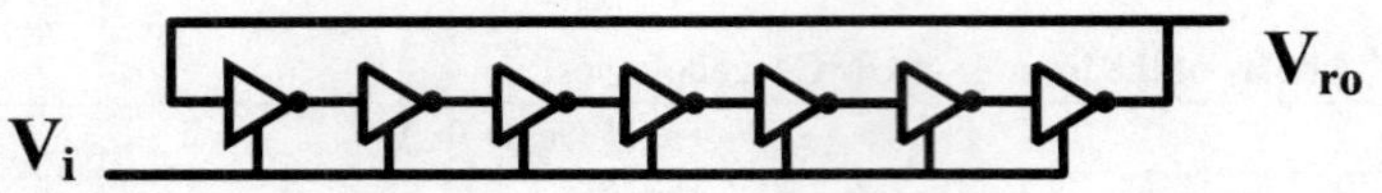

Fig. 2 Ring oscillator as V-to-ϕ integrator

oscillators can generate high precision and high linearity delay, their resolution is limited to the inverter delay (τ_P) [6].

The ring oscillator shown in Fig. 2 can be seen as an ideal voltage-to-phase (V-to-ϕ) integrator with theoretically infinite dc gain at low frequency and low supply voltages [12]. As the process scales down, the analog integrators such as G_m-C integrator or op-amp-RC integrator, which uses OTAs as the building block suffer from high non-linearity as their gain is limited to $g_m r_o$, since the output impedance of the operational amplifier drastically deteriorates. Moreover, the reduction of supply voltage significantly affects the voltage swing, therefore, making it very difficult to improve the gain of the analog integrators. The efficiency of the analog integrators, which mostly relies on operational amplifiers, is highly compromised. Also, analog integrators are power-hungry. In contrary to the above integrators, the CMOS inverter-based ring oscillator that scales well with technology and can act as an integrator in the phase domain. Also, these oscillators perform signal processing in the time-domain, so there is no issue of shrinking of voltage headroom and no performance loss due to scaling-induced imperfections. The ring oscillator integrator (ROI) consists of a ring oscillator with inverter delay stages, as shown in Fig. 2. The relationship between the output oscillator frequency (ω_{ro}) and input voltage (V_i) is given by the oscillator gain (A_{ROI}) and can be written as

$$A_{ROI} = \frac{d\omega_{ro}(t)}{dV_i(t)} \tag{1}$$

where A_{ROI} is the gain of the ring oscillator integrator. Since, we know that

$$\phi_{ro}(t) = \int w_{ro}(t)\,dt \tag{2}$$

Substituting (1) in (2), we get

$$\phi_{ro}(t) = A_{ROI} \times \int V_i(t)\,dt \tag{3}$$

Taking the Laplace Transform of the above equation, then the transfer function will be

$$H_{ROI}(s) = \frac{\phi_{ro}(s)}{V_i(s)} = \frac{A_{ROI}}{s} \tag{4}$$

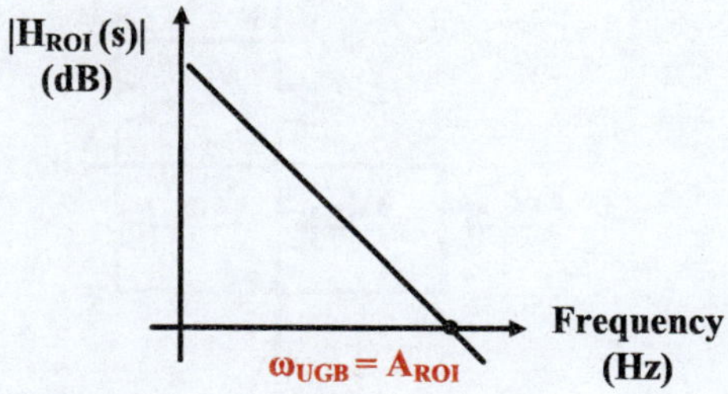

Fig. 3 Bode plot for the ring oscillator integrator (ROI)

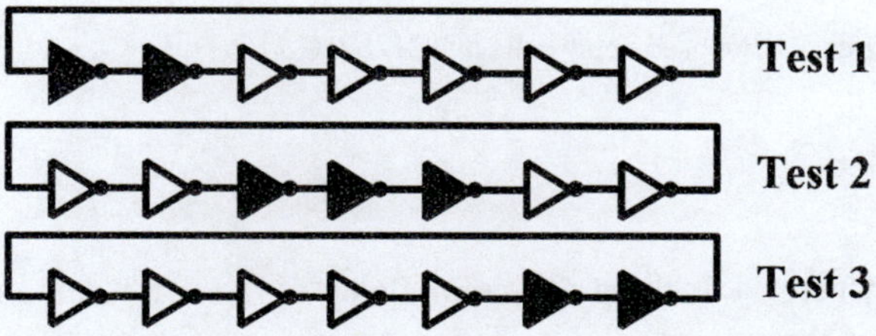

Fig. 4 Barrel shifting of ROI delay cells for dynamic element matching

From the above equation, it is evident that the transfer function of the output phase to the input voltage is inversely proportional to frequency. Hence, the ring oscillator integrator (ROI) can achieve theoretically infinite dc gain at low frequency irrespective of the variations in the device parameters or non-idealities due to the reduction of the output impedance of the transistor as noticed in the analog integrators. Thus, it acts as a lossless integrator, with transfer function independent of the transistor parameters. The modulus of the transfer function in the dB scale is plotted in Fig. 3. In the ring oscillator integrator (ROI), the mismatch between the delay-cells is first-order shaped. It achieves the dynamic element matching (DEM) by following the barrel shifting algorithm for delay cell selection [6, 7], as shown in Fig. 4. So the ring oscillator integrator (ROI) can be considered as the basic building block for the time-domain signal processing, which scales well with the CMOS technology.

3 Multipath Delay Cell

The delay cell is a crucial component in the design of the ring oscillator. The delay generated by adjusting the control voltage of the delay generator should appropriately be defined for the delay stages. There are some factors, such as supply and substrate voltages that affect the precision and linearity of the delay cell. The supply and substrate voltage give rise to the output jitter or phase-noise that will hamper the efficiency of the oscillator. Hence, the circuit should be less susceptible to the supply and substrate-induced noise. It can be done by using the coupled ring oscillator [8, 13], which shows better phase noise for a particular oscillation frequency as compared to the traditional ring oscillator. The oscillation period (T_{osc}) of the conventional ring oscillator is defined as

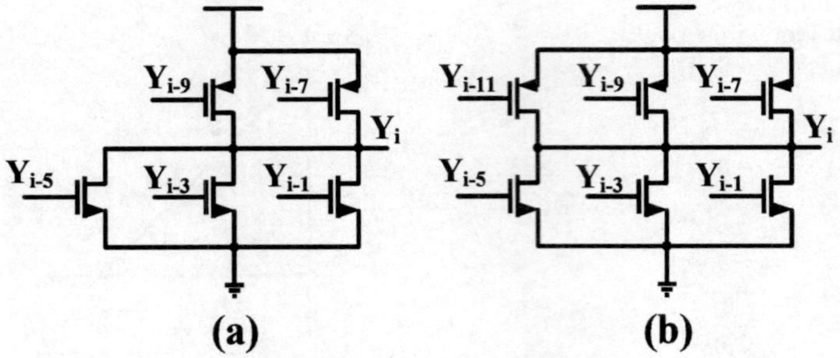

Fig. 5 Proposed delay cells **a** DC1, **b** DC2

$$T_{osc} = 2N \times \tau_P \tag{5}$$

where τ_P is the propagation delay of the delay cell and N is the number of stages of the ring oscillator. The minimum measurable time (T_{LSB}) of $\Sigma\Delta$ TDC can be expressed as [7]

$$T_{LSB} = \frac{T_{osc}}{2N} = \tau_P \tag{6}$$

From the above equation, it is clear that the T_{LSB} is limited to τ_P. Hence, the multipath technique is employed to reduce the effective propagation delay of the oscillator, which takes the inputs from not only previous output, but also the past outputs [6, 7]. For the ring oscillator to operate correctly, the number of inverters in the loop should be odd and preferably be prime. It is observed that the odd-prime number of delay cell in the loop gives better phase noise [6, 7].

Multipath ring oscillators are a selective type of ring oscillator used to increase the performance of the traditional ring oscillator. As already described that in the multipath technique, the oscillator accepts input from the previous output, as well as from the past outputs. In the conventional ring oscillator, each stage of the delay cell changes its state only when the last stage has changed the state. However, in the multipath delay cell oscillator, each stage tries to change its state well before the complete state changeover of the previous stages depending on the coupling. In order to verify this conjecture, we have proposed two designs, namely delay cell1 (DC1) and delay cell2 (DC2), as shown in Fig. 5. In DC1 architecture, the delay cell takes outputs from the just preceding stage Y_{i-1}, as well as from the other previous stage outputs too, i.e., from Y_{i-3}, Y_{i-5}, Y_{i-7} and Y_{i-9} as shown in the Fig. 5a. In a similar manner for DC2 topology, the delay cell accepts the outputs from Y_{i-1}, Y_{i-3}, Y_{i-5}, Y_{i-7}, Y_{i-9} and Y_{i-11} as illustrated in Fig. 5b. Note that the larger widths are assigned to the MOSFETs which are relatively far from the current delay cell. In DC1 architecture, the MOSFET with Y_{i-1} input is just the previous stage, whereas Y_{i-9} is the farthest one. So, the MOSFET with Y_{i-1} input is assigned with a smaller width, whereas Y_{i-9} is assigned with a larger width for offering a low resistive path. The same strategy is followed for DC2 architecture too.

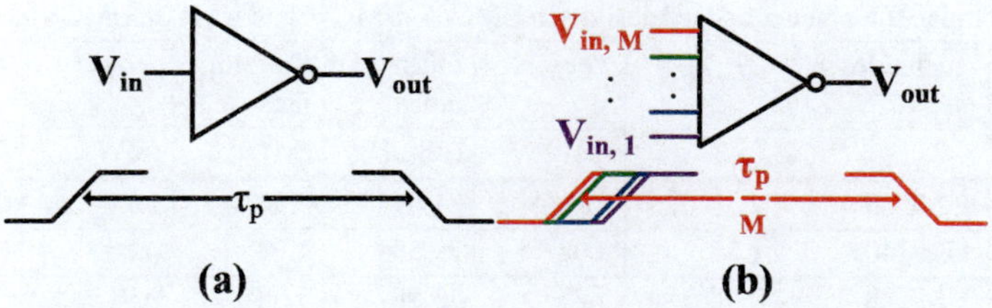

Fig. 6 Propagation delay: **a** Traditional Delay Cell, **b** Multipath Delay Cell

The reduction in the delay of the multipath delay cell can be illustrated intuitively in Fig. 6. Suppose the multipath delay cell has M inputs. The effective propagation delay of the multipath delay cell is decided by the input, which is leading in phase. The $V_{in,M}$ input, which is leading in phase, provides the effective propagation delay of the ring oscillator. The multipath ring oscillator with M number of inputs is equivalent to the M number of coupled ring oscillators. In M number of coupled ring oscillators, the effective propagation delay $\tau_P{}'$ reduces by a factor of M, i.e.

$$\tau_P{}' = \frac{\tau_P}{M} \tag{7}$$

In the proposed designs; DC1 topology, the coupling factor M = 5, whereas for the DC2 topology M = 6. The designer may be tempted to increase the coupling factor for achieving better resolution. In an N-stage ring oscillator, it is advised to choose the coupling inputs to a maximum of $\frac{N-1}{2}$ previous connections; otherwise, the oscillator will oscillate entirely out of the phase. This, in turn, increases the delay of the circuit [4]. For instance, in a 47-stage multipath ring oscillator, the multipath connections that we can connect to the present delay cell would be $\frac{47-1}{2} = 23$ previous outputs connection only.

4 Results and Discussion

The proposed multipath delay cells for $\Sigma\Delta$ TDC are designed and simulated in UMC65, UMC40 and UMC28 CMOS processes using Cadence Virtuoso ADE. The proposed designs are utilised in the design of 47-stage ring oscillators. The performance summary of the multipath delay cell-based coupled oscillators for both DC1 and DC2 topologies in different technologies are compared in Table 2 where $T_{LSB_{MP}}$ is the T_{LSB} with multipath connections, T_{LSB_T} is the T_{LSB} in a conventional manner, P_{MP} is the power consumption with multipath technique, and P_T is the power consumption with the conventional method. For DC1 and DC2 designs, the resolution is improved as we move from UMC65 to UMC28. This demonstrates the advantages of CMOS technology scaling in resolution. The actual value of the coupling factor is nearly matching with the theoretical value and is given by

Table 2 Comparison of the proposed designs in UMC65, UMC40 and UMC28 CMOS technologies

Topology	Technology	$T_{LSB_{MP}}$	T_{LSB_T}	Phase noise	Coupling factor	P_{MP}	P_T
		(ps)	(ps)	(dBc/Hz)	(M)	(mW)	(mW)
DC1	UMC65	2.40	11.86	81.04	4.94	0.71	0.12
	UMC40	1.83	9.02	79.50	4.93	0.85	0.14
	UMC28	0.98	5.27	73.88	5.38	0.91	0.15
DC2	UMC65	2.17	11.99	81.35	5.52	1.04	0.15
	UMC40	1.63	8.94	79.84	5.48	1.27	0.19
	UMC28	0.91	5.37	74.30	5.93	1.30	0.20

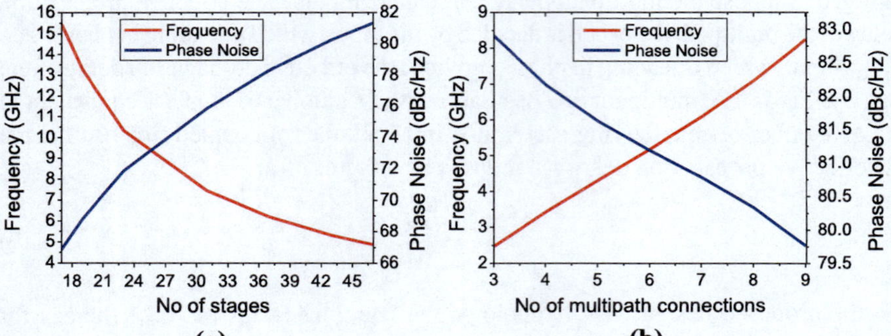

Fig. 7 Frequency and phase noise vs **a** the number of stages **b** the number of multipath connections

$$M = \frac{T_{LSB_T}}{T_{LSB_{MP}}} \tag{8}$$

The theoretical value of the coupling factor (M) for DC1 architecture is 5 and for DC2 architecture, it is 6.

The proposed designs are subjected to the number of stages of parametric simulations. From Fig. 7a, it can be seen that as the number of delay cells (N) in the ring oscillator increases, the oscillation period increases. So, the frequency of oscillation decreases, which improves the phase noise around 1 MHz offset from the carrier. The proposed designs are subjected to the number of multipath connections parametric simulation. From Fig. 7b, as the number of multipath connections in the oscillator increases, the frequency of oscillation will increase, whereas the phase noise around 1 MHz offset from the carrier decreases.

From the above extensive parametric simulations, we can draw the following conclusions. As the number of stages in the ring oscillator increases, the oscillation frequency comes down. Whereas the number of multipath connections increases, the oscillation frequency also increases. The phase-noise is inversely proportional to the operating frequency. The designer has to trade-off between the number of stages and the number of multipath connections, for achieving the desired operating frequency and phase-noise.

5 Conclusion

The coupled ring oscillators for $\Sigma\Delta$ TDCs have been demonstrated in this paper using the multipath techniques in UMC65, UMC40 and UMC28 CMOS process technologies. The ring oscillator integrator (ROI) is projected as the basic building block for the time-domain signal processing, which scales well with the CMOS technology. The qualitative analysis on the multipath technique for improving the resolution of $\Sigma\Delta$ TDC has been established. The effective propagation delay was reduced below τ_P by the factor of the number of multipath connections (M), without the requirement of array oscillators. The effective propagation delay for DC1 architecture has been reduced to $\frac{\tau_P}{5}$, whereas for DC2 architecture, it is $\frac{\tau_P}{6}$.

Acknowledgements The authors would like to extend their heartfelt thanks to Dr. P. V.Ananda Mohan (Technology Advisor, CDAC, Bangalore) for valuable suggestions and comments.

References

1. Henzler S (2010) Time-to-Digital Converters, Springer Series in Advanced Microelectronics, vol 29. Springer, Netherlands, Dordrecht
2. Yuan, F.: CMOS Time-Mode Circuits and Systems. CRC Press (nov 2015)
3. Cao, Y., Leroux, P., De Cock, W., Steyaert, M.: A 0.7mW 13b temperature-stable MASH $\Delta\Sigma$ TDC with delay-line assisted calibration. In: IEEE Asian Solid-State Circuits Conference 2011. pp. 361–364. Jeju (nov 2011)
4. Krishna, R.S.S.M.R., Mal, A.K., Mahapatra, R.: CMOS time-mode smart temperature sensor using programmable temperature compensation devices and $\Delta\Sigma$ time-to-digital converter. Analog Integrated Circuits and Signal Processing (sep 2019)
5. Sung-Mo (Steve) Kang: CMOS Digital Integrated Circuits Analysis & Design. McGraw-Hill (2019)
6. Straayer, M.Z., Perrott, M.H.: A Multi-Path Gated Ring Oscillator TDC With First-Order Noise Shaping. IEEE Journal of Solid-State Circuits **44**(4), 1089–1098 (apr 2009)
7. Krishna, R.S.S.M.R., Mal, A.K., Mahapatra, R.: All MOS Noise-shaped Time-Mode Temperature Sensor. Integration, the VLSI Journal **65**, 74–80 (mar 2019)
8. Maneatis J, Horowitz M (1993) Precise delay generation using coupled oscillators. IEEE Journal of Solid-State Circuits 28(12):1273–1282
9. Lee Seog-Jun, Kim Beomsup, Lee Kwyro (1997) A novel high-speed ring oscillator for multiphase clock generation using negative skewed delay scheme. IEEE Journal of Solid-State Circuits 32(2):289–291
10. Matsumoto, A., Sakiyama, S., Tokunaga, Y., Morie, T., Dosho, S.: A Design Method and Developments of a Low-Power and High-Resolution Multiphase Generation System. IEEE Journal of Solid-State Circuits **43**(4), 831–843 (apr 2008)

11. Yu, W., Kim, K., Cho, S.: A 148fsrms integrated noise 4MHz bandwidth all-digital second-order $\Delta\Sigma$ time-to-digital converter using gated switched-ring oscillator. In: Proceedings of the IEEE 2013 Custom Integrated Circuits Conference. pp. 1–4. San Jose, CA (sep 2013)
12. Drost, B., Talegaonkar, M., Hanumolu, P.K.: Analog Filter Design Using Ring Oscillator Integrators. IEEE Journal of Solid-State Circuits **47**(12), 3120–3129 (dec 2012)
13. Krishna, R.S.S.M.R., Mal, A.K., Mahapatra, R.: Time-Domain Smart Temperature Sensor Using Current Starved Inverters and Switched Ring Oscillator-Based Time-to-Digital Converter. Circuits, Systems, and Signal Processing (aug 2019)

Low-Hardware Digit-Serial Sequential Polynomial Basis Finite Field GF(2^m) Multiplier for Trinomials

Siva Ramakrishna Pillutla and Lakshmi Boppana

Abstract Finite field GF(2^m) multipliers are employed in practical applications such as elliptic curve cryptography (ECC) and Reed-Solomon encoders. Digit-level finite field multipliers are best suitable for applications that require low-hardware implementation while operating at speeds that conform to today's high data rates. With the emergence of Internet of Things (IoT), many resource-constrained devices such as IoT edge devices came into proliferate usage. To secure these constrained devices, ECC must be implemented in these devices with low-hardware complexity. Hence, it requires to design efficient digit-serial finite field multipliers since the performance of ECC greatly depends on the performance of the finite field multiplier employed. Many efficient designs for digit-serial finite field multipliers are presented in the literature to achieve better area and time complexities. In this paper, we present an area-efficient sequential digit-serial finite field multiplier for trinomials. The hardware and time complexities of the proposed multiplier are estimated for GF(2^{409}) and compared with the similar multipliers available in the literature. The comparison shows that the proposed multiplier achieves lower hardware complexity. Therefore, the proposed multiplier is attractive for cost-effective high-speed applications such as IoT edge devices.

Keywords Internet of Things · Elliptic curve cryptography · Finite field arithmetic · Digit-level multiplier

S. R. Pillutla (✉) · L. Boppana (✉)
National Institute of Technology, Warangal, Telangana, India
e-mail: srk100p@student.nitw.ac.in

L. Boppana
e-mail: lakshmi@nitw.ac.in

T. Laxminidhi et al. (eds.), *Advances in Communications, Signal Processing, and VLSI*, Lecture Notes in Electrical Engineering 722,
https://doi.org/10.1007/978-981-33-4058-9_36

1 Introduction

Internet of Things (IoT) connects many physical things that are quite different from conventional computing systems to the network. These physical things are equipped with constrained devices, namely, IoT end-devices/IoT edge devices, to enable them to participate in communication over the network. These constrained IoT devices must be cost-effective while suitable for today's high data speed applications. Enabling security in IoT edge devices is crucial to thwarting many network-based attacks [1]. Some of the security features can be achieved by using public key cryptosystems such as elliptic curve cryptography (ECC) and RSA (Rivest-Shamir-Adleman). ECC with its relatively shorter key-size and relatively less computational requirements is best suitable for constrained devices (IoT edge devices) to implement security services such as digital signature and key-exchange [2]. Since arithmetic operations in ECC heavily involve over finite fields GF(2^m), efficient finite field arithmetic implementations, particularly multiplication, result in a considerable performance improvement in ECC [3]. Hence, the performance of applications that employs ECC for security implementation can be enhanced by designing an efficient finite field multiplier. For IoT edge devices, designing a bit-serial multiplier results in low cost. However, this multiplier is too slow for today's high data speed applications. Designing a parallel multiplier requires very high hardware that results in more cost, which is not desirable for IoT end-devices/IoT edge devices since a typical domestic application needs tens of IoT devices. Digit-serial multipliers, which inherently provide area-delay trade-off and can also make use of full available data bus width of an IoT device, are best suitable for IoT edge devices.

A finite field GF(2^m) has 2^m elements, where the field elements can be represented using various bases such as dual basis, normal basis, polynomial basis, and redundant basis. Polynomial basis is one of the bases recommended by many standard institutes such as National Institute of Standards and Technology (NIST), and designs based on this basis are more regular and modular. In polynomial basis representation, every field element is a reduced modulo an irreducible polynomial. There are various types of irreducible polynomials such as trinomials, pentanomials, all one polynomials, and general irreducible polynomials. Standard institutes recommend sparse polynomials such as trinomials and pentanomials as they result in low-hardware and time complexities. Hence, fields defined over trinomials are more suitable for constrained devices.

Many digit-level finite field multipliers over trinomials are proposed in the literature to achieve better area and time complexities. In [4], two digit-level multipliers based on most significant bit (MSB)/least significant bit (LSB) first algorithms were presented. In this paper, the authors have shown that irreducible polynomials that have low hamming weight and low second highest degree result in complexity reduction. In [5], a bit-parallel word serial multiplier for GF(2^{233}) over trinomials was presented. This multiplier has a parallel partial product generator followed by an accumulation unit. In [6], a high-throughput hardware efficient digit-serial architecture was presented. In this multiplier, by using T flip flops in the accumulation unit authors achieved lower critical path delay as well as low-hardware complexity. In [7], a shifted polynomial basis (SPB) digit-serial multiplier using the proposed (b, 2)-way Karatsuba decomposition was presented to achieve sub-quadratic space complexity. A digit-serial area-efficient multiplier employing a new factoring technique was proposed in [8] to achieve power reduction. In our paper, we propose a new area-efficient polynomial basis digit-level multiplier whose structure comprises of a parallel multiplier followed by an accumulation unit. The parallel multiplier is based on the approach proposed in [9] for a parallel multiplier for all trinomials. This approach when applied for a class of trinomials, $x^m + x^n + 1$ where $n \leq m/2$, gives low-hardware implementation [10]. This class of trinomials also includes two of the five NIST recommended irreducible polynomials suggested for ECC. The proposed area-efficient digit-serial multiplier is suitable for edge devices used in IoT applications.

The organization of the paper is as follows. Section 2 gives preliminaries regarding polynomial basis digit-serial multiplication. Section 3 introduces the mathematical formulations for the proposed digit-serial multiplication and presents the proposed architecture. In addition, a comparison of area and time complexities of the proposed multiplier with the existing similar multipliers is also presented in this section. Conclusions are presented in Sect. 4.

2 Preliminaries

A finite field GF(2^m) has 2^m elements where each element is represented with a polynomial of degree at most $(m-1)$ over GF(2). An mth degree polynomial over which the field GF(2^m) is defined is called the irreducible polynomial $P(x)$ of the field, given by $P(x) = x^m + \sum_{j=m-1}^{1} p_j x^j + 1$, where all the $p_j \in$ GF(2). The polynomial basis can be defined with the set $(x^{m-1}, x^{m-2}, \ldots, x^2, x, 1)$, where x is the root of the irreducible polynomial $P(x)$ of the field.

Let A and B be two arbitrary field elements, given by

$$
\begin{aligned}
A(x) &= \sum_{j=0}^{m-1} a_j x^j \\
&= a_{m-1}x^{m-1} + a_{m-2}x^{m-2} + \cdots + a_1 x^1 + a_0
\end{aligned} \tag{1}
$$

and

$$
\begin{aligned}
B(x) &= \sum_{j=0}^{m-1} b_j x^j \\
&= b_{m-1}x^{m-1} + b_{m-2}x^{m-2} + \cdots + b_1 x^1 + b_0,
\end{aligned} \tag{2}
$$

where all a_j and $b_j \in$ GF(2). Let $D(x)$ be the product of $A(x)$ and $B(x)$. Conventionally, $D(x)$ is obtained by first multiplying the two polynomials A and B followed by modulo reduction using irreducible polynomial $P(x)$. Thus,

$$D(x) = A(x)B(x) \bmod P(x) \tag{3}$$

Two popular digit-level schemes presented in the literature to evaluate $D(x)$ are MSD (most significant digit) first scheme and LSD (least significant digit) first scheme. Let s be the number of digits and l be the width of each digit, then operand B can be written as

$$B = \sum_{j=0}^{s-1} B_j x^{jl} \tag{4}$$

where

$$B_j = \begin{cases} \sum_{t=0}^{l-1} b_{lj+t}x^t, & \text{for } 0 \le j \le s-2 \\ \sum_{t=0}^{m-1-l(s-1)} b_{lj+t}x^t, & \text{for } j = s-1 \end{cases}$$

and $s = \lceil \frac{m}{l} \rceil$. Then, $D(x)$ can be obtained as

$$\begin{aligned} D = (B_0A + B_1(Ax^l \bmod P(x)) + \\ B_2(Ax^lx^l \bmod P(x)) + \cdots + \\ B_{s-1}(Ax^{l(s-2)}x^l \bmod P(x))) \bmod P(x) \end{aligned} \tag{5}$$

for the LSD scheme, and

$$\begin{aligned} D = ((\ldots(((AB_{s-1} \bmod P(x))x^l + AB_{s-2}) \\ \bmod P(x))x^l + \cdots \\ \ldots)x^l + AB_0) \bmod P(x) \end{aligned} \tag{6}$$

for the MSD scheme.

3 Proposed Digit-Serial Multiplier

3.1 *Mathematical Formulation*

Let $P(x) = x^m + x^n + 1$, where $1 \le n \le \lceil \frac{m}{2} \rceil$, be an irreducible trinomial polynomial over which the field GF(2^m) is defined. Let $A(x) = \sum_{j=0}^{m-1} a_j x^j$ and $B'(x) = \sum_{j=0}^{l-1} b'_j x^j$ be two elements, where $l \le m$. Let $C(x)$ denote the product of polynomials A and B' as $C(x) = \sum_{j=0}^{m+l-2} c_j x^j = AB'$. This product expression $C(x) = AB'$ can be expressed using a $(m+l-1) \times l$ matrix M as follows.

$$\begin{bmatrix} c_0 \\ c_1 \\ \vdots \\ c_{l-1} \\ c_l \\ \vdots \\ c_{m-1} \\ c_m \\ \vdots \\ c_{m+l-2} \end{bmatrix} = \begin{bmatrix} a_0 & 0 & 0 & \cdots & 0 & 0 \\ a_1 & a_0 & 0 & \cdots & 0 & 0 \\ \vdots & \vdots & \vdots & \ddots & \vdots & \vdots \\ a_{l-1} & a_{l-2} & a_{l-3} & \cdots & a_1 & a_0 \\ a_l & a_{l-1} & a_{l-2} & \cdots & a_2 & a_1 \\ \vdots & \vdots & \vdots & \ddots & \vdots & \vdots \\ a_{m-1} & a_{m-2} & a_{m-3} & \cdots & a_{m-l+1} & a_{m-l} \\ 0 & a_{m-1} & a_{m-2} & \cdots & a_{m-l+2} & a_{m-l+1} \\ \vdots & \vdots & \vdots & \ddots & \vdots & \vdots \\ 0 & 0 & 0 & \cdots & 0 & a_{m-1} \end{bmatrix} \times \begin{bmatrix} b'_0 \\ b'_1 \\ b'_2 \\ \vdots \\ b'_{l-1} \end{bmatrix}$$

The product polynomial $C(x)$ includes the terms whose degree is more than $m-1$ and these terms can be modulo reduced using the identity $x^m = x^n + 1$. From this identity we can have $x^{m+i} = (x^n + 1)x^i = x^{n+i} + x^i$, where the range of i is assumed to be in the range $0 \le i \le (l-2)$. Also, assume $l \le \lceil m/2 \rceil$, then we have $n + i \le n + l - 2 \le m/2 + \lceil m/2 \rceil - 2 < m$. Thus each term in the product polynomial $C(x)$ whose degree is $(m+i)$ can be reduced to a polynomial of at most degree $(m-1)$ with two terms, $x^{n+i} + x^i$. By this modulo reduction, each $(m+i)$th row of matrix M for $0 \le i \le (l-2)$ is added to the ith and $(n+i)$th rows of it.

Let Q be a $m \times l$ matrix which is obtained from the matrix M, after the modulo reduction process applied. Let matrix Q be decomposed into the sum of three $m \times l$ matrices X, Y, and Z, such that $Q = X + Y + Z$. These three matrices X, Y, and Z can be defined as follows.

$$X = \begin{bmatrix} a_0 & 0 & 0 & \cdots & 0 & 0 \\ a_1 & a_0 & 0 & \cdots & 0 & 0 \\ \vdots & \vdots & \vdots & \ddots & \vdots & \vdots \\ a_{l-2} & a_{l-3} & a_{l-4} & \cdots & a_0 & 0 \\ a_{l-1} & a_{l-2} & a_{l-3} & \cdots & a_1 & a_0 \\ \vdots & \vdots & \vdots & \ddots & \vdots & \vdots \\ a_{m-1} & a_{m-2} & a_{m-3} & \cdots & a_{m-l+1} & a_{m-l} \end{bmatrix}$$

$$Y = \begin{bmatrix} 0 & a_{m-1} & a_{m-2} & \cdots & a_{m-l+2} & a_{m-l+1} \\ 0 & 0 & a_{m-1} & \cdots & a_{m-l+3} & a_{m-l+2} \\ \vdots & \vdots & \vdots & \ddots & \vdots & \vdots \\ 0 & 0 & 0 & \cdots & 0 & a_{m-1} \\ 0 & 0 & 0 & \cdots & 0 & 0 \\ \vdots & \vdots & \vdots & \ddots & \vdots & \vdots \\ 0 & 0 & 0 & \cdots & 0 & 0 \end{bmatrix}$$

$$Z = \begin{bmatrix} 0 & 0 & 0 & \cdots & 0 & 0 \\ \vdots & \vdots & \vdots & \ddots & \vdots & \vdots \\ 0 & 0 & 0 & \cdots & 0 & 0 \\ 0 & a_{m-1} & a_{m-2} & \cdots & a_{m-l+2} & a_{m-l+1} \\ 0 & 0 & a_{m-1} & \cdots & a_{m-l+3} & a_{m-l+2} \\ \vdots & \vdots & \vdots & \ddots & \vdots & \vdots \\ 0 & 0 & 0 & \cdots & 0 & a_{m-1} \\ \vdots & \vdots & \vdots & \ddots & \vdots & \vdots \\ 0 & 0 & 0 & \cdots & 0 & 0 \end{bmatrix} \begin{matrix} \text{0th row} \\ \\ \\ \\ n\text{th row} \\ \\ (n+l-2)\text{th row} \\ \\ \\ \end{matrix}$$

Matrix Z is equivalent to a matrix that can be obtained by shifting matrix Y down by n rows and filling the first n rows with zeros. By employing similar method presented in [9], any ith row of matrix Q can be obtained with simple rewiring of the nth row, Q_n, of the matrix Q. The row Q_n can be computed as

$$Q_n = \begin{cases} (a_n a_{n-1} \ldots a_0 a_{m-1} a_{m-2} \ldots a_{m-l+n+1}) + (0 a_{m-1} \ldots a_{m-l+1}), & \text{if } n \le l \\ (a_n a_{n-1} \ldots a_{n-l+1}) + (0 a_{m-1} \ldots a_{m-l+1}), & \text{if } n > l \end{cases}$$

To compute Q_n, it requires $(l-1)$ two-input XOR gates, and a delay of T_X, where T_X is a delay of a two-input XOR gate. Since Q is an $m \times l$ matrix, $Q \cdot B'$ needs lm AND gates, $(l-1)m$ XOR gates, and $T_A + \lceil \log_2^l \rceil T_X$ delays, where $B' = (b_0, b_1, \ldots, b_{l-1})^t$. After formulating this method of computing the multiplication for $C(x)$ of an m-bit element with a l-bit element where $l \le \lceil m/2 \rceil$, the multiplication of any two arbitrary field elements is considered now as follows.

Let $A(x) = \sum_{i=0}^{m-1} a_i x^i$ and $B(x) = \sum_{i=0}^{m-1} b_i x^i$ be two arbitrary field elements. Let $D(x) = \sum_{i=0}^{m-1} d_i x^i = AB \bmod P(x)$ be the product of the elements A and B. The computation of $D(x)$ can be performed as follows.

Let the element $B(x)$ is partitioned into s digits where each digit of size l bits. Then, we have $s = \lceil \frac{m}{l} \rceil$. It follows

$$B(x) = \sum_{j=0}^{l-1} b_j x^j + \sum_{j=l}^{2l-1} b_j x^j + \cdots + \sum_{j=(s-1)l}^{sl-1} b_j x^j$$
$$= \sum_{j=0}^{l-1} b_j x^j + x^l \sum_{j=0}^{l-1} b_{l+j} x^j + +x^{2l} \sum_{j=0}^{l-1} b_{2l+j} x^j \cdots + x^{(s-1)l} \sum_{j=0}^{l-1} b_{(s-1)l+j} x^j, \tag{7}$$

where all b_js for $j \geq m$ are zero. Now, the product $D(x)$ can be computed as

$$D(x) = A(x)B(x) \bmod P(x) = A(x) \sum_{j=0}^{l-1} b_j x^j \bmod P(x) +$$
$$x^l A(x) \sum_{j=0}^{l-1} b_{l+j} x^j \bmod P(x) + x^{2l} A(x) \sum_{j=0}^{l-1} b_{2l+j} x^j \bmod P(x) + \cdots$$
$$\cdots + x^{(s-1)l} A(x) \sum_{j=0}^{l-1} b_{(s-1)l+j} x^j \bmod P(x)$$

$$= T_0 \bmod P(x) + x^l T_1 \bmod P(x) + x^{2l} T_2 \bmod P(x) + \cdots$$
$$\cdots + x^{(s-1)l} T_{s-1} \bmod P(x)$$

$$= (\ldots((T_{s-1} x^l \bmod P(x) + T_{s-2}) x^l \bmod P(x) + T_{s-3}) x^l \bmod P(x) + \ldots$$
$$\ldots + T_1) x^l \bmod P(x) + T_0, \tag{8}$$

where

$$T_i = A(x) \sum_{j=0}^{l-1} b_{il+j} x^j \bmod P(x). \tag{9}$$

The computation of T_i can be performed using the procedure shown to compute $C(x)$. In the computation of $C(x)$, note that the value of digit-size, l, is taken at most half the value of field order, m. It is acceptable for the constrained devices since the data bus width of these devices is typically 8/16/32 bits only. As per today's security requirements, a field order of at least 233 is required. Hence, the selected l range is quite applicable to today's security requirements for Wireless Sensor Network (WSN) nodes and IoT end-nodes/edge devices.

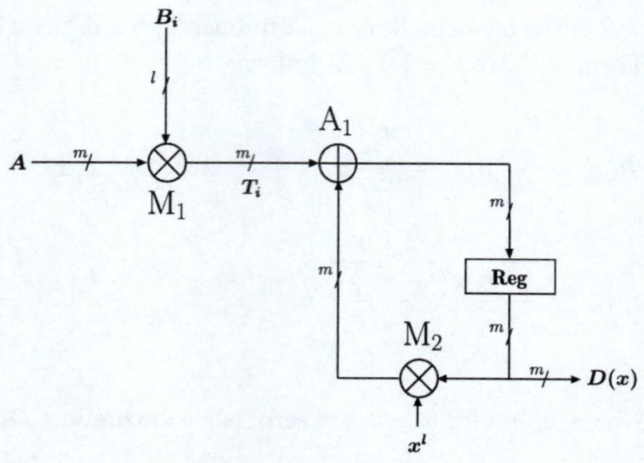

Fig. 1 The proposed structure of the digit-serial multiplier

3.2 *Proposed Structure of the Multiplier*

Based on the proposed formulations, a conceptual block diagram of the digit-serial multiplier is shown in Fig. 1. The structure shown in Fig. 1 realizes the expression given in Eq. 8. Node M1 is a partial parallel $m \times l$ multiplier that multiplies an m-bit element with an l-bit element. It realizes the computation of T_i as given in Eq. 9. Node A_1 performs the additions that are involved in the computation of the expression in Eq. 8. Similarly, Node M_2 performs the interleaved multiplications of partial output product with x^l that are involved in the computation of the expression in Eq. 8. The multiplicand A is made available throughout the computation, while multiplier B enters the structure digit-wise starting from most significant digit (MSD). The structure produces the required multiplication result after a delay of s clock cycles.

3.3 *Area and Time Complexities*

In this section, the area and time complexities of the proposed multiplier are obtained and compared with the existing similar multipliers. The node M_1 which computes T_i performs a similar computation presented for computing $Q \cdot B'$. Hence, it requires lm AND gates and $(l-1)(m+1)$ XOR gates. The node A_1 requires m XOR gates while the node M_2 requires l XOR gates. The structure also requires two m-bit registers, one at the input to register multiplicand A while another as output register, *Reg*. The delays of the nodes M_1, A_1, and M_2 are $T_A + (\lceil \log_2^l \rceil + 1)T_X$, T_X, and T_X, respectively. The critical path of the structure is $T_A + (\lceil \log_2^l \rceil + 2)T_X$. The area and time complexities for the proposed multiplier and the other similar multipliers [4–8] are presented in Tables 1 and 2, respectively.

The analytical comparisons presented in Tables 1 and 2 can be better understood by considering a specific field order m and a specific digit-size l. By selecting the

Table 1 Comparison of area complexities for GF(2^m)

Design	XOR	AND	Register
[4]	$lm + 3l$	lm	$2m + l$
[5]	$lm + (l^2 + l)/2$	lm	$2m + l$
[6]	$lm + (l^2 + l)/2$	lm	$2m + l$
[7]	$69/20m^{\log_4^6} - 1/4m^{\log_4^2} - 11/5$	$m^{\log_4^6}$	$2m - 1$
[8]	$lm + l^2/2 + 3l/2 - 1$	lm	$2m$
[Proposed]	$lm + (2l - 1)$	lm	$2m$

Table 2 Comparison of time complexities for GF(2^m)

Design	Critical path	Latency (clock cycles)
[4]	$T_A + (\lceil \log_2^{2l+1} \rceil)T_X$	$s + 2$
[5]	$T_A + (\lceil \log_2^l \rceil + 2)T_X$	s
[6]	$T_A + (\lceil \log_2^l \rceil + 2)T_X$	$s + 1$
[7]	$T_A + (1 + 3\log_4^m)T_X$	$s + 1$
[8]	$T_A + (\lceil \log_2^l \rceil + 2)T_X$	$s + 1$
[Proposed]	$T_A + (\lceil \log_2^l \rceil + 2)T_X$	s

Table 3 Area and time complexities comparison for GF(2^{409}) over $x^{409} + x^{87} + 1$ with $l = 8$

Design	Area (in terms of number of NAND gate equivalents)	Latency (clock cycles)	Critical path delay (ns)	Delay (ns)	Area-delay product (NAND equivalents × delay (ns))
[4]	13,780	54	0.27	14.58	200,913
[5]	13,804	52	0.23	11.96	165,096
[6]	13,804	53	0.23	12.19	168,271
[7]	23,299	53	0.59	31.27	728,560
[8]	13,788	53	0.23	12.19	168,076
[Proposed]	13,732	52	0.23	11.96	164,235

field to be GF(2^{409}) over an irreducible polynomial $x^{409} + x^{87} + 1$ with a digit-size $l = 8$, the complexities presented in Tables 1 and 2 are computed and presented in Table 3.

We have 65 nm standard library statistics to estimate the time and area requirements. With this technology, the NAND gate equivalents for XOR gate, AND gate, and register are assumed to be 2, 1.25, and 3.75 [11]. The delays for XOR gate and AND gate are assumed to be 0.04 and 0.03 [11]. It is observed from Table 3, the proposed multiplier requires marginally less hardware when compared with other similar multipliers. It is also observed that the proposed multiplier achieves low area-delay product as well. Hence, the proposed digit-serial sequential multiplier is suitable for IoT edge devices which typically have a bus width of 8/16/32 bits.

4 Conclusions

In this paper, a new formulation for the digit-serial finite field multiplication over a class of trinomials and its hardware structure are presented. The proposed multiplier achieves low hardware and a reduction in area-delay product when compared with the multipliers available in the literature. This area-efficient sequential digit-level multiplier is suitable for constrained devices such as IoT edge devices.

References

1. Li S, Da Xu L, Zhao S (2015) The internet of things: a survey. Inform Syst Front 17(2):243–259
2. Suárez-Albela M, Fraga-Lamas P, Fernández-Caramés T (2018) A practical evaluation on RSA and ECC-based cipher suites for IoT high-security energy-efficient fog and mist computing devices. Sensors 18(11):3868
3. Lim CH, Hwang HS (2000) Fast implementation of elliptic curve arithmetic in GF(p^n). In: International workshop on public key cryptography. Springer, pp 405–421
4. Song L, Parhi KK (1998) Low-energy digit-serial/parallel finite field multipliers. J VLSI Signal Process Syst Signal Image Video Technol 19(2):149–166
5. Tang W, Wu H, Ahmadi M (2005) VLSI implementation of bit-parallel word-serial multiplier in GF(2^{233}). In: The 3rd international IEEE-NEWCAS conference. IEEE, pp 399–402
6. Meher P (2007) High-throughput hardware-efficient digit-serial architecture for field multiplication over GF(2^m). In: 2007 6th international conference on information, communications and signal processing. IEEE, pp 1–5
7. Lee CY, Yang CS, Meher BK, Meher PK, Pan JS (2014) Low-complexity digit-serial and scalable SPB/GPB multipliers over large binary extension fields using (b, 2)-way Karatsuba decomposition. IEEE Trans Circuits Syst I Regul Pap 61(11):3115–3124
8. Namin SH, Wu H, Ahmadi M (2016) Low-power design for a digit-serial polynomial basis finite field multiplier using factoring technique. IEEE Trans Very Large Scale Integr (VLSI) Syst 25(2), 441–449
9. Sunar B, Koc CK (1999) Mastrovito multiplier for all trinomials. IEEE Trans Comput 48(5):522–527
10. Choi Y, Chang KY, Hong D, Cho H (2004) Hybrid multiplier for GF(2^m) defined by some irreducible trinomials. Electron Lett 40(14):852–853
11. El-Razouk H, Reyhani-Masoleh A (2015) New bit-level serial GF(2^m) multiplication using polynomial basis. In: 2015 IEEE 22nd symposium on computer arithmetic. IEEE, pp 129–136

Author Index

A
Addepalli, Tathababu, 23
Alex, Zachariah C., 179
Anil Kumar, Chevella, 63
Anitha Sheela, Kancharla, 63
Anitha, V. R., 23
Anjaneyulu, L., 219
Arun Kumar, G., 213

B
Barma, Minakshi Deb, 323
Biswas, Bijit, 83, 213
Boppana, Lakshmi, 401

C
Chhabra, I. M., 123

D
Deb, Sumita, 323
Dikshit, Amrita, 391

G
Gai, Hemanth Kumar, 199
Gopala Swamy, B., 263

H
Hanumantha Rao, T. V. K., 241
Hunagund, P. V., 379

J
Jaffino, G., 227
Jagannadha Swamy, T., 1
Jatoth, Ravi Kumar, 75
Jeslani, Mahima, 335
Jose, J. Prabin, 227

K
Kalpana, Naidu, 199, 335
Kamaraju, M., 249
Karthik Sairam, S., 273
Kaushik, C., 91
Kayala, Vamshi Krishna, 35
Kodali, Prakash, 35
Komati, Shravya Krishna, 315
Kotte, Sowjanya, 49
Krishna, R. S. S. M. R., 391
Krishnasree, Vasagiri, 91
Kumar, Amgothu Ravi, 199
Kumar, G. Arun, 83
Kumar, Gopisetti Indra Sai, 11
Kumar, L. Sobhan, 123
Kumar, R. Manoj, 301

L
Lokam, Anjaneyulu, 115

M
Mahapatra, Rajat, 391
Mal, Ashis Kumar, 391
Mallavarapu, Sandhya, 115
Mallesh, Ch. Yajna, 91
Mallikarjuna Rao, Prudhvi, 349
Misra, Rohit, 293
Mondal, Biplob, 83, 213

T. Laxminidhi et al. (eds.), *Advances in Communications, Signal Processing, and VLSI*,
Lecture Notes in Electrical Engineering 722,
https://doi.org/10.1007/978-981-33-4058-9

Muralidhar, P., 273
Musala, Sarada, 283

N
Nayak, Jagannath, 103
Neelam, Aruna Kumari, 283
Neelima, K., 357

P
Pandya, Rahul Jashvantbhai, 293
Pardhasaradhi, B., 263
Patri, Sreehari Rao, 103, 123
Paul, Satyajit, 323
Pillutla, Siva Ramakrishna, 401
Prasad, MVKS, 103
Prasad, S. Siva, 139
Prathima, S., 179

R
Raghunadh, M. V., 153
Rajasekhar, K., 11
Ramamurthy, G., 1
Rao, C. B. Rama, 139
Rao, L. A. V. Sanhith, 123
Rao, N. Bheema, 153
Ravikanti, Swetha, 219

S
Sahitya, G., 91
Sai Kumar, Aruru, 241
Sambamurthy, N., 249
Santosh, Pavada, 349
Sathya, Vanlin, 199
Satyanarayana, M., 373
Sheela, K. Anitha, 373
Shiddanagouda, F. B., 379
Soma, Prathap, 75
Sravanthi, Chundru, 75
Sridevi, P. V., 301
Srihari, P., 263
Srilakshmi, Palaparthi, 75
Sripati Acharya, U., 263
Subhas, C., 357
Sudhakar, P., 373
Sundaram, M., 227
Sushma, C., 179
Swamy, K. Ayyappa, 179

T
Tatiknoda, Gayathri Devi, 283
Thummala, Kalyani, 283
Tyagi, Akash, 83, 213

V
Vakula, Damera, 315
Vani, R. M., 379
Venkata, Manoj Katta, 165
Vulligaddala, Veeresh Babu, 165